Selected Titles in This Series

23 **Myrna H. Wooders, Editor,** Topics in mathematical economics and game theory. Essays in honor of Robert J. Aumann, 1999

22 **Graciela Chichilnisky, Editor,** Topology and markets, 1999

21 **Shigui Ruan, Gail S. K. Wolkowicz, and Jianhong Wu, Editors,** Differential equations with applications to biology 1998

20 **Peter A. Fillmore and James A. Mingo, Editors,** Operator algebras and their applications II, 1998

19 **William G. Dwyer, Stephen Halperin, Richard Kane, Stanley O. Kochman, Mark E. Mahowald, and Paul S. Selick (Editor-in-Chief), Editors,** Stable and unstable homotopy, 1998

18 **Panos M. Pardalos and Henry Wolkowicz, Editors,** Topics in semidefinite and interior-point methods, 1998

17 **Joachim J. R. Cuntz and Masoud Khalkhali, Editors,** Cyclic cohomology and noncommutative geometry, 1997

16 **Victor P. Snaith, Editor,** Algebraic K-theory, 1997

15 **Stephen P. Braham, Jack D. Gegenberg, and Robert J. McKellar, Editors,** Sixth Canadian conference on general relativity and relativistic astrophysics, 1997

14 **Mourad E. H. Ismail, David R. Masson, and Mizan Rahman, Editors,** Special functions, q-series and related topics, 1997

13 **Peter A. Fillmore and James A. Mingo, Editors,** Operator algebras and their applications, 1997

12 **Dan-Virgil Voiculescu, Editor,** Free probability theory, 1997

11 **Colleen D. Cutler and Daniel T. Kaplan, Editors,** Nonlinear dynamics and time series: Building a bridge between the natural and statistical sciences, 1997

10 **Jerrold E. Marsden, George W. Patrick, and William F. Shadwick, Editors,** Integration algorithms and classical mechanics, 1996

9 **W. H. Kliemann, W. F. Langford, and N. S. Namachchivaya, Editors,** Nonlinear dynamics and stochastic mechanics, 1996

8 **Larry M. Bates and David L. Rod, Editors,** Conservative systems and quantum chaos, 1996

7 **William F. Shadwick, Perinkulam Sambamurthy Krishnaprasad, and Tudor Stefan Ratiu, Editors,** Mechanics day, 1996

6 **Anna T. Lawniczak and Raymond Kapral, Editors,** Pattern formation and lattice gas automata, 1996

5 **John Chadam, Martin Golubitsky, William Langford, and Brian Wetton, Editors,** Pattern formation: Symmetry methods and applications, 1996

4 **William F. Langford and Wayne Nagata, Editors,** Normal forms and homoclinic chaos, 1995

3 **Anthony Bloch, Editor,** Hamiltonian and gradient flows, algorithms and control, 1994

2 **K. A. Morris, Editor,** Control of flexible structures, 1993

1 **Michael J. Enos, Editor,** Dynamics and control of mechanical systems: The falling cat and related problems, 1993

Fields Institute Communications

The Fields Institute for Research in Mathematical Sciences

Topics in Mathematical Economics and Game Theory

Essays in Honor of Robert J. Aumann

Myrna H. Wooders
Editor

American Mathematical Society
Providence, Rhode Island

The Fields Institute for Research in Mathematical Sciences

The Fields Institute is named in honour of the Canadian mathematician John Charles Fields (1863–1932). Fields was a visionary who received many honours for his scientific work, including election to the Royal Society of Canada in 1909 and to the Royal Society of London in 1913. Among other accomplishments in the service of the international mathematics community, Fields was responsible for establishing the world's most prestigious prize for mathematics research—the Fields Medal.

The Fields Institute for Research in Mathematical Sciences is supported by grants from the Ontario Ministry of Education and Training and the Natural Sciences and Engineering Research Council of Canada. The Institute is sponsored by McMaster University, the University of Toronto, the University of Waterloo, and York University and has affiliated universities in Ontario and across Canada.

2000 *Mathematics Subject Classification.* Primary 91Axx, 91Bxx.

Library of Congress Cataloging-in-Publication Data

Topics in mathematical economics and game theory : essay in honor of Robert J. Aumann / Myrna H. Wooders, editor.

p. cm. — (Fields Institute Communications, ISSN 1069-5265 ; v. 23)

Includes bibliographical references.

ISBN 0-8218-0525-8 (alk. paper)

1. Game theory. 2. Economics, Mathematical. I. Aumann, Robert J. II. Wooders, Myrna Holtz. III. Series.

QA269.T67 1999

519.3—dc21 99-046513

CIP

∞ The paper used in this book is acid-free and falls within the guidelines
established to ensure permanence and durability.
This publication was prepared by The Fields Institute.
Visit the AMS home page at URL: http://www.ams.org/

10 9 8 7 6 5 4 3 2 1 04 03 02 01 00 99

Contents

Preface

Since the publication of the Theory of Games by John von Neumann and Oskar Morgenstern, game theory has played a continually increasing role in economics. It has also been of growing importance in other sciences, including biology, political science and psychology. Several scientists made early seminal contributions and have continued to be leaders in the field. We have in mind especially John Harsanyi, Lloyd Shapley, Martin Shubik, Robert Aumann, and Reinhard Selten (listed in order of age). In addition to his contributions to game theory and economics, Robert Aumann also made a number of significant contributions to mathematics, including his thesis,

"Asphericity of Alternating Knots", Annals of Mathematics 64 (1956), pp. 374-392 and also:

"Integrals of Set-Valued Functions", Journal of Mathematical Analysis and Applications 12 (1965), pp. 1-12.

"Measurable Utility and the Measurable Choice Theorem", in La Decision, Editions du Centre National de la Recherche Scientifique, 1969, pp. 15-26, containing Aumann's measurable selection theorem,

"A Variational Problem Arising in Economics", Journal of Mathematical Analysis and Applications 11 (1965), pp.488-503 (with Micha Perles), and

"Orderable Set Functions and Continuity III: Orderability and Absolute Continuity", SIAM Journal on Control and Optimization 15 (1977), pp. 156-162 (with Uri Rothblum).

Finally, no listing, however partial, of Aumann's contributions to mathematics can omit

"Values of Non-Atomic Games", Princeton University Press, 1974, xi + 333 pp. (with Lloyd S. Shapley),

which has led to large numbers of articles (one appearing in this volume).

Auman's publications in game theory began with

"Acceptable Points in General Cooperative n-Person Games", in Contributions to the Theory of Games IV, Annals of Math. Study 40, Princeton University Press, 1959, pp. 287-324.

In this insightful and thought-provoking work Aumann introduced the important concept of "acceptable points" for a game. In a letter to the editor of this volume, responding to a request to reprint this paper, Bob Aumann wrote:

> "There's a soft spot in my heart for it: It is my first published paper in game theory proper. At the time it was published, I didn't realize that the set of "acceptable points" is simply the beta-core. This realization – or better, reformulation – is intimately related to the development of the NTU theory, which came a year or two later."

There are soft spots in the hearts of many game theorists for Bob Aumann, so we are especially pleased to reprint "Acceptable Points in General Cooperative n-Person Games" in this volume.

For a complete list of Aumann's publications through 1995, see Game and Economic Theory: Selected Contributions in Honor of Robert J. Aumann, Sergiu Hart and Abraham Neyman, Editors, University of Michigan Press, June 1995.

A number of the papers in the current volume, or related works, were presented at the Fields Institute Conference "New Directions in the Theory of Markets and Games", organized by Kenneth Arrow, Sergiu Hart, Andreu Mas-Colell, Jean-Francois Mertens, Walter Trockel, and Myrna Wooders (Chair), held in honor of Robert Aumann in 1995. The program of the conference concludes the volume.

In view of Aumann's contributions to mathematics and also the social sciences, it is most appropriate that this Fields Institute Communication volume be in his honor. It is with great pleasure that the contributors and the editor dedicate this volume to Bob Aumann.

Myrna H. Wooders
Editor

October 1998

Fields Institute Communications
Volume **23**, 1999

Acceptable Points in General Cooperative n-Person Games

Robert J. Aumann
Department of Mathematics
The Hebrew University
Jerusalem, Israel
raumann@math.huji.ac.il

1 Introduction

In this paper, we introduce a new theory for n-person games. Like many previous theories for n-person games, it is based on the search for a reasonable "steady state" in the "supergame,"[1] i.e., the game each play of which consists of a long sequence of plays of the original game. We use two approaches: The usual approach, in which a single play is studied, and an attempt is made to find an n-tuple of strategies that intuitively speaking might be acceptable for a steady state in the supergame (Sections 4 and 5); and an approach based on a mathematical analysis of the supergame itself (Sections 6 and 7). The results we obtain by the two approaches turn out to be essentially equivalent (Sections 9 and 10).

We will consider here only cooperative games, i.e., games in which coalitions are permitted; side payments will be forbidden.[2] However, the ideas introduced here also apply to non-cooperative games. Subsequent papers will consider non-cooperative as well as cooperative games.

We will not explicitly consider games involving chance; however, everything we say here carries over with no essential change to games involving chance.

Section 2 describes the notation we will use throughout. The reader may omit this section at first and use it only for reference. Section 8 is devoted to the proof of lemmas needed in the subsequent sections; it makes strong use of the approachability-excludability theory of Blackwell [**4**]. Sections 11, 12, and 13 are devoted to some miscellaneous remarks and counterexamples.

2 Notation

Throughout the paper, we will be concerned with an n-person game G. N will denote the set of all players; the individual players will be denoted by the positive

This paper originally appeared in Contributions to the Theory of Games IV, Annals of Mathematics Studies, Vol. 40, edited by R. D. Luce and A. W. Tucker, pp. 287–324, ©1959 by Princeton University Press, Princeton. It has been reproduced with the permission of the author and the publisher.

[1]The name is due to Luce and Raiffa [**6**].

[2]Of course it is known (see [**1**]) that the cooperative game with side payments is a special case of the cooperative game without side payments.

integers $1, \ldots, n$. If x^i is an object defined for each $i \in N$, and if $B \subset N$, then x^B will denote the set of x^i where $i \in B$; it will be called a B-vector. When $B = N$, we will often omit the superscript and the prefix; that is, we will write x instead of x^N, and will refer to x simply as a vector rather than as an N-vector. If X is a set of vectors, we denote by X^B the set of all x^B where $x \in X$. When we speak of a subset B of N, i.e., of a $B \subset N$, we will mean a non-null subset, unless we specifically include the null subset. All B-vector equalities or inequalities will be taken to hold term by term.

We will be concerned only with random variables taking a fixed finite number of values, and possibly with infinite sequences of such random variables. If x is a random variable ranging over a finite set Z, then the probability distribution of x may be considered as a function y from Z to the reals satisfying the relations:

$$y(z) \geq 0, \quad z \in Z \tag{2.1}$$

and

$$\sum_{z \in Z} y(z) = 1. \tag{2.2}$$

We can then write

$$y = \sum_{z \in Z}^{*} y(z) z \tag{2.3}$$

where the * indicates that the sum is merely symbolic and is not meant to be an ordinary sum. $y(z)$ is simply the probability that a random variable distributed according to y will take the value z.

The set of all probability distributions on Z will be denoted by $C(Z)$. The set of all pure strategies p^i for player i will be denoted by P^i. P is then the space of pure strategy vectors p. $C(P)$ is the space of **correlated** strategy vectors. A correlated strategy vector is a probability distribution on the set of all pure strategy **vectors**, just as a mixed strategy is a probability distribution on individual strategies. In general, to make use of a correlated strategy vector, the players have to agree to consult the **same** random device in choosing the pure strategies they will play. Correlated strategy vectors are thus not usable in non-cooperative games. We will usually abbreviate $C(P)$ to C. The prefix $c-$ in front of a word will stand for "correlated."

Let X^B be a space of B-vectors, and let $B = \cup_{j=1}^{k} B_j$ be a partition of B into subsets. If for $j = 1, \ldots, k$, we have

$$y^{B_j} \in C(X^{B_j}),$$

then $(y^{B_1}, \ldots, y^{B_k})$ denotes the joint probability distribution of $y^{B_1}, \ldots, y^{B_k}$.

If Z_1 and Z_2 are finite sets and $y \in C(Z_1 \times Z_2)$, then $y \mid Z_1$ (sometimes also written $y \mid C(Z_1)$) will be the distribution given by

$$y \mid Z_1 = \sum_{z_1 \in Z_1}^{*} z_1 \sum_{z_2 \in Z_2} y(z_1, z_2).$$

If $B_1 \subset B_2$ and $y^{B_2} \in C(X^{B_2})$, then we will simply write y^{B_1} instead of $y^{B_2} \mid X^{B_1}$. This applies to all symbols **except** γ. Thus γ^{B_1} is **not** the same as $\gamma^{B_2} \mid P^{B_1}$. ($\gamma^B$ will be defined on $C^B(P)$.)

Let Z_1 and Z_2 be finite sets, and suppose we have a function f that takes Z_1 into Z_2. If nothing is said to the contrary, we will assume that the definition of f

is extended linearly; that is, f is a function from $C(Z_1)$ to $C(Z_2)$ defined by

$$f(y) = \sum_{z_1 \in Z_1}^{*} f(z_1)y(z_1).$$

If for each $i \in N$, we have a function f^i defined on X^i, then we may define a **vector function** f on X by means of

$$f(x) = (f^1(x^1), \ldots, f^n(x^n)) \tag{2.4}$$

for all $x \in X$. It is important to remember, though, that the inverse process does not in general work. A vector function f defined on X does not yield a unique definition of a function f^i defined on X^i.

In general, when we have defined f^i on X^i, we will use the function f on X in the above sense.

For $p \in P$, the payoff to the ith player will be denoted by $E^i(p)$. If $c \in C, H(c)$ will be the mean of the distribution $E(c)$. $E(c)$ is called the **payoff distribution** when c is played, $H(c)$ the **expected** payoff when c is played. If $p \in P$, we have $H(p) = E(p)$.

The null set will be denoted by $\emptyset$.

We will be concerned with partitions of N. Let $\mathcal{P}_N$ denote the set of all subsets of N. Formally, a partition of N is a vector d whose components are members of $\mathcal{P}_N$, and which satisfies the condition

$$j \in d^i \Leftrightarrow d^i = d^j, \quad i, j \in N.$$

The set of all partitions of N will be denoted by D. The **constant** partition d_N is defined by

$$d_N^i = N, \quad i \in N. \tag{2.5}$$

The **identity** partition d_e is defined by

$$d_e^i = \{i\}, \quad i \in N.$$

Denote by R the set of all vectors r, each of whose components r^i is a member of $\mathcal{P}_N$ containing i. Clearly $D \subset R$, and $D^i = R^i$. For a given d, the number of distinct d^i will be called $n(d)$. We choose a set of representatives $i_1, \ldots, i_{n(d)}$, one out of each distinct d^i. If c^B is a c-strategy B-vector, define

$$e(c^B) = B.$$

(The e stands for exponent; B is called the **exponent** of c^B.) Denote by T the set of all vectors t, each of whose components is some correlated strategy B-vector (not necessarily the same B for each member of N), for which

$$i \in e(t^i), \quad i \in N.$$

Let

$$d^i(t) = \{j \mid t^j = t^i\}, \quad i \in N, \quad t \in T. \tag{2.6}$$

Then $d(t)$ is a partition of N, and we have

$$d^i(t) \subset e(t^i), \quad i \in N.$$

Define

$$c(t) = \left(\left(t^{i_1}\right)^{d^{i_1}(t)}, \ldots, \left(t^{i_{n(d(t))}}\right)^{d^{i_{n(d(t))}}(t)} \right) \tag{2.7}$$

where $i_1, \ldots, i_{n(d(t))}$ are a set of representatives for $d(t)$.

In discussing games in extensive form, we will call the vertices of the game tree that are not moves by the name "terminal" rather than "play". This is necessary to avoid confusion, as we will be dealing with repeated plays of the same game, in the more common sense of the word.

As a final remark, we mention that everything we have stated probabilistically can be restated, where necessary, in the language of measure theory, and rigorously justified in that language. All sets we use can be proved to be measurable.

3 General cooperative games

We use the word "general" in the title of the paper and of this section in order to emphasize that, unlike the cooperative games of von Neumann–Morgenstern, the games to be considered here exclude side payments. On the other hand, full and free cooperation is permitted between the players, in the sense that they may communicate with each other freely prior to each play and form any coalitions they may consider convenient. The agreements under which the coalitions are chosen are to be considered enforceable. (However, see Section 13 for further discussion of this point.) We will usually omit the word "general" and refer simply to a cooperative game. This is in accordance with the usage introduced by Nash [**1**].

In applications, the prohibition on side payments may be the result either of a physical barrier to side payments (in economic applications this might take the form of certain provisions in an anti-trust law) or of the lack of a common unit of measurement for the payoff. As has been pointed out by Raiffa [**8**], if there is a common unit of measurement for the payoff but there exists some physical barrier to the making of side payments, then the theory need only be invariant under linear transformations of the payoff function that do not destroy the common unit of measurement; i.e., under transformations of the form

$$H' = aH + b \tag{3.1}$$

where both a and b are scalar multiples of the vector $(1, \dots, 1)$ (a by a positive factor, b by an arbitrary factor). If, however, there is no common unit of measurement for the payoff, then the theory should be invariant under independent linear transformations of the payoff function; i.e., under transformations of the form (3.1), where a is an arbitrary positive vector and b is an arbitrary vector. In fact, the theory presented in this paper is invariant under the wider class of transformations, so that the results may be used in either kind of application.

Mathematically, the cooperative game is given by a vector payoff function that is assumed to be linear on the space C of correlated strategy vectors. The cooperative game differs from the non-cooperative game only in that the use of correlated strategy vectors is permitted in the former, but not in the latter.

This points up a reason for studying cooperative games even if our a priori interest lies in the field of non-cooperative games only. Consider a long sequence of plays (we call this a **superplay**) of the game

	1	2
1	$(4, 0)$	$(0, 0)$
2	$(0, 0)$	$(0, 4)$

.

One of the most "natural" ways for the players to proceed in such a superplay would be to alternate between playing $(1,1)$ and $(2,2)$. This would tend to happen even in the non-cooperative case, by a sort of "silent gentlemen's agreement". However, the resulting payoff, $(2,2)$, is obtainable only by a correlated strategy vector, not by mixed strategies. Thus if we wish to arrive at results that are valid for the supergame by considering a single play, we must permit ourselves the use of correlated strategy vectors. This amounts to considering cooperative games.

4 Acceptable points

Let c be a correlated strategy vector in G. Obviously, c will be a candidate for a possible steady state in the supergame unless some subset B of the players can, by concerted action, increase their payoff. Here it is not sufficient that the players of B be able to increase their payoffs by changing their strategies while the players of $N-B$ maintain their strategies as they are. For the players of $N-B$ will not in general maintain their strategies fixed in the face of a change on the part of B, but will also change **their** strategies in order to meet the new conditions; thus B's glory would be short-lived indeed. In order to rule out c as a candidate for a possible steady state, we have to be sure that the players of $N-B$ cannot prevent the players of B from improving their payoff by concerted action. Of course, we must require that **all** the members of B obtain more at the new point than they do at c, for they must all have an incentive to cooperate in the venture.

Definition 4.1 *Let $c_0 \in C$. c_0 is said to be c-acceptable if there is no $B \subset N$ such that for each $c^{N-B} \in C^{N-B}$, there is a $c^B \in C^B$ for which*

$$H^B(c^B, c^{N-B}) > H^B(c_0).$$

We write "c-acceptable" instead of "acceptable" because in subsequent papers of this series, we will be defining a slightly different kind of acceptability, and we will use prefixes to distinguish them. When in heuristic discussion in the sequel we use the word "acceptable" without a prefix, then it is to be taken to mean "acceptable" with an arbitrary prefix (but one that is fixed throughout the discussion). For the purposes of this paper alone, "acceptable" can be taken to mean the same thing as "c-acceptable".

The set of all c-acceptable c-strategy vectors is denoted by A_c. Note that the notion of acceptability is a "global" one. In other words, whether or not a point is acceptable depends only on the payoff at that point, not on the point itself. This is also easy to accept from the intuitive standpoint; as long as he is getting the same payoff, it makes little difference to a player which strategy he is playing.

Definition 4.2 *A payoff vector h is said to be c-acceptable, if for some $c \in A_c$, we have*

$$H(c) = h.$$

It will be convenient in the sequel to have available the following trivial restatement of 4.2.

Lemma 4.3 *A payoff vector h is c-acceptable if and only if for each $B \subset N$, there is a $c^{N-B} \in C^{N-B}$, such that for all $c^B \in C^B$, there is an $i \in B$ for which*

$$H^i(c^B, c^{N-B}) \geq h^i.$$

5 Two-person, zero-sum games

A sine qua non of theories of n-person games is that for two-person, zero-sum games, they reduce to the von Neumann theory. Because of the global nature of our theory, we can only expect that every acceptable payoff vector yield to each player the value of the game to that player. This is indeed the case.

Theorem 1 *In a two-person, zero-sum game G, a payoff vector h is c-acceptable if and only if*

$$h^1 = v, \tag{5.1}$$

where v is the value of G to player 1.

Proof Let h be c-acceptable. From Lemma 4.3 applied to $B = \{1\}$, we deduce that there is a $c_0^2 \in C^2$, such that for all $c^1 \in C^1$, we have

$$H^1(c^1, c_0^2) \leq h^1.$$

Hence

$$\max_{c^1 \in C^1} H^1(c^1, c_0^2) \leq h^1.$$

Hence

$$\min_{c^2 \in C^2} \max_{c^1 \in C^1} H^1(c^1, c^2) \leq h^1. \tag{5.2}$$

C^1 and C^2 are the mixed strategy spaces for players 1 and 2 respectively. Hence the left side of (5.2) is v, and we deduce

$$v \leq h^1. \tag{5.3}$$

The value of G to player 2 is $-v$. Proceeding as in the proof of (5.3), we obtain

$$-v \leq h^2. \tag{5.4}$$

But

$$h^2 = -h^1. \tag{5.5}$$

Combining (5.4) and (5.5), we obtain

$$v \geq h^1. \tag{5.6}$$

Combining (5.3) and (5.6), we obtain (5.1).

Conversely, assume (5.1). Then (5.3) and (5.2) follow at once. But for $B = \{1\}$, (5.2) is precisely the condition for c-acceptability given by Lemma 4.3. Similarly we establish the condition for $B = \{2\}$. To establish it for $B = \{1, 2\}$, note that for **any** (c^1, c^2), we have

$$H^1(c^1, c^2) + H^2(c^1, c^2) = 0 = h^1 + h^2,$$

whence we must have either

$$H^1(c^1, c^2) \leq h^1$$

or

$$H^2(c^1, c^2) \leq h^2.$$

Thus h is c-acceptable.

6 Strategies and payoffs in the supergame

In the notion of supergame that will be used in this paper, each superplay consists of an **infinite** number of plays of the original game G. On the face of it, this would seem to be unrealistic, but actually it is more realistic than the notion in which each superplay consists of a fixed finite (large) number of plays of G. In the latter notion, the fact that the players **know** when they have arrived at the last play becomes the decisive factor in the analysis, overshadowing all other considerations.[3] This is unnatural, because in the usual case, the players can always anticipate that there will still take place in the future an indefinite number of plays. This is especially true in political and economic applications, and even holds true for the neighborhood poker "club". Of course when looked at in the large, nobody really expects an infinite number of plays to take place; on the other hand, after each play we do expect that there will be more. A. W. Tucker has pointed out that this condition is mathematically equivalent to an infinite sequence of plays, so that is what our notion of supergame will consist of.

Roughly speaking, we will be interested in "behavior" strategies in the supergame; that is, we will be interested in strategy vectors that determine the c-strategy vector to be played on each play as a function of the information possessed by the players about the outcomes of the previous plays. Before we pass to the formal definition of a supergame c-strategy vector, let us examine in detail the procedure followed during the play of a cooperative supergame. The first question to be answered is: What information about the outcome of each previous play does each player i actually possess? The most he can know is exactly what pure strategy vector was actually played; the least he can be sure of knowing, in every game, is the strategy he himself played and the payoff he himself received. In between is a whole range of possibilities, depending on the rules of G. The situation can be concisely described by means of a function u^i that takes P onto some identification space of P. We define u^i so that for $p_1, p_2 \in P, u^i(p_1) = u^i(p_2)$ if and only if player i cannot tell after a play of G, whether the vector p_1 or the vector p_2 has been played. The minimum knowledge that each player i has, in accordance with our remarks above, insures us that if

$$u^i(p_1) = u^i(p_2),$$

then

$$p_1^i = p_2^i$$

and

$$H^i(p_1) = H^i(p_2).$$

Every game G has associated with it an information vector u satisfying these conditions. In particular, these conditions assure us of the existence of functions ϕ^i and ψ^i which map $u^i(P)$ onto P^i and $H^i(P)$ respectively. Some relations connecting the vector functions H, ϕ, ψ, and u are the following:

$$\begin{aligned} \phi \circ u &= \text{identity} \\ \psi \circ u &= H \\ \psi &= H \circ \phi. \end{aligned}$$

The information vector just defined is not included in the classical formulation of the extensive form of a game, as given, say, in [**3**]. That is because the state of

[3]See, for example, the excellent analysis of the supergame of the game known as the "Prisoner's Dilemma" that appears in 5.5 of [**6**].

information at the **end** of a game has heretofore been of little interest, as the payoff in no way depends on it. In a number of consecutive plays of a game, however, it is seen to be of vital importance. Incidentally, inclusion of this concept in the extensive structure of a game necessitates a revision in the notion of equivalence as formulated in [**5**]. We hope to treat the subject in more detail in a subsequent paper. For the present, we content ourselves with the remark that the information vector may be included in the extensive description of the game by associating with each player i, an information partition on the set of all terminals. In a game of perfect information, this partition would be the identity partition for each i.

All that we have said in this section up to now is quite general, applying to the non-cooperative as well as to the cooperative case. In the latter case, the information function does not exhaust the information that a player has to work with. Prior to each play, consultation takes place among the various players to determine the constitution of the coalitions and the correlated strategy B-vector that each coalition B will use. Formally, each player designates the coalition B to which he wishes to belong, and the correlated strategy B-vector he wishes B to play. Those players who can agree among each other form coalitions and play the agreed upon c-strategy B-vector.[4]

Naturally, although each player knows nothing about the coalitions in which the other players participate, he does remember the coalition B of which he himself was a member on previous plays.

The formal definition is as follows:

Definition 6.1 *A supergame c-strategy f^i for player i is an infinite sequence of functions*

$$f_0^i, f_1^i, \dots, f_k^i, \dots$$

where f_k^i is a function from

$$((U_1^i \times R_1^i) \times \dots \times (U_k^i \times R_k^i))$$

to T^i; here U_j^i is a copy of $u^i(P)$ and R_j^i is a copy of R^i, for $j = 1, \dots, k$. The vector f is called a supergame c-strategy vector.

In the sequel, we will denote $U_j^i \times R_j^i$ by J_j^i. J_j^i represents the information about the jth play available to player i.

To define the payoff to a supergame c-strategy vector, we must first define the payoff to a given super**play**, as a function of its individual constituent plays. For this purpose, we use the arithmetic mean of the payoffs (in the limiting case, this amounts to the first Cesáro sum); Blackwell [**4**] also used the arithmetic mean under similar circumstances. Heuristically, one can argue both for and against this method; in my mind the heuristic advantages outweigh the disadvantages. Other methods have been proposed (see [**6**]) for obtaining the superplay payoff from the payoffs to its constituent plays; the most prominent seems to be the discounting of future payoffs back to the present, using a fixed rate of interest. This has the disadvantage of putting an unnaturally heavy stress on the beginning of the supergame. Nevertheless, it would be interesting to see an analysis based on this payoff notion, especially for certain kinds of economic applications.

A more perplexing problem in defining the payoff to a supergame c-strategy vector f is that of somehow combining the various possible superplays that might occur if f is played. The first impulse is to use some notion of **expected** payoff. We

---

[4]If $B_1 \subset B_2$ and every member of B_1 proposed (B_2, c^{B_2}) while the members of $B_2 - B_1$ proposed something else, then the members of B_1 form a coalition and play c^{B_1}.

shall indeed define such a notion, but it is not an appropriate criterion for the players in choosing their supergame c-strategy. The chief reason for our being interested in expected payoffs for **individual** plays is that in a long sequence of plays, the law of large numbers turns an expected payoff into an **actual** payoff. A payoff for the whole superplay that is only **expected**, but not eventually approached, will not satisfy most players. Thus if a player has joined a coalition on the strength of a high expectation and is being disappointed in this expectation, he will soon express his disappointment by resigning from the coalition, regardless of his originally high expectation. For a clear example of the pitfalls of expected payoffs, see Section 12. What is required is some kind of criterion based on a law of large numbers. We give a restricted definition of this kind in this section, but reasoning of this type in general will dominate the remainder of the paper.

Let us first return to the question of expected payoff. For each $t \in T$, define

$$s(t) = (u(c(t)),\ \ d(t)). \tag{6.2}$$

Let f be a supergame c-strategy vector. For each $k \geq 1$, we may define a member $z_k(f) = z_k$ of $C(J_1 \times \ldots \times J_k)$ recursively as follows:

$$z_1 = s(f_0) \tag{6.3}$$

$$z_k = \sum^{*}_{y \in J_1 \times \ldots \times J_{k-1}} z_{k-1}(y)(y, s(f_{k-1}(y))),\ \ k > 1. \tag{6.4}$$

z_k is the probability distribution of possible outcomes of plays up to the kth. Define a member $x_k(f) = x_k$ of $C(J_k)$ as follows:

$$x_k = s(f_{k-1}(z_{k-1})),\ k \geq 1. \tag{6.5}$$

x_k is the probability distribution of outcomes of the kth play. Finally, define $E_k = E_k(f)$ and $H_k = H_k(f)$ by

$$E_k = E(c(f_{k-1})z_{k-1}))) \tag{6.6}$$

and

$$H_k = H(c(f_{k-1}(z_{k-1}))). \tag{6.7}$$

E_k represents the probability distribution of possible payoff vectors on the kth play; H_k is the expected payoff for the kth play.

If

$$H(f) = \lim_{k \to \infty} \frac{1}{k} \sum_{r=1}^{k} H_r(f) \tag{6.8}$$

exists, then f is said to be **summable in the mean**.

Let $v = (v_1, \ldots, v_k, \ldots)$ be a sequence of random variables in which v_k is distributed according to the distribution $x_k(f)$. These random variables are not in general independent. Instead of saying that the random variables v_k are distributed according to $x_k(f)$, we will sometimes say that the random variable v is distributed according to f. The probability of a statement concerning a random variable v distributed according to f will usually be denoted Prob_f.

Definition 6.9 *A supergame c-strategy vector f is said to be* **summable** *if it is summable in the mean and if a sequence of random variables distributed according to $E_k(f)$ obeys the strong law of large numbers.*[5]

If v is an infinite sequence $(v_1, \ldots, v_k, \ldots)$ where $v_k \in J_k$ for all k, define

$$H_k(v) = \psi(v_k \mid U_k) \quad (= H(\phi(v_k \mid U_k))) \tag{6.10}$$

and

$$S_k(v) = \frac{1}{k} \sum_{j=1}^{k} H_j(v). \tag{6.11}$$

Note that if v is a random variable distributed according to f, then $H_k(v)$ is a random variable distributed according to $E_k(f)$.

Lemma 6.12 *A necessary and sufficient condition that a supergame c-strategy vector f be summable is that there exist a vector $H(f)$ such that for each vector $\varepsilon > 0$, we have*

$$\lim_{k\to\infty} \text{Prob}_f(\mid S_r(v) - H(f) \mid \geq \varepsilon \text{ for some } r \geq k) = 0.$$

Furthermore, this $H(f)$ is then identical with the $H(f)$ defined by (6.8).

The proof, which is straightforward, will be omitted.

7 Strong equilibrium points

When we come to investigate what supergame c-strategy vectors f are liable actually to occur, we are faced with a problem somewhat different from the one we discussed in Section 4. There one could always look forward to future plays in order to punish deviations by the other players. In the supergame, however, there is only one superplay. Thus the players will have to make sure the first time that deviations won't pay. This amounts to saying that no coalition B can increase the payoff of all its members by substituting a different supergame c-strategy B-vector g^B while $N - B$ **maintains** the supergame c-strategy $(N - B)$-vector f^{N-B}. It is clear that such an f would be very strongly stable, since no **coalition** could have any incentive to move away from it, especially if we demand that B cannot even increase its payoff with positive probability, no matter how small. This is a much stronger version of what corresponds in the supergame to an ordinary equilibrium point.

The formal definition is as follows:

Definition *Let f be a summable supergame c-strategy vector. f is a* **strong equilibrium** c**-point** *if there is no $B \subset N$ for which there is a supergame c-strategy vector g satisfying*

$$g^{N-B} = f^{N-B} \tag{7.1}$$

and

$$\text{there is a } B\text{-vector } \varepsilon^B > 0 \text{ for which}$$
$$\lim_{k\to\infty} \text{Prob}_g(S_r^B(v) \geq H^B(f) + \varepsilon^B \text{ for some } r \geq k) > 0. \tag{7.2}$$

(The limit exists because the sequence involved is monotone decreasing.)

[5] See [**2**], p. 207. The law is there stated for one dimensional random variables; the extension to n dimensions is straightforward.

The set of all strong equilibrium c-points is denoted by S_c.

It is necessary to use the form $S_r^B(v) \geq H^B(f)+\varepsilon^B$ rather than $S_r^B(v) > H^B(f)$ because substitution of the latter statement would make Condition 7.2 true even for $f = g$. We wish to include only those cases in which substitution of g for f will yield B an advantage that at least does not tend to the vanishing point, doubtful and rare as it may be.

Condition 7.2 is very weak, which means that the concept of strong equilibrium c-point is a very strong one (i.e., comparatively few supergame c-strategy vectors are strong equilibrium c-points). On the other end of the spectrum, we could replace (7.2) by:

$$\begin{gathered}\text{There is a } B\text{-vector } \varepsilon^B > 0 \text{ for which} \\ \lim_{k\to\infty} \text{Prob}_g(S_r^B(v) \geq H^B(f) + \varepsilon^B \text{ for all } r \geq k) = 1.\end{gathered} \tag{7.3}$$

Denote by $\tilde{S}_c$ the set of supergame c-strategy vectors obtained when (7.3) is substituted for (7.2) in the definition of strong equilibrium c-point.

It is easy to see that

$$(7.3) \Rightarrow (7.2),$$

whence it follows that

$$S_c \subset \tilde{S}_c$$

and

$$H(S_c) \subset H(\tilde{S}_c). \tag{7.4}$$

We will demonstrate the opposite inequality as well. This shows that the notion of strong equilibrium point does not really depend on the strength of the Condition 7.2.

8 Application of approachability and excludability theory

For two-person, zero-sum games with vector payoffs, Blackwell [**4**] has defined the notion of "approachability" or "excludability" of a set in the payoff space with a supergame strategy employed by one or the other of the players. Blackwell does not use the information vector u which we introduced in Section 6, but apart from that, our Definition 6.1 applied to two-person, zero-sum games reduces to his definition of supergame strategy (for games not involving chance). Most of his definitions and results can be used here without any changes being necessary due to our use of the information vector.

Let us consider the two-person, zero-sum game with vector payoffs that is obtained from our game G by considering the coalition B to be the first player and the coalition $N - B$ the second player, the payoff to the first player being given by H^B.

Lemma 8.1 *Let $\Lambda(f, \varepsilon^B)$ denote the closed convex set consisting of all B-vectors λ^B for which*

$$\lambda^B \geqq H^B(f) + \varepsilon^B.$$

Then if there is an $\varepsilon^B > 0$ such that $\Lambda(f, \varepsilon^B)$ is approachable with g^B, then (7.3) holds.

Proof This lemma follows at once from the definition of approachability; see Section 1 of [**4**]. Our games do not involve chance, so the elements of the game matrix should be considered as B-vectors rather than probability distributions on

B-vectors. We have abbreviated "approachable in the game matrix" to "approachable"; this will be done throughout the sequel.

Lemma 8.2 *If there is an $\varepsilon^B > 0$ such that, for each $c^{N-B} \in C^{N-B}$, there is a $c^B \in C^B$ for which*

$$H^B(c^B, c^{N-B}) \geq H^B(f) + \varepsilon^B,$$

then there is a g^B for which (7.3) holds.

Proof The condition given in the lemma is equivalent to the statement that for each $c^{N-B} \in C^{N-B}, H^B(C^B, c^{N-B})$ intersects $\Lambda(f, \varepsilon^B)$. In accordance with Theorem 3 of [**4**], this in turn is equivalent to the statement that there is a g^B with which $\Lambda(f, \varepsilon^B)$ is approachable. Applying (8.1), we obtain the conclusion.

The information situation in [**4**] is slightly different from ours. In our application, the requirements of [**4**] may be stated as follows: Both B and $N - B$ know the payoffs to B on each previous play; nothing else is known. In our situation, B knows its own previous payoffs, but $N - B$ need not know what the payoffs to B were; on the other hand, both B and $N - B$ may have information given by the information function, that they do not have in [**4**]. An analysis of the proof of Theorem 3 of [**4**] shows that these differences do not affect the truth of the theorem, so that we may still apply it here.

To see this intuitively, note that the only difference that could interfere with the approachability of $\Lambda(f, \varepsilon^B)$, with g^B would be the additional information that $N - B$ has in our case that it does not have in [**4**]. Examination of the proof of Theorem 3 of [**4**] indicates that g_k^B depends only on the previous payoffs, and can be defined for **any** set of previous payoffs without affecting the truth of the theorem; thus no amount of additional information can possibly help $N - B$ in preventing B from approaching $\Lambda(f, \varepsilon^B)$ with g^B.

Lemma 8.3 *If there is a $c^{N-B} \in C^{N-B}$ (called γ^{N-B}) such that for each $c^B \in C^B$, there is an $i \in B$ for which*

$$H^i(c^B, \gamma^{N-B}) \leq H^i(f),$$

then for every B-vector $\varepsilon^B > 0$, and every supergame c-strategy vector β for which for all $(v_1, \ldots, v_k)$, we have

$$\beta_k^i(v_1, \ldots, v_k) = \gamma^{N-B}, \quad i \in N - B, \quad k \geq 0, \tag{8.4}$$

we have

$$\lim_{k \to \infty} \text{Prob}_\beta(S_r^B(v) \geq H^B(f) + \varepsilon^B \text{ for some } r \geq k) = 0.$$

Proof The hypothesis of the lemma is equivalent to the assertion that for each $\varepsilon^B > 0, \Lambda(f, \varepsilon^B)$ fails to intersect $H(C^B, \gamma^{N-B})$. Applying Theorem 3 of [**4**], we conclude that for each $\varepsilon^B > 0, \Lambda(f, \varepsilon^B)$ is excludable with the supergame c-strategy $(N - B)$-vector given by (8.4). From the definition of excludability (Section 1 of [**4**]) it then follows that there is a number $\delta > 0$ such that

$$\lim_{k \to \infty} \text{Prob}_\beta \quad (\text{For all } r \geq k, \text{ there is an } i \in B \text{ such that}$$

$$S_r^i(v) < H^i(f) + \varepsilon^i - \delta) = 1,$$

whence

$$\lim_{k \to \infty} \text{Prob}_\beta(S_r^B(v) \geq H^B(f) + \varepsilon^B \text{ for some } r \geq k) = 0.$$

9 The main theorem: first half

We will now show that our two approaches culminating in the definition of c-acceptability (4.1) and strong equilibrium (7.1, 7.2, and 7.3) actually yield results that are essentially the same. More precisely, for every c-acceptable strategy vector c, there is a strong equilibrium c-point (according to either 7.2 or 7.3) in supergame c-strategies that has the same payoff as c; and conversely. Thus c-acceptable points for a single play are seen to correspond exactly to strong equilibrium c-points in the supergame. The main theorem can be concisely stated by means of the equation

$$H(A_c) = H(S_c) \; (= H(\tilde{S}_c)).$$

This section is devoted to the proof of the relation

$$H(\tilde{S}_c) \subset H(A_c).$$

The proof rests heavily on Lemma 8.2. In addition to (8.2), the main fact needed is that if B cannot be prevented from obtaining more than h^B, then they cannot be prevented from obtaining a fixed amount more than h^B either. This is shown in Lemma 9.1.

Lemma 9.1 *Let h be a vector and B be a subset of N. If for each $c^{N-B} \in C^{N-B}$, there is a $c^B \in C^B$ for which*

$$H^B(c^B, c^{N-B}) > h^B,$$

then there is a positive B-vector ε^B such that for each $c^{N-B} \in C^{N-B}$, there is a $c^B \in C^B$ for which

$$H^B(c^B, c^{N-B}) \geq h^B + \varepsilon^B.$$

Proof We first remark that if $G(x, y)$ is a continuous real-valued function of two variables x and y ranging over compact sets X and Y respectively, then the function defined by

$$F(y) = \max_{x \in X} G(x, y)$$

is continuous on Y.

Now define a function F on C^{N-B} by

$$F(c^{N-B}) = \max_{c^B \in C^B} \min_{i \in B} (H^i(c^B, c^{N-B}) - h^i).$$

By the above remark, F is continuous. Since C^{N-B} is compact, F takes on its minimum at a point c_0^{N-B} in C^{N-B} By the hypothesis of the lemma,

$$F(c_0^{N-B}) > 0.$$

Setting

$$\varepsilon^1 = F(c_0^{N-B})$$

for all $i \in B$, we obtain the conclusion of the lemma.

Theorem 2 $H(\tilde{S}_c) \subset H(A_c)$.

Proof Let f be a summable supergame c-strategy vector; set

$$h = H(f).$$

Suppose

$$h \notin H(A_c), \tag{9.2}$$

i.e., suppose there is a B for which the hypothesis of Lemma 9.1 holds. Then the conclusion of Lemma 9.1, which is the same as the hypothesis of Lemma 8.2, holds.

Hence we deduce the conclusion of Lemma 8.2, i.e., there is a g^B for which (7.3) holds. Setting

$$g^{N-B} = f^{N-B},$$

we obtain a B and a g obeying (7.1) and (7.3), whence

$$f \notin \tilde{S}_c. \tag{9.3}$$

As we can go through the same argument for any f satisfying $h = H(f)$, we may deduce from (9.3) that

$$h \notin H(\tilde{S}_c). \tag{9.4}$$

We have shown that (9.2) $\Rightarrow$ (9.4), whence we can deduce that

$$H(\tilde{S}_c) \subset H(A_c),$$

which is what we set out to prove.

10 The main theorem: second half

This section is devoted to the proof of the relation

$$H(A_c) \subset H(S_c);$$

together with Theorem 2, this yields the main theorem. The train of thought of the proof is somewhat as follows:

Suppose h to be a c-acceptable payoff. It is easy to set up a supergame c-strategy vector f whose payoff is h. But we must also incorporate in f a foolproof system for the punishment of any deviators; f must make sure that crime does not pay. The machinery for accomplishing this is at hand; by the definition of c-acceptability, for each B there is a c-strategy $(N-B)$-vector γ^{N-B} whose use prevents B from obtaining more than h^B. It remains only to **find** the culprits. To do this, we note that any players who at one time or another deviated from f, were at the time of their deviation certainly not included in the same coalition as the "orthodox" players. Since each player knows who was in his coalition on all previous plays, the deviators are thus easily spotted. The set B_k of players who deviated on some play up to the $(k+1)$st is monotone increasing with k. Every time it grows, i.e., every time another player joins the ranks of the deviators, the remaining players revise their strategy to punish at least one of the **new** set of deviators. Since N is finite, there must be a set B that includes all the deviators and that is actually attained by B_k after a finite number of plays. The use of the strategy γ^{N-B} will then make certain that crime does not pay for at least one of the deviators. The proof is complicated by the fact that B_k and B usually depend on which pure strategies were chosen from among those considered by the correlated strategy vectors used on previous plays.

The strong equilibrium point f just described is one of "unrelenting ferocity"[6] against offenders. It exhibits a zeal for meting out justice that is entirely oblivious to the sometimes dire consequences to oneself or to the other faithful — i.e., those who have not deviated. There are other, more reasonable strong equilibrium points, which give deviating players a chance to return to the fold. These are important in connection with the general non-cooperative game; I will return to them in a subsequent paper on that topic. Here it is simpler and just as efficacious to use the strong equilibrium point of unrelenting ferocity.

[6]The term is due to Luce and Raiffa [**6**].

Theorem 3 $H(A_c) \subset H(S_c)$.

Proof Let $h \in H(A_c)$, and suppose $\gamma^N \in A_c$ is such that

$$H(\gamma^N) = h. \tag{10.1}$$

Then by (4.3), for each $B \subset N$ there is a $c^{N-B} \in C^{N-B}$, which we will call γ^{N-B}, such that for each $c^B \in C^B$, there is an $i \in B$ for which

$$H^i(c^B, \gamma^{N-B}) \leq h^i. \tag{10.2}$$

Now define a supergame c-strategy vector f as follows:

$$\begin{aligned} f_0^i &= \gamma^N, \quad i \in N \\ f_k^i(v_1^i, \ldots, v_k^i) &= \gamma^{v_k^i|R^i}, \; k > 0, \quad i \in N, \quad v_j^i \in J_j^i, \quad j \leq k. \end{aligned} \tag{10.3}$$

We must prove that the payoff to f is h and that f is a strong equilibrium c-point. It is convenient to break the proof into a number of lemmas. However, these lemmas will depend on previous assumptions and formulae, and are valid only within the context of this proof. They will be numbered like ordinary formulae used within the proof.

Lemma 10.4 *For $k > 0$ and $i \in N$, we have*

$$f_k^i(z_k) = \gamma^N.$$

Proof The proof is by induction on k. For $k = 0$, (10.4) follows at once from (10.3). Assume (10.4) for a given k Applying (2.6) and (2.7), we obtain

$$c(f_k(z_k)) = \gamma^N$$

and

$$d(f_k(z_k)) = d_N.$$

Hence, by (6.2),

$$s(f_k(z_k)) = (u(\gamma^N), d_N). \tag{10.5}$$

Now by (6.4) and the linearity of f_k^i we have

$$\begin{aligned} f_{k+1}^i(z_{k+1}) &= \sum_{y \in J_1 \times \ldots \times J_k}^{*} z_k(y) f_{k+1}^i(y, s(f_k(z_k))) \\ &= \sum_{y \in J_1 \times \ldots \times J_k}^{*} z_k(y) f_{k+1}^i(y, (u(\gamma^N), d_N)) && \text{(by (10.5))} \\ &= \sum_{y \in J_1 \times \ldots \times J_k}^{*} z_k(y) f_{k+1}^i(y^i, (u^i(\gamma^N), N)) && \text{(by (2.4) and (2.5))} \\ &= \sum_{y \in J_1 \times \ldots \times J_k}^{*} z_k(y) \gamma^N && \text{(by (10.3))} \\ &= \gamma^N \sum_{y \in J_1 \times \ldots \times J_k} z_k(y) \\ &= \gamma^N. && \text{(by (2.2))} \end{aligned}$$

This completes the induction and the proof of the lemma.

Lemma 10.6 *For $k, j \geq 0$ and $k \neq j, c(f_k(z_k))$ is statistically independent of $c(f_j(z_j))$.*

Proof By (10.3), the choice of strategies after each play depends only on the coalitions previously chosen. Thus the sequence of partitions of N into coalitions is strictly determined. For a given play, the choice of c-strategy B-vectors depends only on this sequence (which is, in fact, a sequence of constant partitions), and **not** on the strategies previously chosen. This completes the proof.

Lemma 10.7 *f is summable, and $H(f) = h$.*

Proof Applying (2.7) to (10.4), we obtain

$$c(f_k(z_k)) = \gamma^N,$$

whence by (6.6)

$$E_k = E(c(f_k(z_k))) = E(\gamma^N). \tag{10.8}$$

By Lemma 10.6, the strong law of large numbers[7] applies to each component of $c(f_k(z_k))$, hence to $c(f_k(z_k))$ itself, and hence also to E_k. By (10.8) and (6.7), we have

$$\begin{aligned} H_k &= H(\gamma^N) \\ &= h. \qquad \text{(by (10.1))} \end{aligned}$$

Hence

$$\begin{aligned} H(f) &= \lim_{k\to\infty} \frac{1}{k} \sum_{r=1}^{k} H_r \qquad \text{(by (6.8))} \\ &= \lim_{k\to\infty} \frac{1}{k}\, kh \\ &= \lim_{k\to\infty} h \\ &= h. \end{aligned}$$

By Definition 6.9, the proof is complete.

Lemma 10.9 *If $v_k \mid R$ is a partition and $k > 0$, then*

$$j \in e(f_k^i(v_1, \ldots, v_k)) \Rightarrow f_k^j(v_1, \ldots, v_k) = f_k^i(v_1, \ldots, v_k).$$

Proof From (10.3) and the left side of the conclusion, we deduce that $j \in v_k^i \mid R^i$. Hence since $v_k \mid R$ is a partition, it follows that

$$v_k^i \mid R^i = v_k^j \mid R^j,$$

and then the right side of the conclusion follows at once from (10.3).

In order to show that f is a strong equilibrium c-point, we will first assume that it is not.

Assumption 10.10 *There is a $B \subset N$ (which we will call B^*) for which there is a supergame c-strategy vector g for which*

$$g^{N-B^*} = f^{N-B^*} \tag{10.11}$$

---

[7]See [**2**], p. 207, The Kolmogoroff Criterion, or p. 208, the Theorem.

and there is a B^*-vector $\varepsilon^{B^*} > 0$ for which

$$\lim_{k\to\infty} \text{Prob}_g(S_r^{B^*}(v) \geq H^{B^*}(f) \ + \varepsilon^{B^*} \ \text{ for some } r \geq k) > 0. \tag{10.12}$$

It will be our task to show that (10.10) leads to an absurdity.

Let $v = (v_1, \dots, v_k, \dots)$ be an arbitrary sequence for which $v_k \in J_k, k \geq 1$. Let $B(v)$ denote the set of i for which there is a $k_i \geq 0$ such that

$$g^i_{k_i}(v_1, \dots, v_{k_i}) \neq f^i_{k_i}(v_1, \dots, v_{k_i}). \tag{10.13}$$

It follows from (10.11) and (10.13) that

$$B(v) \subset B^*. \tag{10.14}$$

Definition 10.18 *For each v and for $k \geq 0$, let $B_k(v)$ denote the set of i for which there is a k_i such that*

$$0 \leq k_i \leq k$$

and such that (10.13) holds. Define

$$B_{-1}(v) = \emptyset. \tag{10.19}$$

For each v, there is clearly a $k(v)$ such that

$$B_k(v) = B(v), \quad k \geq k(v). \tag{10.20}$$

Furthermore, it is not difficult to see that (10.21): For each $v, B_k(v)$ is monotone increasing with k. We will in the sequel restrict use of the symbol $k(v)$ to the **first** $k(v)$ satisfying (10.20).

Definition 10.22 *We say that v occurs with* **positive probability** *if for each $k > 0$,*

$$z_k(g^\cdot(v_1, \dots, v_k) > 0;$$

that is, if for each $k > 0$, and under the assumption that the supergame c-strategy vector g is played, $(v_1, \dots, v_k)$ occurs with positive probability.

Lemma 10.23 *If v occurs with positive probability, then for each $k > 0$, $v_k \mid R$ is a partition.*

Proof By (6.4) and (10.22), $(v_1, \dots, v_k)$ occurs with positive coefficient in

$$(v_1, \dots, v_{k-1}, s(g_{k-1}(v_1, \dots, v_{k-1}))).$$

Hence v_k occurs with positive coefficient in $s(g_{k-1}(v_1, \dots, v_{k-1}))$. Applying (6.2), we conclude that $v_k \mid R$ occurs with positive coefficient in $d(g_{k-1}(v_1, \dots, v_{k-1}))$. Since by definition, $d(g_{k-1}(v_1, \dots, v_{k-1}))$ is a probability combination of a number of partitions of N, it follows that $v_k \mid R$ must be a partition, which completes the proof.

Lemma 10.24 *Let v occur with positive probability, and assume that (10.10) holds. For $k \geq 0$ and $i \in N - B_k(v)$, we have*

$$g^i_k(v_1, \dots, v_k) = \gamma^{N-B_{k-1}(v)}.$$

Proof We will write B instead of $B(v)$, and for each k, B_k instead of $B_k(v)$. The proof is by induction on k. First let $k = 0$. If $i \in N - B_0$, then it follows from Definition 10.18 that there is no k_i satisfying $k_i = 0$ and (10.13). In other words, for $i \in N - B_0$, we have

$$g^i_0 = f^i_0.$$

Hence by (10.3), we have for $i \in N - B_0$,

$$\begin{aligned} g_0^i &= \gamma^N \\ &= \gamma^{N-B_{-1}} \qquad \text{(by (10.19))}, \end{aligned}$$

which establishes (10.24) for $k = 0$.

Next suppose we have established (10.24) for some $k \geq 0$ and all preceding k. For $i \in B_k - B_{k-1} = (N - B_{k-1}) \cap B_k$, we must have, by Definition 10.18, that (10.13) holds for $k_i = k$. Hence

$$g_k^i(v_1, \ldots, v_k) \neq f_k^i(v_1, \ldots, v_k), \quad i \in B_k - B_{k-1}. \tag{10.25}$$

Now if $k = 0$, it follows from (10.19), (10.25), and (10.3) that

$$g_0^i \neq \gamma^{N-B_{-1}}, \quad i \in B_0 - B_{-1}. \tag{10.26}$$

If $k > 0$, we have for $j \in B_k - B_{k-1}$ that

$$\begin{aligned} j \in N - B_{k-1} &= e(g_k^i(v_1, \ldots, v_k)), \quad i \in N - B_k \quad \text{(by induction hypothesis)} \\ &= e(f_k^i(v_1, \ldots, v_k)), \quad i \in N - B_k \qquad \text{(by (10.18))}. \end{aligned}$$

Applying Lemma 10.9 and Lemma 10.23, we obtain

$$f_k^i(v_1, \ldots, v_k) = f_k^j(v_1, \ldots, v_k), k > 0, \ i \in N - B_k, j \in B_k - B_{k-1}. \tag{10.27}$$

From (10.18) and the induction hypothesis, we obtain

$$f_k^i(v_1, \ldots, v_k) = g_k^i(v_1, \ldots, v_k) = \gamma^{N-B_{k-1}}, \ i \in N - B_k, \ k > 0.$$

Combining this with (10.27), we obtain

$$f_k^j(v_1, \ldots, v_k) = \gamma^{N-B_{k-1}}, j \in B_k - B_{k-1}, k > 0,$$

i.e.,

$$f_k^i(v_1, \ldots, v_k) = \gamma^{N-B_{k-1}}, i \in B_k - B_{k-1}, k > 0.$$

From this and (10.25) we deduce, for $k > 0$, that

$$g_k^i(v_1, \ldots, v_k) \neq \gamma^{N-B_{k-1}}, i \in B_k - B_{k-1}; \tag{10.28}$$

the same result for $k = 0$ is given by (10.26). Hence (10.28) holds for $k \geq 0$.

We certainly have

$$g_k^i(v_1, \ldots, v_k) \neq \gamma^{N-B_{k-1}}, i \in B_{k-1},$$

for otherwise $i \notin e(g_k^i(v_1, \ldots, v_k))$, contradicting the membership of g_k^i in T^i. Combining this with the induction hypothesis and (10.28), we obtain

$$g_k^i(v_1, \ldots, v_k) \begin{cases} = \gamma^{N-B_{k-1}}, & i \in N - B_k \\ \neq \gamma^{N-B_{k-1}}, & i \in B_k. \end{cases}$$

It follows that

$$d^i(g_k(v_1, \ldots, v_k)) = N - B_k, \quad i \in N - B_k,$$

and in particular, since $N - B_{k+1} \subset N - B_k$, we have

$$d^i(g_k(v_1, \ldots, v_k)) = N - B_k, \quad i \in N - B_{k+1}. \tag{10.29}$$

Now for $i \in N - B_{k+1}$, we have by Definition 10.18 that

$$g_{k+1}^i(v_1, \ldots, v_{k+1}) = f_{k+1}^i(v_1, \ldots, v_{k+1}) = \gamma^{v_{k+1}^i | R^i}. \tag{10.30}$$

By reasoning similar to that used in the proof of (10.23), we obtain that $v^i_{k+1} \mid R^i$ occurs with positive coefficient in $d^i(g_k(v_1, \dots, v_k))$. Combining this with (10.29) and (10.30), we obtain that for $i \in N - B_{k+1}$

$$g^i_{k+1}(v_1, \dots, v_{k+1}) = \gamma^{N-B_k}.$$

This completes the induction and establishes (10.24).

From (10.24) and (10.20) we deduce that if v occurs with positive probability, then

$$g^i_k(v_1, \dots, v_k) = \gamma^{N-B(v)}, \quad i \in N - B(v), \quad k > k(v),$$

whence

$$c(g_k(v_1, \dots, v_k)) = (\gamma^{N-B(v)}, c^{B(v)}(g_k(v_1, \dots, v_k))), \quad k > k(v).$$

Setting

$$c^B(g_k(v_1, \dots, v_k)) = \gamma^B_k, B \subset N, \ k > k(v),$$

we obtain

$$c(g_k(v_1, \dots, v_k)) = (\gamma^{B(v)}_k, \gamma^{N-B(v)}), \quad k > k(v). \tag{10.31}$$

Lemma 10.32 *For each $m \geq 0$, there is a positive integer k, such that for all v, r, and $h_1, \dots, h_m$ for which*

$$r \geq k \tag{10.33}$$

and

$$h_j \in H(P), \ \ j = 1, \dots, m \tag{10.34}$$

and

$$S^{B^*}_r(v) \geq H^{B^*}(f) + \varepsilon^{B^*} \tag{10.35}$$

we have

$$r > m \tag{10.36}$$

and

$$\frac{1}{r}\left(\sum_{j=1}^{m} h^{B^*}_j + \sum_{j=m+1}^{r} H^{B^*}_j(v)\right) \geq H^{B^*}(f) + \frac{1}{3}\varepsilon^{B^*}. \tag{10.37}$$

Proof First of all, we may choose

$$k > m,$$

so that (10.36) is satisfied. Next, choose

$$k \geq \max_{i \in B^*} \frac{3m}{\varepsilon^i} \max_{p \in P} H^i(p). \tag{10.38}$$

We then have

$$\left| S_r^{B^*}(v) - \tfrac{1}{r}\left(\sum_{j=1}^{m} h_j^{B^*} + \sum_{j=m+1}^{r} H_j^{B^*}(v)\right)\right|$$

$$= \left| \tfrac{1}{r}\left(\sum_{j=1}^{r} H_j^{B^*}(v)\right) - \tfrac{1}{r}\left(\sum_{j=1}^{m} h_j^{B^*} + \sum_{j=m+1}^{r} H_j^{B^*}(v)\right)\right| \qquad \text{(by (6.11))}$$

$$= \tfrac{1}{r}\left|\sum_{j=1}^{m}(H_j^{B^*}(v) - h_j^{B^*})\right|$$

$$\le \tfrac{1}{k}\sum_{j=1}^{m}\left|H_j^{B^*}(v) - h_j^{B^*}\right| \qquad \text{(by (10.33))}$$

$$\le \tfrac{1}{k}\sum_{j=1}^{m} 2 \max_{p\in P} H^{B^*}(p) \qquad \text{(by (10.34))}$$

$$= \tfrac{2m}{k} \max_{p\in P} H^{B^*}(p)$$

$$\le \tfrac{2}{3}\varepsilon^{B^*} \qquad \text{(by (10.37))}$$

Combining this with (10.35), we obtain (10.37). This completes the proof.

Corollary 10.39 *For each $m \ge 0$, the following holds: For all sufficiently large k, we have that for all v, r, and $h_1, \dots, h_m$ for which (10.33), (10.34), and (10.35) hold, (10.36) and (10.37) follow.*

Proof Because if (10.32) is true for a given k, it is certainly true for all larger k.

Lemma 10.40 *For each $m \ge 0$ and $B \subset N$, including the null set, we have*

$$\lim_{k\to\infty} \operatorname{Prob}_g((S_r^{B^*}(v) \ge H^{B^*}(f) + \varepsilon^{B^*}$$

for some $r \ge k$) and $k(v) = m$ and $B(v) = B) = 0$.

Proof First let $B = \emptyset$. Then for the v we are considering (inside Prob_g), we have

$$B(v) = \emptyset. \qquad (10.41)$$

Since $B_k(v)$ is monotone increasing with k and always contained in $B(v)$, it follows from (10.41) that for all k,

$$B_k(v) = \emptyset.$$

Applying (10.24), we obtain for $k \ge 0$ and $i \in N$,

$$g_k^i(v_1, \dots, v_k) = \gamma^N. \qquad (10.42)$$

Let θ be a supergame c-strategy vector given for all v by

$$\theta_k^i(v_1, \dots, v_k) = \gamma^N, k \ge 0, i \in N. \qquad (10.43)$$

From (10.42) and (10.43), we obtain

$$\begin{aligned}
\lim_{k\to\infty} \quad & \text{Prob}_g((S_r^{B^*}(v) \geq H^{B^*}(f) + \varepsilon^{B^*} && \text{for some } r \geq k) \\
& && \text{and } k(v) = m \\
& && \text{and } B(v) = B). \\
& = \lim_{k\to\infty} \text{Prob}_\theta((S_r^{B^*}(v) \geq H^{B^*}(f) + \varepsilon^{B^*} && \text{for some } r \geq k) \\
& && \text{and } k(v) = m \\
& && \text{and } B(v) = B) \\
& \leq \lim_{k\to\infty} \text{Prob}_\theta(S_r^{B^*}(v) \geq H^{B^*}(f) + \varepsilon^{B^*} && \text{for some } r \geq k) \\
& \leq \lim_{k\to\infty} \text{Prob}_\theta(S_r^i(v) \geq H^i(f) + \varepsilon^i && \text{for some } r \geq k) \\
& = \lim_{k\to\infty} (1 - \text{Prob}_\theta(S_r^i(v) < H^i(f) + \varepsilon^i && \text{for all } r \geq k) \\
& = 1 - \lim_{k\to\infty} \text{Prob}_\theta(S_r^i(v) < H^i(f) + \varepsilon^i && \text{for all } r \geq k) \\
& \leq 1 - \lim_{k\to\infty} \text{Prob}_\theta(| \ S_r^i(v) - H^i(f) \ | < \varepsilon^i && \text{for all } r \geq k),
\end{aligned} \tag{10.44}$$

where i is an arbitrary member of B^*.

Now when v is distributed according to θ, we deduce from (10.43) and (6.10) that the $H_j^i(v)$ are independent random variables whose mean is $H^i(\gamma^N)$. Applying (6.11), (10.1), (10.7), and the strong law of large numbers, we obtain that the second term of the right side of (10.44) is 1, whence we deduce that the left side vanishes. This completes the proof in the case $B = \emptyset$.

Next, suppose $B \neq \emptyset$. From (10.14) we obtain that $B(v) \subset B^*$. Thus if $B \not\subset B^*$, then $\text{Prob}_g(B(v) = B) = 0$, and our lemma is already proved. Thus we may assume without loss of generality that

$$B \subset B^*. \tag{10.45}$$

For $j = 1, \ldots, m$, suppose the random variable w_j^B to be distributed according to an arbitrary but fixed member of $C^B(J_j^B)$, which we will call γ_j^B, while w_j^{N-B} is distributed according to γ^{N-B}. Set

$$h_j = H(w_j), \qquad j = 1, \ldots, m. \tag{10.46}$$

Applying Corollary 10.39, we obtain

$$\begin{aligned}\lim_{k\to\infty} \quad &\mathrm{Prob}_g((S_r^{B^*}(v) \geq H^{B^*}(f) + \varepsilon^{B^*} && \text{for some } r \geq k) \\ & && \text{and } k(v) = m \\ & && \text{and } B(v) = B) \\ &\leq \lim_{k\to\infty} \mathrm{Prob}\left(\left(\frac{1}{r}\left(\sum_{j=1}^{m} h_j^{B^*} + \sum_{j=m+1}^{r} H_j^{B^*}(v)\right) \right.\right. && \geq H^{B^*}(f) + \tfrac{1}{3}\varepsilon^{B^*} \\ & && \left.\text{for some } \ r \geq k\right) \\ & && \text{and } \ k(v) = m \\ & && \left.\text{and } \ B(v) = B\right),\end{aligned} \tag{10.47}$$

where $v_{m+1}, \ldots, v_r$ are distributed according to $x_{m+1}(g), \ldots, x_r(g)$, while h_j is given by (10.46). If for $j = k+1, \ldots, r$, we will let w_j be a random variable whose distribution is $(\gamma_j^B, \gamma^{N-B})$, where the w_j^B may be dependent on w_ℓ with $\ell < j$ but the w_j^{N-B} must be independent of all w_ℓ with $\ell \neq j$, then we may conclude from (10.31) and (6.10) that the right side of (10.47) is

$$\begin{aligned}&\leq \lim_{k\to\infty} \mathrm{Prob}\left(\frac{1}{r} \sum_{j=1}^{r} H^{B^*}(w_j) \geq H^{B^*}(f) + \frac{1}{3}\varepsilon^{B^*} \quad \text{for some } \ r \geq k\right) \\ &\leq \lim_{k\to\infty} \mathrm{Prob}\left(\frac{1}{r} \sum_{j=1}^{r} H^{B}(w_j) \geq H^{B}(f) + \frac{1}{3}\varepsilon^{B} \quad \text{for some } \ r \geq k\right)\end{aligned} \tag{10.48}$$

by (10.45). Now there is certainly a supergame c-strategy vector β for which

$$c(\beta_{j-1}(z_{j-1}(\beta))) = (\gamma_j^B, \gamma^{N-B}), \quad j \geq 1,$$

whence the right side of (10.48)

$$= \lim_{k\to\infty} \mathrm{Prob}_\beta(S_r^B(v) \geq H^B(f) + \frac{1}{3}\varepsilon^B \quad \text{for some} \quad r \geq k). \tag{10.49}$$

Applying (10.2), (10.7) and Lemma 8.3, we obtain that the right side of (10.49) vanishes. Combining this with (10.48) and (10.47), we obtain the lemma.

Lemma 10.50

$$\lim_{k\to\infty} \mathrm{Prob}_g(S_r^{B^*}(v) \geq H^{B^*}(f) + \varepsilon^{B^*} \text{ for some } r \geq k) = 0.$$

Proof Letting $\bigcup$ stand for mutually exclusive disjunction, we have

$$\begin{aligned}
1 &= \text{Prob}_g\left(\bigcup_{m=0}^{\infty} k(v) = m\right) \\
&= \sum_{m=0}^{\infty} \text{Prob}_g(k(v) = m) \\
&= \sum_{m=0}^{\infty} \text{Prob}_g(k(v) = m \text{ and } \bigcup_{B\subset N} (B(v) = B)) \\
&= \sum_{m=0}^{\infty} \sum_{B\subset N} \text{Prob}_g(k(v) = m \text{ and } B(v) = B).
\end{aligned} \tag{10.51}$$

Let $Q(v, k)$ stand for the expression

$$S_r^{B^*}(v) \geq H^{B^*}(f) + \varepsilon^{B^*} \text{ for some } r \geq k.$$

Then

$$\begin{aligned}
&\lim_{k\to\infty} \text{Prob}_g(Q(v, k)) \\
&\quad = \lim_{k\to\infty} \text{Prob}_g(Q(v, k) \text{ and } \bigcup_{m=0}^{\infty} (k(v) = m) \text{ and } \bigcup_{B\subset N} (B(v) = B)) \\
&\quad = \lim_{k\to\infty} \sum_{m=0}^{\infty} \sum_{B\subset N} \text{Prob}_g(Q(v, k) \text{ and } k(v) = m \text{ and } B(v) = B) \\
&\quad = \sum_{m=0}^{\infty} \sum_{B\subset N} \lim_{k\to\infty} \text{Prob}_g(Q(v, k) \text{ and } k(v) = m \text{ and } B(v) = B)
\end{aligned}$$

(since the series is dominated by the series on the right side of (10.51))

$$\begin{aligned}
&= \sum_{m=0}^{\infty} \sum_{B\subset N} 0 \qquad \text{(by Lemma 10.40)} \\
&= 0.
\end{aligned}$$

This completes the proof of Lemma 10.50.

Lemma 10.50 contradicts (10.12) and thus establishes the truth of Theorem 3.

Corollary 4 $H(A_c) = H(S_c) = H(\tilde{S}_c)$.

Proof Combine (7.4), Theorem 2 and Theorem 3.

11 Existence of acceptable points

All two-person games, zero-sum or not, have acceptable points. The proof, which is not difficult, will be given in a subsequent paper of this series devoted exclusively to the application of this theory to two-person games. Unfortunately, though, when we go beyond two-person games, we find games in which there are no acceptable points. It is instructive to examine an example of such a game.

G is a three-person, zero-sum game. The three persons form a community and at regular intervals hold elections for mayor and vice-mayor. The mayor draws a salary of 2; the vice-mayor draws a salary of 1; the remaining player pays the salaries.

Let h be an arbitrary payoff vector. Since G is zero-sum, not all the players can have a positive payoff. Without loss of generality, let

$$h^1 \le 0. \tag{11.1}$$

Now the minimum payoff to 1, even in pure strategies, is -3; hence since G is zero-sum, it follows that

$$h^2 + h^3 \le 3. \tag{11.2}$$

From (11.2) it follows that at least one of 2 and 3 gets at most $1\frac{1}{2}$; so we may write without loss of generality

$$h^2 \le 1\frac{1}{2}. \tag{11.3}$$

Now by agreeing to vote for 2 for mayor and for 1 for vice-mayor, 1 and 2 can both increase their payoffs; and there is nothing that 3 can do to prevent this.

Actually, the game G just described does not **deserve** an acceptable point. In a long sequence of plays of G, a steady state will never be reached; no sooner do we start settling down to one, than two of the players will see that it is relatively disadvantageous for them, and will certainly not be willing to agree to it for the remainder of the superplay. The game is **inherently unstable**. The same argument holds for all games not possessing acceptable points. Thus in seeking a steady state for the supergame, it would seem that we may restrict ourselves to games possessing acceptable points. These are called **stable** games.

12 The expected payoff

In order to see why the notion defined by (6.8) is essentially misleading, let us consider the three-person game G of perfect information which is given in extensive form in Figure 1.

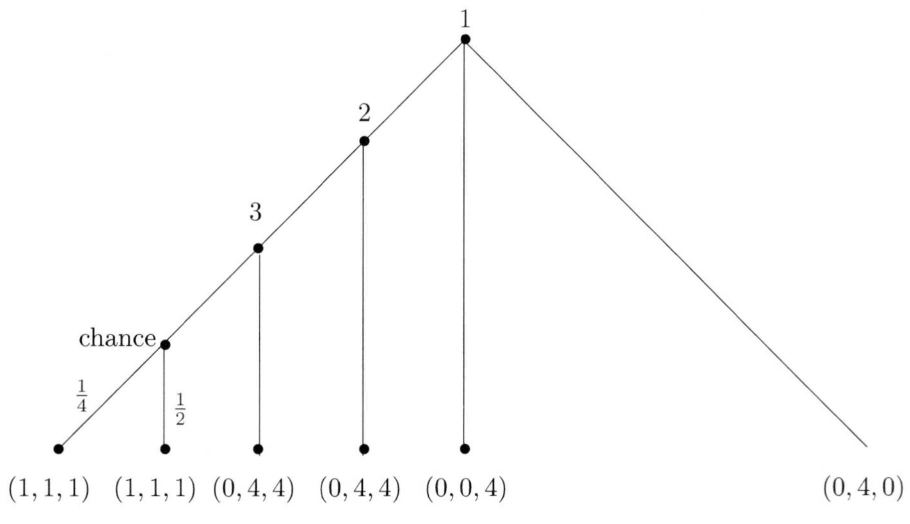

Figure 1 The payoff vector at each terminal gives the payoff to Players 1, 2, and 3 in that order

The payoff vector at each terminal gives the payoff to players $1, 2$, and 3 in that order.

Player 1 has three strategies, ℓ^1, c^1, and r^1 (left, center, and right). The other players have two strategies each; they will be denoted by ℓ^j and r^j, where $j = 0, 2$, and 3. Chance[8] is denoted by 0.

The point (ℓ^1, ℓ^2, ℓ^3), yielding a payoff of $(1, 1, 1)$, is acceptable. No player or group of players can obtain a higher payoff. However, if 1 plays the supergame c-strategy given by (10.3), there is a supergame c-strategy $(2, 3)$-vector that will yield an **expected** payoff of $(0, 2, 2)$ rather than $(1, 1, 1)$. This supergame c-strategy $(2, 3)$-vector, which we will call the "plan", may be described as follows: On the first play, both players make the "orthodox" choice, i.e., they concur with player 1 in choosing (ℓ^1, ℓ^2, ℓ^3). On all subsequent plays, we distinguish two cases:

1) If chance played ℓ^0 on the first play, then player 2 makes the "orthodox" choice, i.e., that given by (10.3), while player 3 plays r^3.
2) If chance played r^0 on the first play, then player 3 makes the "orthodox" choice, while player 2 plays r^2.

According to (10.3), player 1 will respond to this treachery with a wounded roar and proceed to wreak dire vengeance. In case 1 he will finish off player 3 by playing r^1; in case 2 he will play c^1, thus avenging himself on the treacherous player 2. In each case, he is convinced that the orthodox player has been loyal all along, and therefore will not feel too unhappy about his getting 4 while player 1 himself gets nothing. The expected payoff is $(0, 2, 2)$, an improvement for both 2 and 3. For 2 and 3, this sounds like a good set-up, at least on paper. Who wouldn't want an expected payoff of 2 instead of 1? Let's see how it works out in practice.

Imagine that chance has played ℓ^0 on the first play. For the second play, player 1, all unsuspecting, suggests (ℓ^1, ℓ^2, ℓ^3), and player 2 concurs, according to plan. Player 3 is now faced with the choice of going along with the plan and playing r^3, thus bringing down upon himself the eternal wrath of player 1, or of concurring in the choice of (ℓ^1, ℓ^2, ℓ^3), thus possibly making player 2 angry at him, but still assuring himself of a steady income of 1. It won't take him long to make his choice. He has nothing to gain and everything to lose by going along with the plan; he will get nothing but 0 for the rest of eternity for his pains, while player 2 enjoys a steady payoff of 4. Player 3 will not now be consoled by the fact that if chance had played the other way, **he** would have been getting 4 while player 2 would have been getting 0; in no sense does this constitute an incentive for him to follow the plan as things stand now. The point is that in a single play of a game, it may be worthwhile for each of the parties to plan to co-operate, even if it hurts, because in future plays, **he** may be the beneficiary. But in a superplay, there **are** no future superplays; **this is it**; by going along with the plan, player 3 will be ruining himself not just for one play, but for always. It just isn't worthwhile, and no player would do it. Similar remarks apply to player 2. The plan does yield a higher expected payoff, but will never be carried out.

It is possible to define strong equilibrium c-points using an appropriate formula involving expected payoffs in place of (7.2). If this is done, the first half of the main

[8]It may be objected that we have excluded chance from consideration in this paper. Actually, chance is used in this example only to provide, as conveniently and cheaply as possible, a random device on which players 2 and 3 will be able to peg their choices. We could eliminate chance at the cost of complicating the example considerably. In order to exhibit the principle involved as clearly as possible, we have retained chance; but it is important to remember that it does not form an essential part of the example.

theorem, the analogue of Theorem 2, remains true. However, the second half, the analogue of Theorem 3, fails. The game G that we have been discussing provides the counterexample. If we start out with $h = (1,1,1)$, then not only does the f given by (10.3) fail, as we have shown, but it can be shown that there is then no strong equilibrium c-point (in the sense of expected payoffs) whose payoff is h.

13 Enforceability of agreements

There are two kinds of agreements involved in a cooperative supergame. One is the formal agreement that is contracted prior to a given play by the members of a coalition B, who agree to play a certain correlated strategy B-vector on that play. This agreement expires immediately after the play in question is completed; its validity does not extend to any subsequent plays. The other is the informal "gentlemen's agreement" in which the players undertake informally to make certain choices or join certain coalitions on each of a number of plays, or even for the entire superplay; they do not form a part of the formal mathematical structure introduced in Section 6. These are the agreements that are involved in the supergame strategy vectors that we have used in our proofs. When in heuristic discussions above we referred to "orthodoxy" or "deviationism", it was the second kind of agreement to which we were referring. To distinguish between the two kinds of agreements, we will call the first formal, the second informal.

As mentioned in Section 3, together with the formal agreement there comes a formal enforcement mechanism. Put completely precisely, if for the kth play the players choose the vector $t \in T$, then each $i \in N$ **must** become a member of $d^i(t)$, and the coalition $d^i(t)$ **must** then play $(t^i)^{d^i(t)}$. If the reader wishes, he may consider that once the choice of t has been made, the remainder of the play is no longer in the hands of the individual players, but in the hands of an umpire. Such an extreme interpretation is unnecessary for practical purposes; we refer to it only to make clear the intuitive meaning of formal enforcement.

On the other hand, the informal agreement can be enforced in no such formal way. Violation of such an agreement can be "prevented" only by the threat or implied threat of retaliation on subsequent plays. The exact nature of this "pseudo-enforceability" is of course the crux of what we have been investigating in the foregoing sections.

In the light of these remarks, it is an interesting fact that if in the supergame we abandon formal enforceability for the formal agreements, the main theorem remains true. The possibility of retaliation is sufficient to enforce even the formal agreements, without any necessity for formal enforcement apparatus. We have not proved this statement in the foregoing; the structure introduced in Section 6 was constructed with formal enforceability of formal agreements in mind. A structure permitting the violation of formal agreements would be somewhat more complicated.[9] However, upon a little reflection, the reader will be able to convince himself of the truth of the statement, at least intuitively.

Of course, when considering just a single play rather than the supergame, as in the definition of acceptability, then there are only formal agreements and these are formally enforceable.

---

[9]Roughly speaking, each play would be described by a choice of a $t \in T$ as in Section 6, followed by a chance choice of a pure strategy vector in accordance with $c(t)$, followed by some move permitting the players to renege on their commitments if they wish.

References

[1] Nash, J. F., "Non-cooperative games," Annals of Mathematics **54** (1951), pp. 286-295.

[2] Feller, W., An Introduction to Probability Theory and its Applications, John Wiley and Sons, 1950.

[3] Kuhn, H. W., "Extensive games and the problems of information," Annals of Mathematics Study No. 28 (Princeton, 1953), pp. 193-216.

[4] Blackwell, D., "An analog of the minimax theorem for vector payoffs," Pacific J. Math. **6** (1956), pp. 1-8.

[5] Thompson, F. B., "Equivalence of games in extensive form," The RAND Corporation, Research Memorandum 759, 1952.

[6] Luce, R. D., and Raiffa, H., Games and Decisions, John Wiley and Sons, 1957.

[7] von Neumann, J., and Morgenstern, O. Theory of Games and Economic Behavior, Princeton, 1944, 3rd ed., 1953.

[8] Raiffa, H., "Arbitration schemes for generalized two-person games," Annals of Mathematics Study No. 28 (Princeton, 1953), pp. 361-387.

Non-cooperative Game Theory

Fields Institute Communications
Volume **23**, 1999

Young Measure Techniques for Existence of Cournot-Nash-Walras Equilibria

Erik J. Balder
Mathematical Institute
University of Utrecht
3508A Utrecht, the Netherlands
balder@math.ruu.nl

Abstract. A new approach to existence of Walrasian equilibria for competitive markets is presented. It is based on a quasi-variational inequality for Young measures, the solution of which, when purified by a standard purification argument, constitutes the Walrasian equilibrium for the competitive market. This continues and extends an earlier approach by the present author to the existence of Cournot-Nash equilibrium distributions and ordinary Nash equilibria for games with a measure space of players.

1 Introduction

In [**9, 12, 10**] a new, effective approach was given to existence of various Nash equilibria in game theory and mathematical economics. It is based upon a reformulation of the equilibrium problem as a *variational inequality* for Young measures, that is, transition probabilities which have been topologized in a way that originated in the work of L. C. Young. The Nash equilibrium then corresponds canonically to the Young measure solution of the variational inequality or to some purification thereof. Incidentally, the natural importance of Young measures for these problems comes from the fact that they form the canonical mixed strategy profiles. The success of the approach can be measured by the power of its central Nash equilibrium existence result [**10**, Theorem 2.1]: Partly on its own and partly by the use of purification, it was shown to have as its applications quite general results of the following kinds: existence of mixed Nash equilibria in ordinary games with a finite or countably infinite number of players, existence of Cournot-Nash equilibrium distributions in anonymous games with a measure space of types or players (thus considerably generalizing earlier such results by A. Mas-Colell [**21**], M. A. Khan and A. Rustichini [**19, 20**]), and existence of pure equilibria in games with a continuum of players, as pioneered by D. Schmeidler [**25, 24**].

In this paper an extension is presented of the above mentioned central existence result in [**10**]. This time, it is derived from a *quasi-variational inequality* for Young

1991 *Mathematics Subject Classification.* Primary 90C40, 14E20; Secondary 46E25, 20C20.

measures. It covers, *inter alia*, R. Aumann's seminal existence result in [**4**] for a Walrasian equilibrium in a competitive economy with a continuum of agents. The potential for further generalizations of this new approach to Walrasian equilibria will be the subject of further work. We mention that many other applications of Young measures, including several to other subjects in mathematical economics, can be found in [**11**].

2 Main result

Let $(T, \mathcal{T}, \mu)$ be a finite measure space of *players* (or *types* – the precise interpretation is irrelevant for the purposes of this paper). Let S be a metrizable Suslin space of *actions*, and let $D \subset T \times S$ be a $\mathcal{T} \times \mathcal{B}(S)$-measurable subset of $T \times S$. Recall that a metric space is *Suslin* if it is the continuous image of a Polish (i.e., metrizable, separable and complete) space [**15**, III]; $\mathcal{B}(S)$ stands for the Borel σ-algebra on S; by $\mathcal{D}$ we denote the σ-algebra $D \cap (\mathcal{T} \times \mathcal{B}(S))$, induced on D by the product σ-algebra $\mathcal{T} \times \mathcal{B}(S)$.

Assumption A.1 *$D_t := \{s \in S : (t, s) \in D\}$ is compact for every $t \in T$.*

Let P, the *abstract set of prices*, be a subset of a Hausdorff locally convex space and let $V : D \times P \to [-\infty, +\infty)$ be an abstract $\mathcal{D} \times \mathcal{B}(P)$-measurable *budget function*: Given the price p, each player $t \in T$ can only choose those actions $s \in D_t$ which obey the inequality $V(t, s, p) \leq 0$. The corresponding budget set $\{s \in D_t : V(t, s, p) \leq 0\}$ will be denoted by $B_{t,p}$.

Assumption A.2 *P is a compact, metrizable and convex set.*

Assumption A.3

(i) *$V_t := V(t, \cdot, \cdot)$ is continuous on $D_t \times P$ for every $t \in T$.*
(ii) *$V(t, s, \cdot)$ is concave on P for every $(t, s) \in D$.*
(iii) *There exists $\phi : T \to \mathbf{R}$, μ-integrable, with $\sup_{s \in D_t, p \in P} |V(t, s, p)| \leq \phi(t)$ for every $t \in T$.*

Assumption A.4 *$B_{t,p} := \{s \in D_t : V(t, s, p) \leq 0\} = \mathrm{cl}\,\{s \in D_t : V(t, s, p) < 0\}$ for every $t \in T$, $p \in P$.*

A *Young measure* (alias transition probability) from T into S is a mapping $\delta : t \mapsto \delta_t$ from T into the set $M_1^+(S)$ of all probability measures on $(S, \mathcal{B}(S))$, which is such that $t \mapsto \delta_t(B)$ is $\mathcal{T}$-measurable for every $B \in \mathcal{B}(S)$. Basic properties of such transition probabilities can be found in [**1**, 2.6] and [**23**, III.2]. We denote the set of all such Young measures by $\mathcal{R}$; the set consisting of all $\delta \in \mathcal{R}$ such that $\delta_t(D_t) = 1$ for every $t \in T$ is denoted by $\mathcal{R}_0$. Observe that each $\delta \in \mathcal{R}_0$ can be considered as a mixed action profile: if the profile $\delta \in \mathcal{R}_0$ is realized, then every player $t \in T$ chooses the probability measure δ_t as his/her mixed action on D_t. An equivalence relationship on $\mathcal{R}$ is obtained by defining any δ and δ' in $\mathcal{R}$ to be *a.e.-equivalent* if $\delta(t) = \delta'(t)$ for μ-almost every $t \in T$.

Following [**10**], let Y be a metrizable Suslin space of *profile abstracts* and let $e_t : \mathcal{R} \to Y$ be player t's *mixed externality*: she/he can only discern the mixed action profile δ in the form of the abstract $e_t(\delta)$.[1] Namely

Assumption A.5

(i) *$t \mapsto e_t(\delta)$ is $\mathcal{T}$-measurable for every $\delta \in \mathcal{R}$.*

[1] After this paper was written, it was shown in [**13**] that the present framework can be simplified considerably, with $Y := \mathcal{R}$ and $e_t :=$ identity on $\mathcal{R}$. See [**13**] for a detailed discussion.

(ii) *For every* $t \in T$, $e_t : \mathcal{R} \to Y$ *is continuous with respect to the narrow topology on* $\mathcal{R}$.
(iii) $e_t(\delta) = e_t(\delta')$ *for every* $t \in T$ *and every a.e.-equivalent pair* δ *and* δ' *in* $\mathcal{R}$.

Recall from [**6, 7**] that the *narrow* topology on $\mathcal{R}$ is the weakest topology for which all functionals $\delta \mapsto \int_B [\int_S c(x)\delta_t(ds)]\mu(dt)$, $B \in \mathcal{T}$, $c \in \mathcal{C}_b(S)$, are continuous. Here $\mathcal{C}_b(S)$ stands for the set of continuous bounded real-valued functions on S.

Let $U : D \times P \times Y \to \mathbf{R}$ be a $\mathcal{D} \times \mathcal{B}(P \times Y)$-measurable *utility function*: If the price $p \in P$ and the mixed action profile $\delta \in \mathcal{R}_0$ are realized, then the expected payoff for each player $t \in T$ is given by

$$\Pi_t(\delta) := \int_{D_t} U(t, s, p, e_t(\delta))\delta_t(ds),$$

assuming that the integral exists.

Assumption A.6 $U_t := U(t, \cdot, \cdot, \cdot)$ *is continuous on* $D_t \times P \times Y$ *for every* $t \in T$.

Theorem 2.1 (main existence result) *Under Assumptions A.1–A.6 there exist a mixed action profile* $\delta^* \in \mathcal{R}_0$ *and a price* $p^* \in P$ *such that the following hold:*
(a) $\delta_t^*(\mathrm{argmax}_{s \in B_{t,p^*}} U_t(s, p^*, e_t(\delta^*))) = 1$ *for* μ*-a.e.* $t \in T$,
(b) $\sup_{p \in P} \int_T [\int_{D_t} V_t(s, p)\delta_t^*(ds)]\mu(dt) \leq 0$.

Remark 2.2 *Theorem 2.1 generalizes the principal Theorem 2.1 of* [**10**], *where the abstract price set P and budget function V do not appear. There are two small technical differences: (1) Assumption A.6 is slightly stronger than the corresponding assumptions 2.4 and 2.6 in* [**10**]; *however, as underlined by Remark 3.3, Assumption A.6 can be relaxed in the same way when P and V are absent. (2) Assumption A.5(iii) does not occur in* [**10**]; *this is indicative of a small difference in the proofs: Ky Fan's inequality* [**3**], *used in* [**9, 10**], *holds in a non-Hausdorff setting* [**16**, pp. 500-501], *but at present it is not known to the author if Kakutani's theorem, used in the proof of Theorem 3.1, can be extended to a non-Hausdorff setting as well.*[2]

3 Proof of the main result

The proof of the main result turns around the following existence result for quasi-variational inequalities [**2, 5, 22**].

Theorem 3.1 *Let C be a compact convex and nonempty subset of a Hausdorff locally convex topological vector space. Let* $b : C \times C \to [-\infty, +\infty]$ *be such that*

$$b \text{ is upper semicontinuous,} \tag{3.1}$$

$$b(x, \cdot) \text{ is quasiconcave for each } x \in C. \tag{3.2}$$

Also, let $F : C \to 2^C$ *be a multifunction with convex and nonempty values, such that*

$$F \text{ has a closed graph.} \tag{3.3}$$

$$a : x \mapsto \sup_{x' \in F(x)} b(x, x') \text{ is lower semicontinuous.} \tag{3.4}$$

[2]By the previous footnote, Assumption A.5(iii) can indeed be removed. Moreover, after this paper was written, a non-Hausdorff version of Theorem 3.1 was obtained by the author in [**13**].

Then there exists $x^* \in C$ *such that* $x^* \in F(x^*)$ *and* $b(x^*, x^*) \geq b(x^*, x)$ *for all* $x \in F(x^*)$.

Proof Since C is compact, the upper semicontinuity of a multifunction is obviously equivalent to the closedness of its graph. For $x \in C$ define $G(x)$ to be the closed convex and nonempty set of all $x' \in F(x)$ such that $b(x, x') \geq a(x)$. It is clear from the assumptions that G has a closed graph; hence, $G : C \to 2^C$ is upper semicontinuous. An application of Kakutani's theorem [**5**] then gives the existence of $x^* \in C$ which is a fixed point of G, after which the desired properties of x^* follow from the definitions of G and a. □

It would appear that the following result does not figure exactly as stated in the literature. Of course, it is well-known in less tailored formats; cf. [**18**, Proposition 11, p. 28].

Proposition 3.2 *Let* Q *and* Z *be metric spaces and let* $u, v : Q \times Z \to \mathbf{R}$ *be such that*

$$u \textit{ is lower semicontinuous,}$$

$$v(q, \cdot) \textit{ is upper semicontinuous on } Z \textit{ for every } q \in Q,$$

$$\{z \in Z : v(q, z) \leq 0\} = \text{cl}\,\{z \in Z : v(q, z) < 0\} \textit{ for every } q \in Q.$$

Then $w : Q \to (-\infty, +\infty]$, *defined by*

$$w(q) := \sup_{z \in Z, v(q,z) \leq 0} u(q, z),$$

is lower semicontinuous on Q.

Proof Let $q_k \to q_0$ in Q; we shall demonstrate $\alpha := \liminf_k w(q_k) \geq w(q_0)$. First, we can assume that $w(q_k) \to \alpha$, rather than concentrating our proof from now on to a suitable subsequence. Let $r < w(q_0)$ be arbitrary; there exists $z_0 \in Z$ such that $u(q_0, z_0) > r$ and $v(q_0, z_0) \leq 0$. By the last of the conditions, the latter inequality implies that there exists a sequence (z_k) in Z such that $v(q_0, z_k) < 0$ for all k and $z_k \to z_0$. By the second condition, this implies that for every $j \in \mathbf{N}$ there exists $k_j \in \mathbf{N}$ such that $v(q_{k_j}, z_j) < 0$, and $k_{j+1} > k_j$ can always be arranged. Now $\alpha = \lim_j w(q_{k_j})$ by the first step of the proof, so it follows that $\alpha \geq \liminf_j u(q_{k_j}, x_j) \geq u(q_0, x_0) > r$, using the given lower semicontinuity of u. □

Remark 3.3 *Obviously, if* $v \equiv 0$ *then, instead of the three conditions of Proposition 3.2, it is enough to require that* $u(\cdot, z)$ *be lower semicontinuous on* Q *for every* $z \in Z$.

In the notation which anticipates the application of Theorem 3.1, we set $C := \mathcal{R}_0 \times P$ and define $b : C \times C \to \mathbf{R}$ by

$$b(\delta, p; \eta, q) := b_1(\delta, p, \eta) + b_2(\delta, q).$$

Here

$$b_1(\delta, p, \eta) := \int_T \left[\int_{D_t} \arctan U_t(s, p, e_t(\delta))\eta_t(ds) \right] \mu(dt),$$

$$b_2(\delta, q) := \int_T \left[\int_{D_t} V_t(s, q)\delta_t(ds) \right] \mu(dt).$$

Observe that b_1 and b_2 are well-defined by our measurability conditions for D, U, e, V, by the integrable boundedness of V (Assumption A.3(iii)) and $\tilde{U}$ (hence the arctangent transformation) and by basic properties of transition probabilities [**1**, 2.6], [**23**, III.2].

Lemma 3.4 *$b : C \times C \to \mathbf{R}$ meets conditions (3.1) and (3.2).*

Proof To verify (3.1), we observe that, by virtue of Assumptions A.3(i), (iii), A.5 and A.6, the upper semicontinuity (here actually continuity) of b_1 and b_2 follows from Theorems 2.2, 2.5 in [**7**],[3] completely similar to the proof of [**9**, Lemma 5(i)]. As for (3.2), observe that $b(\delta, p; \cdot, \cdot)$ is obviously concave for each $(\delta, p) \in C$ (use Assumption A.3(ii)). □

Next, we define $F : C \to 2^C$ by $F(\delta, p) := F'(p) \times P$, where

$$F'(p) := \{\eta \in \mathcal{R} : \eta_t(B_{t,p}) = 1 \text{ for } \mu\text{-a.e. } t \in T\}.$$

Lemma 3.5 *$F : C \to 2^C$ has convex nonempty values and meets (3.3).*

Proof Let $p \in P$ be arbitrary. Convexity of the set $F'(p)$ is obvious. To verify nonemptiness of $F'(p)$, let (d_i) be a Castaing representation of the multifunction $t \to D_t$; i.e., all $d_i : T \to S$ are measurable functions and $D_t = \text{cl}\,\{d_i(t) : i \in \mathbf{N}\}$ for μ-a.e. $t \in T$. Then by Assumptions A.3, A.4 we have for any $s \in S$ that $s \in B_{t,p}$ if and only if for every $k \in \mathbf{N}$ there exists $i \in \mathbf{N}$ such that s and $d_i(t)$ have a distance of at most k^{-1} and $V(t, d_i(t), p) < 0$. Thus, it is easily seen that the graph of $t \mapsto B_{t,p}$ is $\mathcal{T} \times \mathcal{B}(S)$-measurable, so by the von Neuman-Aumann measurable selection theorem [**14**, III.2] there exists a measurable function $f : T \to S$ for which $f(t) \in B_{t,p}$ for μ-a.e. $t \in T$. Consequently, the Dirac Young measure ϵ_f belongs to $F'(p)$. Next, we show that the graph of F' is closed. For $p \in P$, $\eta \in \mathcal{R}_0$ we obviously have $\eta \in F'(p)$ if and only if $I_A(\eta, p) \leq 0$ for all $A \in \mathcal{T}$. Here

$$I_A(\eta, p) := \int_A \left[\int_{D_t} V_t^+(s, p)\eta_t(ds) \right] \mu(dt)$$

with $V^+ := \max(V, 0)$. By Assumption A.3(i), $1_A(t)V_t^+$ is lower semicontinuous (actually continuous) on $D_t \times P$ for every t, so it follows by Theorems 2.2, 2.5 of [**7**], completely similar to the proof of upper semicontinuity in Lemma 3.4, that each functional I_A is lower semicontinuous (actually continuous) on $\mathcal{R} \times P$. So the graph of the multifunction F' is closed, which implies that the graph of F is also closed. □

Let $a : C \to \mathbf{R}$ be given by

$$a(\delta, p) := \sup_{(\eta,q) \in F(\delta,p)} b(\delta, p; \eta, q).$$

Lemma 3.6 *a can be decomposed as $a(\delta, p) = a_1(\delta, p) + a_2(\delta)$, with*

$$a_1(\delta, p) := \int_T \int_{D_t} \sup_{s \in B_{t,p}} \tilde{U}_t(s, p, e_t(\delta))\mu(dt), \;\; a_2(\delta) := \sup_{q \in P} b_2(\delta, q).$$

Proof By additive separation it is evident that $a(\delta, p) = \sup_{\eta \in \mathcal{R}_0} b_1(\delta, p, \eta) + \sup_{q \in P} b_2(\delta, q)$. This explains the second term $a_2(\delta)$. The first term $a_1(\delta, p)$ follows by a measurable selection argument, completely similar to the proof of the first claim in the proof of Proposition 3 in [**9**, p. 349]. □

[3] All results in [**7**] are stated for S a metrizable Lusin space. By the trick introduced in [**8**], they extend to – and beyond – the present case with S metrizable Suslin.

Lemma 3.7 *a meets (3.4).*

Proof We use Lemma 3.6. Define $\tilde{w} : T \times P \times Y \to \mathbf{R}$ by $\tilde{w}(t,p,y) := \sup_{s \in B_{t,p}} \tilde{U}_t(s,p,y)$. First, we claim that $\tilde{w}$ may be supposed $\mathcal{T} \times \mathcal{B}(P \times Y)$-measurable. Indeed, let (d_i) be the Castaing representation of $t \to D_t$ that was used in the proof of Lemma 3.5. By the measurability and continuity conditions for U and V we have

$$\tilde{w}(t,p,y) = \sup_i \{\tilde{U}(t,d_i(t),p,y) : V(t,d_i(t),p) \le 0\}$$

for all (t,p,y) in the product of a set of μ-measure 1 and $P \times Y$, and the expression on the right is clearly $\mathcal{T} \times \mathcal{B}(P \times Y)$-measurable.

Secondly, $u := \tilde{w}(t,\cdot,\cdot)$ is lower semicontinuous on $P \times Y$ for each t; with $v := V_t$, $Q := P \times Y$, and $Z := D_t$ this is a direct application of Proposition 3.2. Thus, we may once more apply Theorems 2.2 and 2.5 of [**7**], and this gives the desired lower semicontinuity of a_1 on C, in view of Lemma 3.6. Finally, in the proof of Lemma 3.4 continuity of b_2 on C was established, so lower semicontinuity of a_2 on $\mathcal{R}_0$ follows immediately. The decomposition of a in Lemma 3.6 then shows that a is lower semicontinuous on C. □

Proof (of Theorem 2.1). We apply Theorem 3.1 in a quotient setting to ensure the Hausdorff property. This quotient corresponds to the notion of a.e.-equivalence that was introduced on $\mathcal{R}$; see [**7**] and the proof of Proposition 4 in [**9**] for some details (see also our comments concerning the use of Assumption A.5(iii) in Remark 2.2). Above, we checked the validity of the conditions (3.1)–(3.4), and the remaining conditions of Theorem 3.1 are easily verified: In view of Assumption A.1, (the quotient of) $\mathcal{R}_0$ is compact for (the quotient of) the narrow topology by Theorem 2.3 in [**7**]; see also [**8**]. Hence, $C := \mathcal{R}_0 \times P$ is compact by Assumption A.2. Further, $\mathcal{R}_0$ and its quotient are obviously convex, and the nonemptiness of $\mathcal{R}_0$ has already been tacitly asserted in the proof of Lemma 3.5.

Therefore, we may apply Theorem 3.1, which gives the existence of a pair (δ^*, p^*) in C such that

$$b_1(\delta^*, p^*, \delta^*) = a_1(\delta^*, p^*) \text{ and } b_2(\delta^*, p^*) = a_2(\delta^*),$$

where Lemma 3.6 has already been used. The first identity implies immediately that for μ-a.e. $t \in T$

$$\int_{D_t} \tilde{U}_t(s, p^*, e_t(\delta^*))\delta^*(t)(ds) = \int_{D_t} \sup_{s \in B_{t,p^*}} \tilde{U}_t(s, p^*, e_t(\delta^*)),$$

giving (a) in Theorem 2.1 by monotonicity of the arctangent transformation. The second identity follows from

$$b_2(\delta^*, p^*) := \int_T \left[\int_{D_t} V_t(s,p^*)\delta^*(t)(ds) \right] \mu(dt) \le 0,$$

since $\delta^*(t)(B_{t,p^*}) = 1$ for μ-a.e. $t \in T$. □

4 Some applications

As mentioned, Theorem 2.1 essentially generalizes the main result of [**10**], so a number of applications can of course be found in [**10**]. As for new applications of Theorem 2.1, we limit ourselves in this paper to establishing the connection with the seminal existence result of R. J. Aumann [**4**]. Forthcoming work by the author will present a more complete account, with new, more refined applications, including those which mix aspects of differential information and equilibrium distributions [**9**] with the competitive market features of this paper.

To derive Aumann's result, it is enough to demonstrate that Theorem 2.1 implies the main existence theorem of J. Greenberg et al. [**17**, p. 33]; indeed, by a standard truncation argument they show that the former result follows from theirs [**17**, Remark 1].

Let $X_1 : T \to 2^{\mathbf{R}^l_+}$ and $X_2 : T \to 2^{\mathbf{R}^l}$ be multifunctions with measurable graphs, the *consumption* and *production* correspondence of the model in [**17**]. Let Δ be the unit simplex in $\mathbf{R}^l$ and let U_c : graph $X_1 \times \Delta \to \mathbf{R}$ be a measurable *(consumption) utility function.* Also, let $\omega : T \to \mathbf{R}^l$ be a measurable selector of X_1, the *initial endowment profile.*

Assumption B.1

(i) *$X_1(t)$ and $X_2(t)$ are compact for every $t \in T$.*
(ii) *there exists integrable $\phi : T \to \mathbf{R}$ with $\sup_{x_1 \in X_1(t)} |x_1| \le \phi(t)$ and $\sup_{x_2 \in X_2(t)} |x_2| \le \phi(t)$.*

Assumption B.2

(i) *$\omega(t) \in \operatorname{int} X_1(t)$ for every $t \in T$.*
(ii) *$X_1(t)$ is convex for every $t \in T$.*
(iii) *$0 \in X_2(t)$ for every $t \in T$.*

Assumption B.3 *$U_c(t, \cdot, \cdot)$ is continuous on $X_1(t) \times \Delta$ for every $t \in T$.*

Corollary 4.1 *Suppose that the measure space $(T, \mathcal{T}, \mu)$ is nonatomic. Then under Assumptions B.1–B.3 there exist integrable selectors f^* of X_1 and g^* of X_2 (that is, an allocation and a production plan) and an equilibrium price system $p^* \in \Delta$ such that the following hold, with $X_{t,p^*} := \{x_1 \in X_1(t) : x_1 \cdot p^* \le \omega(t) \cdot p^* + \sup_{x_2 \in X_2(t)} x_2 \cdot p^*\}$:*
(a) $f^*(t) \in \operatorname{argmax}_{x_1 \in X_{t,p^*}} U_c(t, x_1, p^*)$ and $g^*(t) \in \operatorname{argmax}_{x_2 \in X_2(t)} x_2 \cdot p^*$,
(b) $\int_T f^*(t)\mu(dt) \le \int_T \omega(t)\mu(dt) + \int_T g^*(t)\mu(dt)$.

Proof In the model introduced in Section 2 we use the following substitutions: $S := \mathbf{R}^l \times \mathbf{R}^l$, $D_t := X_1(t) \times X_2(t)$, $P := \Delta$, Y and e trivial (or absent), $V(t, x_1, x_2, p) := (x_1 - \omega(t) - x_2) \cdot p$, and $U(t, x_1, x_2, p) := U_c(t, x_1, p) + x_2 \cdot p$. The assumptions of Theorem 2.1 are easily seen to hold; observe in particular that Assumption A.4 holds by a well-known property of closed convex sets with nonempty interior. Thus, there exist a Young measure $\delta^* : T \to M_1^+(S)$ and $p^* \in \Delta$ such that

$$\delta^*(t)(\operatorname{argmax}_{(x_1,x_2) \in B_{t,p^*}} [U_c(t, x_1, p^*) + x_2 \cdot p^*]) = 1 \text{ for } \mu\text{-a.e. } t \in T$$

and

$$\sup_{p \in \Delta} \int_T \left[\int_{X_1(t) \times X_2(t)} (x_1 \cdot p - \omega(t) \cdot p - x_2 \cdot p) \delta^*(t)(d(x_1, x_2)) \mu(dt) \right] \le 0.$$

Here $B_{t,p^*} = \{(x_1, x_2) \in X_1(t) \times X_2(t) : (x_1 - \omega(t) - x_2) \cdot p^* \le 0\}$. A moment's reflection shows that the maximum of $U_c(t, x_1, p^*) + x_2 \cdot p^*$ over all (x_1, x_2) in B_{t,p^*} is attained by $(\bar{x}_1, \bar{x}_2) \in B_{t,p^*}$ if and only if $\bar{x}_1 \in M_1(t) := \operatorname{argmax}_{x_1 \in X_{t,p^*}} U_c(t, x_1, p^*)$ and $\bar{x}_2 \in M_2(t) := \operatorname{argmax}_{x_2 \in X_2(t)} x_2 \cdot p^*$. Here X_{t,p^*} is as defined in the statement of the result. Thus, it follows that the first property of (δ^*, p^*) can be restated as

$$\delta^*(t)(M_1(t) \times M_2(t)) = 1 \text{ for } \mu\text{-a.e. } t \in T.$$

Observe that measurability of the graphs of M_1 and M_2 has already been demonstrated in the proof of Lemma 3.7, in view of the fact that the set

$$\operatorname{argmax}_{(x_1,x_2)\in B_{t,p^*}}[U_c(t,x_1,p^*)+x_2\cdot p^*]$$

was shown to be equal to $M_1(t)\times M_2(t)$ above. The second property of (δ^*,p^*) implies

$$\int_T\left[\int_{X_1(t)\times X_2(t)}(x_1-\omega(t)-x_2)\delta^*(t)(d(x_1,x_2))\mu(dt)\right]\leq 0$$

(consider the unit vectors in Δ), which can be rewritten as $z_1-\int_T\omega(t)\mu(dt)-z_2\leq 0$, with $z_i:=\int_T[\int_{X_1(t)\times X_2(t)}x_i\delta^*(t)(d(x_1,x_2))]\mu(dt)$, $i=1,2$.

By a purification argument based upon Lyapunov's theorem (see [**10**, Section 3.4] for details and see [**6**] and [**11**] for additional details), there exist measurable – hence integrable by Assumption B.1(ii) – μ-a.e.-selectors f^* of M_1 and g^* of M_2 (whence (a) follows) such that $z_1=\int_T f^*\,d\mu$ and $z_2=\int_T g^*\,d\mu$ (whence (b)). □

References

[1] Ash, R. B. [1972], *Real Analysis and Probability,* Academic Press, New York.

[2] Aubin, J.-P. [1993], *Optima and Equilibria,* Springer-Verlag, Berlin.

[3] Aubin, J.-P. and Ekeland, I. [1984], *Applied Nonlinear Analysis,* Wiley, New York.

[4] Aumann, R. J. [1966], *Existence of competitive equilibria in markets with a continuum of traders,* Econometrica **34**, 1-17.

[5] Baiocchi, C. and Capelo, A. [1984], *Variational and Quasivariational Inequalities,* Wiley, New York.

[6] Balder, E. J. [1984], *A general approach to lower semicontinuity and lower closure in optimal control theory,* SIAM J. Control Optim. **22**, 570-598.

[7] Balder, E. J. [1988], *Generalized equilibrium results for games with incomplete information,* Math. Oper. Res. **13**, 265-276.

[8] Balder, E. J. [1989], *On Prohorov's theorem for transition probabilities,* Sém. Analyse Convexe **19**, 9.1-9.11.

[9] Balder, E. J. [1991], *On Cournot-Nash equilibrium distributions for games with differential information and discontinuous payoffs,* Econ. Theory **1**, 339-354.

[10] Balder, E. J. [1995], *A unifying approach to existence of Nash equilibria,* Intern. J. Game Theory **24**, 79-94.

[11] Balder, E. J. [1995], *Lectures on Young Measures,* Cahiers du Centre de Recherche de Mathématiques de la Décision (CEREMADE) **9517**, Université Paris-Dauphine, Paris.

[12] Balder, E. J. [1996], *Comments on the existence of equilibrium distributions,* J. Math. Econom. **25**, 307-323.

[13] Balder, E. J. [1997], *On equilibria in continuum games,* Preprint no. 993, Department of Mathematics, University of Utrecht, J. Math. Econom., to appear.

[14] Castaing, C. and Valadier, M. [1977], *Convex Analysis and Measurable Multifunctions,* Lecture Notes in Math. **580**, Springer-Verlag, Berlin.

[15] Dellacherie, C. and Meyer, P.-A. [1975], *Probabilités et Potentiel.* Hermann, Paris (English translation: North-Holland, Amsterdam).

[16] Ding, X. P. and Tan, K.-K. [1990], *Generalized variational inequalities and generalized quasi-variational inequalities,* J. Math. Anal. Appl. **148**, 497-508.

[17] Greenberg, J., Shitovitz, B. and Wieczorek A. [1979], *Existence of equilibria in atomless production economies with price dependent preferences,* J. Math. Econom. **6**, 31-41.

[18] Hildenbrand, W. [1974], *Core and Equilibria of a Large Economy.* Princeton University Press.

[19] Khan, M. A. and Rustichini, A. [1991], *Cournot-Nash equilibrium distributions for games with differential information,* Pitman Research Notes **252** (M.A. Théra and J.-B. Baillon, eds.), Longman, Essex, 245-259.

[20] Khan, M. A. and Rustichini, A. [1993], *Cournot-Nash equilibrium distributions for games with uncertainty and imperfect information,* J. Math. Econom. **22**, 35-59.

[21] Mas-Colell, A. [1984], *On a theorem of Schmeidler,* J. Math. Econom.**13**, 201-206.

[22] Mosco, U. [1975], *Implicit variational problems and quasi-variational inequalities,* Lecture Notes in Math. **543** (J. P. Gossez et al., eds.), Springer-Verlag, Berlin.

[23] Neveu, J. [1965], *Mathematical Foundations of the Calculus of Probability,* Holden-Day, San Fransisco.

[24] Rath, K. P. [1992], *A direct proof of the existence of pure strategy equilibria in games with a continuum of players,* Econ. Theory **2**, 427-433.

[25] Schmeidler, D. [1973], *Equilibrium points of non-atomic games,* J. Statist. Phys. **7**, 295-300.

Fields Institute Communications
Volume **23**, 1999

Modeling Free Choice in Games

Steven J. Brams
Department of Politics
New York University
New York, NY 10003 USA
`brams@is2.nyu.edu`

Abstract. The free and independent choices that game theory assumes are, on occasion, constrained by the game form postulated and its associated rules of play. Thus if a game in normal (or strategic) form has a unique Nash equilibrium in pure strategies, the standard theory presumes rational players will choose it. But changing the rules of play to allow players to think ahead and move and countermove from such an equilibrium may lead to a different equilibrium outcome, according to the "theory of moves" (TOM).

It turns out that of the 78 distinct 2×2 strict ordinal games, there are exactly six, subsumed by a generic "Freedom Game," in which the more far-sighted rationality calculations of TOM induce a player with a dominant strategy to depart from the unique pure-strategy Nash equilibrium associated with it and switch to its dominated strategy, which in turn induces the other player also to switch its strategy. The outcome that results is a "nonmyopic equilibrium" (NME) – based on a set of rules that makes the order of play endogenous and assumes players will not cycle – that benefits one and sometimes both players.

A second set of new rules, which assumes that cycling is possible, affords the players still more freedom of choice. However, this freedom may be limited by the exercise of "moving power," which is the ability of one player to outlast the other player in the move-countermove process. Both sets of rules not only help one understand real-life choices players have actually made – some of which proved surprising – but also facilitiate modeling dynamic processes that, by expanding the possible moves of players, enhance their freedom of choice.

1 Introduction

A tenet of noncooperative game theory is that players are free to make any choices they like, as long as they are permitted by the postulated game form and its associated rules of play. In an extensive-form game defined by a game tree,

1991 *Mathematics Subject Classification.* Primary 90D40; Secondary 90D10.
Thank you to the C. V. Starr Center for Applied Economics at New York University for its support.

for example, a player can choose any branch at a node where it is his or her turn to move next. In a normal (or strategic)-form game defined by a payoff matrix – wherein strategies describe plans of action that may be contingent on the choices of other players – a player can choose any strategy.

This freedom of choice is presumed channeled, however, by the rational calculations of the players. That is, while free to make any choices they please, players are, nevertheless, assumed to make moves, or choose strategies, that lead to better rather than worse outcomes. A major question in game theory is how to define "better" and "worse."

I sidestep this issue by analyzing games in which better and worse are, on the surface, uncontroversial. These are two-person normal-form games of complete information in which one player has a dominant strategy. The other player, anticipating the first player's choice of this strategy, has a best response, which leads to a Nash equilibrium.

It turns out that in some of these games, the player with the dominant strategy can do better by choosing his or her dominated strategy, provided he or she can do so before the other player has the opportunity to move or choose a strategy. Despite the dependence of the outcome on the order of play in such games, most game-theoretic models do not make the order of play endogenous – that is, the subject of rational calculation by the players.[1] Rather, the rules generally prescribe that one player moves first, or the players move simultaneously.

Chicken is perhaps the best-known example of a game in which the order of play matters. If the players make simultaneous choices, as is assumed in the usual 2×2 representation of this game, they may suffer their mutually worst outcome if neither "chickens out" by not swerving. But if one player (say, a male) can convince the other player (a female) that he will not swerve, he will obtain his best payoff, given that she responds optimally by swerving. Her optimal response, however, garners her only her next-worst payoff, so there is a disadvantage to a player's moving second, rather than first, in Chicken.[2] In this paper, I begin by assuming that players make simultaneous choices in a 2×2 "generic game," which I call the Freedom Game.[3] This game is *generic* in the sense that it subsumes several 2×2 strict ordinal games, in which each of the two players has two strategies and is able strictly to rank the resulting four outcomes from best to worst (i.e., there are no ties).

The Freedom Game is defined by two conditions:

---

[1]There are some recent exceptions, including [**13, 14**], [**22**], [**24**], and [**1**]. Typically, these models allow a player in the preplay phase of a game to choose when it will move in the play of the game. However, the choice of when to move applies only to a player's initial strategy choice, whereas the nonmyopic calculations to be developed here assume that players, starting at outcomes, make moves and countermoves that depend on thinking several steps ahead.

[2]For reviews of the literature on the effects of different rules or protocols – including the timing of moves – on game outcomes, see [**15**], [**6**], and [**26**].

[3]Elsewhere I call this game the "Surprise Game" [**7**] – for reasons I will spell out later – but here I give this game a very different interpretation and, hence, a new name. The nomenclature of "Freedom Game" is intended to be ironic, because the standard rules of play in fact restrict the choices that rational players would make in this game, compared with those they would make under two sets of new rules (to be described). While implications of the first set of rules were analyzed in [**7**], their interpretation here is new; both the implications and the interpretation of the second set of rules have not been previously analyzed for the Freedom Game.

(1) one player, whom I assume is Row (R), has a dominant strategy, which guarantees that the game has a unique Nash equilibrium in pure strategies ([**14**], p. 50, Lemma 1);[4]

(2) R has an incentive to depart from his dominant strategy and choose his dominated strategy, given that he has the opportunity to move first in a 2×4 game.[5]

On first blush, this seems surprising, because R's dominated strategy is unconditionally worse than his dominant strategy.[6]

Of the 78 distinct 2×2 strict ordinal games [**21**], R has an incentive to move first in six games, which break down into two classes. In three of the six games (class I), C also does better when R moves first, so it is in her interest to allow this to happen.

In the other three games (class II), there is a conflict of interest between the two players, because C does better moving first herself, or at least maintaining the status quo of simultaneous play in the 2×2 game. Consequently, if who moves first is left endogenous by the rules of play, then the structure of payoffs in the class II games leaves uncertain what will happen. I have incorporated this ambiguity into the theory of moves, or TOM ([**5, 6**]; and [**11**]), which adds a dynamic component to classical game theory. Instead of assuming that players choose strategies in normal-form games, TOM assumes that play begins at outcomes in the matrix, which are called "states" because they may be only temporary in nature. Players then decide whether to move or not move from these states by making rational calculations (to be described later) in order to reach "nonmyopic equilibria" (NMEs).

One theoretical finding of this paper is that the six 2×2 games in which R has an incentive to move first are the only games of the 78 in which a player would, according to TOM, move from a unique dominant-strategy (i.e., for one player) Nash equilibrium. It is in this sense that these equilibria are unstable: each player would, thinking ahead about its rational moves and countermoves, move from the Nash equilibrium to an NME that is a different state.[7]

The one-to-one correspondence between games in which the Nash equilibria and NMEs differ, on the one hand, and dominant-strategy games in which who moves first matters, on the other, illustrates how TOM captures order-of-play effects directly in its concept of an NME. This relieves one of the task of having to ask whether, when the order of play is endogenous, the dominant-strategy Nash

---

[4]If neither player has a dominant strategy, there are either multiple Nash equilibria or no Nash equilibria in pure strategies. For example, Chicken has two pure-strategy Nash equilibria; it also has a mixed-strategy equilibrium if the payoffs are not ordinal (as assumed here) but cardinal utilities. In games like Chicken that have multiple Nash equilibria, there is an equilibrium-selection problem, which is the subject of the so-called refinements literature in game theory that seeks to narrow down ("refine") the set of Nash equilibria to some proper subset, preferably with only one element [**23**]. By contrast, here I seek not to delimit Nash equilibria, since there is only one in the games studied, but instead to determine when a unique dominant-strategy Nash equilibrium will *not* be chosen.

[5]Henceforth, I assume R is male and the column player (C) is female.

[6]To be sure, R's dominated strategy in the 2×2 game is not dominated in the 2×4 game that results when R moves first. In fact, it is R's best response to C's (now-dominant) strategy in the 2×4 game, as I will show in Section 3. But this rationale for R's switching from his dominant to his dominated strategy in a 2×2 game is cumbersome, as compared with making the order of moves endogenous in an explicitly dynamic theory.

[7]Technically, a Nash equilibrium is not a state but the strategies of players that produce this state. For convenience, however, I will usually identify equilibria – both Nash and nonmyopic – by their state (or payoffs at this state) rather than by the strategy pair that yields this state.

equilibrium in the 2×2 game is unstable. It is if and only if the players would move from this equilibrium to a different NME.

But the fragility of Nash equilibria in the six 2×2 games is not just a function of the nonmyopic calculations that lead to NMEs. Under a radically different set of rules of play, these Nash equilibria are no better predictors of where the players might end up.

The key difference between the first set of rules and the second set is that the second set allows the players to cycle. In two of the six games, the outcome that occurs under the second set depends on which, if either, player has "moving power," which roughly means having the ability to force the player without moving power to stop when this player has the next move. In the other four games, the outcome does not depend on which player has moving power.

Remarkably, whether moving power is effective (first two games) or irrelevant (second four games), the outcome induced is never the Nash equilibrium; it may or may not be the NME (depending on who has moving power). Thus, once again, the Nash equilibrium is suspect as a predictor of behavior when players are given free rein to move and countermove from outcomes, not just choose strategies.

2 Freedom *not* to be surprised or forced to concede

The unexpected tie-in of unstable Nash equilibria on the one hand, and alternative rules that predict different outcomes on the other, is more than just a theoretical issue. Not only have these different outcomes actually occurred, but when they did they were the occasion of great surprise. As a case in point, the Japanese surprise attack on Pearl Harbor in 1941, which [**7**] modelled as a class II Freedom Game in which the Japan had an incentive to seize the initiative by attacking-at the expense of the United States – even though its dominant strategy in the simultaneous – play game was not to attack, shocked the world.

Taking the initiative may involve peaceful as well as belligerent actions, as illustrated by the second case analyzed in [**7**], Egyptian president Anwar Sadat's 1977 offer to visit Jerusalem to negotiate a peace treaty directly with the Israelis. The situation Sadat and Israeli prime minister Menachem Begin faced was modeled as a class I Freedom Game in which, unlike the Pearl Harbor game, both players benefited when the player with the dominant strategy (i.e., Sadat) moved first. That abandoning this strategy was risky for Sadat, but not irrational, seems borne out by his initial success yet later assassination.

Both the Japanese attack and Sadat's offer were astonishing events at the time. Such surprises, in my opinion, are well modelled in terms of a player's (perfectly rational) abandonment of a dominant strategy in favor of a dominated strategy, which in turn induces a countermove on the part of the surprised player to a different state, or NME.[8]

In laying bare the rationality of these moves in these cases, TOM helps to make them explicable and, therefore, unsurprising, at least in retrospect. Prospectively, TOM might help decision makers avoid "anticipation failures," caused by incomplete information about what rules are applicable. By making endogenous who moves first, TOM introduces a flexibility into thinking about how different scenarios might arise; by assuming that players think ahead, it provides a calculus for

[8]It is worth noting here that surprise does not arise from incomplete information about player payoffs but rather from incomplete information about the rules of play, or possibly which player has moving power, which will be discussed later.

tracing out the dynamic moves they are likely to make over time, including when play will terminate.

On occasion, however, players may be less able to determine an endpoint than be convinced that they can outlast the other player. Then they might well choose to cycle. Indeed, in two of the six 2×2 games subsumed by the Freedom Game, this is a rational choice: the player who can force the other player to stop – by virtue of having moving power – can induce a better outcome than if the other player has moving power (to be defined). But because the Nash equilibrium will never be chosen in either these or the other four games – whichever player possesses moving power – it is not an outcome to which players with a propensity to cycle would be attracted.

Put another way, the building-block concept of stability in noncooperative game theory, that of a Nash equilibrium, tends to cover up the choices that rational players not only would have good reason to make, but have actually made, in the Freedom Game. In particular, the unique Nash equilibrium in this game is very vulnerable: rational players would abandon this equilibrium under two plausible sets of rules, one of which assumes nonmyopic calculations, the other of which assumes cycling to try to outlast an opponent.

Attempting to run an opponent into the ground, so to speak, is hardly an uncommon strategy. The Reagan and Bush administrations pursued this strategy with a vengeance against the Soviet Union, "winning" the Cold War when the Soviet Union collapsed in 1991. Egypt fought five wars with Israel between 1948 and 1973 before both sides, out of despair and futility at fighting again (Israel remained militarily superior, but Egypt had greater manpower), met at Camp David in 1978 and signed a peace treaty in 1979.

When apparently stable and long-lasting relationships rapidly, and often surprisingly, change, we are frequently at a loss to explain why. The instability of the Nash equilibria in the six 2×2 games subsumed by the Freedom Game arises from the constraints imposed by the rules of play in standard game theory. When these rules are changed so as to give fuller expression to the range of choices players can and do make, the surprise disappears.

Specifically, these rules, and the rationality calculations underlying them, artificially constrain players. Empirically, these rules may be untenable as models of real-world behavior, as the cases previously mentioned illustrate. Philosophically, the situation may be worse: a theory should not proscribe choices that are immediately harmful (e.g., departures from Nash equilibria) but, in the longer term, are rationally justifiable.

Such departures in the Freedom Game, as I will show for the first set of rules to be analyzed, trigger a subsequent departure by the second player that, in the end, may rebound to the benefit of both players. Under the second set of rules, in which cycling can occur, the first player may realize a preferred outcome if he or she – with the kind of stamina that moving power provides – is able to force the other player to stop. When this happens, it may be viewed as a less benign force, abridging the free choice of the victimized player.

In sum, I will, using TOM, suggest alternative sets of rules for modelling free choice in games. One set enables players to think ahead and make moves that do not necessarily lead to immediately better outcomes but ultimately improve the lot of one, and sometimes both, players. Another set allows players to repeat choices by cycling, especially when they think their power to hold out against the other player will lead to a better outcome. Indeed, this power may be beneficial, but instead of expanding the free choice of the players, it contracts the choice of the one forced to concede, illustrating how asymmetrical power relationships can repress freedom.

3 Order of play and unstable Nash equilibria

The Freedom Game (simultaneous-play version) is shown in Figure 1(a), in which R has two strategies, s_1 and s_2, and C has two strategies, t_1 and t_2. The payoffs to the players at the resulting four possible outcomes are given by ordered pairs (x_1, x_2), where x_1 is the payoff to R and x_2 the payoff to C.

Column (C)

Row (R)		t_1	t_2	
	s_1	(a_1, a_2)	(b_1, b_2)	← Dominant strategy
	s_2	(d_1, d_2)	(c_1, c_2)	

Figure 1(a) Freedom Game (Simultaneous Play)

Column (C)

Row (R)		t_1/t_1	t_2/t_2	t_1/t_2	t_2/t_1
	s_1	(a_1, a_2)	(b_1, b_2)	(a_1, a_2)	(b_1, b_2)
	s_2	(d_1, d_2)	(c_1, c_2)	(c_1, c_2)	(d_1, d_2)
					↑ Dominant strategy

Figure 1(b) Freedom Game (R Moves First)

Key: (x_1, x_2) = (payoff to R, payoff to C)
t_i/t_j: choose t_i if R chooses s_1, t_j if R chooses s_2 $(i, j = 1 \text{ or } 2)$
Nash equilibria underscored
Complete ordering for $R: a_1 > d_1 > b_1 > c_1$
Partial ordering for $C: b_2 > a_2$ and $d_2 > c_2$

Figure 1

I show in this section under what conditions the unique dominant-strategy Nash equilibrium in the Freedom Game is vulnerable to order-of-play effects. The conditions for dominance, a Nash equilibrium, and instability are the following:

1. Dominance Without loss of generality, assume that s_1 is R's *dominant strategy*, so

$$a_1 > d_1; \quad b_1 > c_1. \tag{3.1}$$

That is, whatever strategy C chooses (t_1 or t_2), R prefers his payoffs associated with s_1 to those associated with s_2, making s_1 R's unconditionally better strategy.

2. Nash equilibrium Without loss of generality, assume that (b_1, b_2) is the unique *Nash equilibrium* that is associated with R's dominant strategy of s_1, so

$$b_2 > a_2. \tag{3.2}$$

That is, C prefers her payoff at (b_1, b_2) to that which she would receive at (a_1, a_2) if she switched from t_2 to t_1; similarly, R prefers (b_1, b_2) to (c_1, c_2) - and so would not switch from s_1 to s_2 – because of the dominance of s_1 assumed in (3.1) above.

3. Instability To ensure that R benefits by moving first and choosing his *dominated strategy* of s_2, which is unconditionally worse than s_1, it must be the case that R prefers (d_1, d_2) to the simultaneous-play Nash equilibrium of (b_1, b_2):

$$d_1 > b_1. \tag{3.3}$$

Also, to ensure that R will induce (d_1, d_2) rather than (c_1, c_2) when he chooses s_2, C must prefer the former outcome to the latter:

$$d_2 > c_2. \tag{3.4}$$

When inequalities (3.3) and (3.4) are both satisfied, the Nash equilibrium, (b_1, b_2), is said to be *unstable*.

Combining inequalities (3.1) and (3.3) for R, and (3.2) and (3.4) for C, gives a complete ordering of payoffs for R and a partial ordering for C:

$$R : a_1 > d_1 > b_1 > c_1; \quad C : b_2 > a_2 \text{ and } d_2 > c_2. \tag{3.5}$$

The partial ordering for C admits six different complete orderings:

$$\begin{array}{ll} b_2 > a_2 > d_2 > c_2; & d_2 > c_2 > b_2 > a_2; \\ b_2 > d_2 > c_2 > a_2; & d_2 > b_2 > a_2 > c_2; \\ b_2 > d_2 > a_2 > c_2; & d_2 > b_2 > c_2 > a_2. \end{array}$$

That is, these six orderings all preserve the partial ordering of C; because they are the only ones to do so, they are exhaustive.

The single strict ordering of payoffs for R, and the six strict orderings for C, define six strict ordinal games, which are shown in Figure 2. In these games, 4 indicates a best payoff, 3 next best, 2 next worst, and 1 worst. Thus, the higher the number, the greater the payoff; but because these payoffs are ordinal, they indicate only an ordering of outcomes from best to worst, not the degree to which a player prefers one outcome over another. The payoffs to R and C are given as ordered pairs (i, j) in parentheses, where i is the payoff to R and j the payoff to C. The ordered pairs $[u, v]$ in brackets just below the parenthetical pairs will be explained in Section 4.[9]

I have grouped the six games into the two classes, I and II, mentioned in Section 1. In the three class I games, the (d_1, d_2) lower-left outcome induced when R moves first and chooses his dominated strategy of s_2 is *Pareto-superior* – that is, better for both players-to the upper-right Nash equilibrium of (b_1, b_2). In the three class II games, neither (d_1, d_2) nor (b_1, b_2) is Pareto-superior to the other outcome.

---

[9]The numbers shown above each game are those given in the classification schemes of [**6**] and, in parentheses, [**21**].

Case I: (d_1, d_2) Pareto-superior to Nash equilibrium of (b_1, b_2)

27(47)

$(4,1)$	$(2,3)$
$[3,4]$	$[3,4]$
$(3,4)^{c*}$	$(1,2)^r$
$[3,4]$	$[3,4]$

28(48)

$(4,1)$	$(2,2)$
$[3,4]$	$[3,4]$
$(3,4)^{c*}$	$(1,3)^r$
$[3,4]$	$[3,4]$

48(57)

$(4,2)^r$	$(2,3)$
$[4,2]$	$[3,4]$
$(3,4)^c$	$(1,1)$
$[3,4]$	$[3,4]$

Case II: (d_1, d_2) not Pareto-superior to Nash equilibrium of(b_1, b_2)

49(44)

$(4,1)$	$(2,4)$
$[2,4]$	$[3,3]$
$(3,3)^{c*}$	$(1,2)^r$
$[2,4]$	$[2,4]/[3,3]$

50(45)

$(4,3)^{r*}$	$(2,4)$
$[4,3]$	$[4,3]$
$(3,2)^c$	$(1,1)$
$[4,3]/[2,4]$	$[2,4]/[4,3]$

56(56)

$(4,2)^r$	$(2,4)$
$[4,2]$	$[3,3]$
$(3,3)^c$	$(1,1)$
$[2,4]$	$[2,4]/[3,3]$

Figure 2 Six 2×2 Strict Ordinal Games Subsumed by Freedom Game (Simultaneous Play)

Key: (x_1, x_2) = (payoff to R, payoff to C) in original game
$[x_1, x_2]$ = [payoff to R, payoff to C] in anticipation game
4 = best; 3 = next best; 2 = next worst; 1 = worst
c = best state C can induce with moving power
r = best state R can induce with moving power
$*$ = state induced by moving power that is better for both players
Nash equilibria underscored in original and anticipation games (except anticipation games 27 and 28, in which all states are equilibria)
Nonmyopic equilibria (NMEs) circled in original game

Before discussing in more detail differences in these two classes of games, I emphasize what they have in common: the Nash equilibrium of (b_1, b_2) in each class is unstable. Specifically, R has an incentive to choose s_2 first, thereby inducing C – by responding optimally – to choose t_1, resulting in (d_1, d_2) rather than the simultaneous-play Nash equilibrium of (b_1, b_2) . Interestingly enough, if *both* players have dominant strategies in a 2×2 game (e.g., Prisoners' Dilemma), it is never rational for R (or C) to disrupt the unique pure-strategy Nash equilibrium by moving first. To see this, assume, as before, that (b_1, b_2) is the unique Nash equilibrium. If not only R's strategy associated with this equilibrium is dominant, but C's is too, then in addition to inequality (3.2), $c_2 > d_2$, which reverses inequality (3.4) that is necessary for instability.[10]

In class I games, not only would R have an incentive to move first, but C would presumably encourage him to do so in order to benefit herself over what she would obtain at the Pareto-inferior Nash equilibrium, (b_1, b_2). By comparison, this is emphatically not true of the class II games, in which C does better at (b_1, b_2), whereas R does better at (d_1, d_2), which he can induce by moving first.[11]

If R moves first, the game is, technically, not a 2×2 game but a 2×4 game, in which C has four strategies – contingent on what R does – rather than two. C's strategies may be described as follows:

$$t_i/t_j : \quad \text{choose } t_i \text{ if } R \text{ chooses } s_1, t_j \text{ if } R \text{ chooses } s_2 \ (i, j = 1 \text{ or } 2).$$

Thus, for example, t_2/t_1 means that C chooses her second strategy (t_2) if R chooses his first strategy (s_1), and C chooses her first strategy (t_1) if R chooses his second strategy (s_2).

Assume R moves first in the 2×2 Freedom Game of Figure 1(a). The 2×4 game that results is shown in Figure 1(b), with each of the four payoffs in the 2×2 game occurring at two different outcomes in the 2×4 matrix. For example, (a_1, a_2) in the 2×2 game occurs in the 2×4 game when R chooses s_1 and C chooses either t_1/t_1 or t_1/t_2, because the latter two choices both imply that C chooses t_1 when R chooses s_1 in the 2×2 game.

Whereas R has a dominant strategy of s_1 in the 2×2 game, he has no dominant strategy in the 2×4 game. While R's inequality (3.5) shows that s_1 is better for him than s_2 when C chooses t_1/t_1 t_2/t_2, or t_1/t_2, s_2 is better than s_1 when C chooses t_2/t_1. Hence, s_1 is not unconditionally better for R than s_2. On the other hand, t_2/t_1 for C is weakly dominant: it is at least as good, and sometimes better, than any of C's other three strategies – whether R chooses s_1 or s_2 – according to C's inequality (3.5).

Assume C chooses her dominant strategy in the 2×4 Freedom Game in Figure 1(b). Anticipating this choice, R's best response is to choose s_2, resulting in (d_1, d_2), the unique dominant-strategy Nash equilibrium in this game.[12] In other words, there is a reversal or roles when R moves first: whereas R possesses a dominant strategy in the 2×2 game (simultaneous play), C does in the 2×4 game (R moves first), leading to a new Nash equilibrium.

The rational choice of (d_1, d_2) in the 2×4 game reinforces my earlier contention: if the order of moves is endogenous, R has an incentive to move first. By redefining, in effect, what game is being played, R can induce his preferred outcome of (d_1, d_2),

---

[10]Specifically, this reversal eliminates R's incentive to move from the Nash equilibrium, because C will not countermove to (d_1, d_2).

[11]Maoz and Felsenthal in **[19]** argue that R must make a self-binding commitment to induce the Pareto-superior (d_1, d_2) outcome in class I games. But this is unnecessary if the order of play is assumed to be endogenous, because it is in C's interest, too, that R move first. By contrast, a credible commitment is much more called for in class II games, which Maoz and Felsenthal do not analyze, because of the players' conflict of interest over the order of play.

[12]The outcome (b_1, b_2) associated with the choice of s_1 by R and t_2/t_2 by C is also a Nash equilibrium. Thus, (b_1, b_2)'s equilibrium status carries over from the 2×2 game, but its selection in the 2×4 game is dubious: it would require that C choose a weakly dominated strategy, forsaking her dominant strategy of t_2/t_1, associated with the (d_1, d_2) Nash equilibrium, which gives her a higher payoff than (b_1, b_2) in class I games

because this outcome is the unique dominant-strategy Nash equilibrium in the 2×4 game.[13]

C, as I indicated earlier, will have no quarrel with R for moving first in the three class I games in Figure 2, because both players do better at (d_1, d_2) than (b_1, b_2). In the class II games, by comparison, C will be motivated to preempt R: in the resulting 4×2 game (i.e., when C moves first, which is not shown), (b_1, b_2) is the unique dominant-strategy equilibrium – just as it is in the 2×2 game – which C prefers to (d_1, d_2).

The players face a major problem if each thinks he or she can successfully preempt the other. Then R will choose s_2 when C chooses t_2, resulting in (c_1, c_2). This outcome is worse for both players than either (b_1, b_2) or (d_1, d_2) in all class II games, wherein precisely this conflict is likely to occur.

This conflict has somewhat the flavor of Chicken, whereby each player does better at his or preferred equilibrium, but each courts disaster if he or she is not successful in implementing it. But if the players recognize the problem of not coordinating their choices in such a "preemption game" – and there is a compromise outcome that benefits both, as in Chicken, even if it is not as good for each as successful preemption – then perhaps the disaster can be averted.[14]

The analysis so far demonstrates how a preemption problem can crop up, even if a simultaneous-play 2×2 game has a unique dominant-strategy Nash equilibrium, when the order of moves is endogenous. But this analysis required looking separately at 2×2 and 2×4 games that had the same payoff structures – only the order of moves was different.

I next describe a dynamic approach to the analysis of the stability of equilibria, in which the 2×2 game is assumed to be the only game played, but under two very different sets of rules from those of classical game theory. This approach offers new insight into not only why players might depart from an (unstable) Nash equilibrium but also where they might end up and how they might get there. Thereby TOM makes "surprise" moves both less surprising and more explicable in terms of the dynamics of rational choice.[15]

4 The theory of moves (TOM)

The starting point of TOM is a payoff matrix, or *configuration*, in which the order of play is not specified.[16] In fact, players are assumed not even to choose strategies but instead to move and countermove from outcomes, or states, by looking ahead and using "backward induction" to determine the rationality of both their moves and those of an opponent.

[13]It is also a Stackelberg equilibrium [**25**] for the case in which R is assumed to be the leader, by virtue of moving first, and C the follower. But the first-mover's advantage is eliminated if there is even a slight amount of noise, impairing the follower's ability to detect the leader's prior move [**2**]; here I assume this move is perfectly observable.

[14]Game-theoretic models of successful deterrence, based on Chicken, are analyzed in [**4**] and [**10**]; see also [**27**] and [**20**].

[15]For a more general argument, and plethora of examples, that "surprise" depends on the rules of play – specifically, not understanding them or their consequences – see [**12**].

[16]The subsequent description of TOM, which also encompasses rules not discussed here or in later sections that allow for threats, misperception, and the like, is adapted from [**6**].

Because game theory assumes that players choose strategies simultaneously,[17] it does not raise questions about the rationality of moving or departing from outcomes – at least beyond an immediate departure, à la Nash. In fact, however, most real-life games do not start with simultaneous strategy choices but commence at outcomes. The question then becomes whether a player, by departing from an outcome, can do better not just in an immediate or myopic sense but, rather, in an extended or nonmyopic sense. In the case of 2×2 games, TOM postulates four *rules of play*, which describe the possible choices of the players at different stages:

1. Play starts at an outcome, called the *initial state*, which is at the intersection of the row and column of a 2×2 payoff matrix.
2. Either player can unilaterally switch its strategy, and thereby change the initial state into a new state, in the same row or column as the initial state.[18] The player who switches, who may be either R or C, is called player 1 ($P1$).
3. Player 2 ($P2$) can respond by unilaterally switching its strategy, thereby moving the game to a new state.
4. The alternating responses continue until the player ($P1$ or $P2$) whost turn it is to move next chooses not to switch its strategy. When this happens, the game terminates in a *final state*, which is the outcome of the game.

Note that the sequence of moves and countermoves is strictly alternating: first, say, R moves, then C moves, and so on, until one player stops, at which point the state reached is final and, therefore, the outcome of the game.[19]

The use of the word *state* is meant to convey the temporary nature of an outcome, before players decide to stop switching strategies. I assume that no payoffs accrue to players from being in a state unless it is the final state and, therefore, becomes the outcome (which could be the initial state if the players choose not to move from it).

To assume otherwise would require that payoffs be cardinal rather than ordinal, with players accumulating them as they pass through states. I eschew this assumption in part because I think payoffs to players in most real-life games cannot be quantified and summed across the states visited. More significant, payoffs in the games that most interest me depend overwhelmingly on the final state reached, not on how it was reached. In politics, for example, the payoff for most politicians is not in campaigning, which is arduous and costly, but in winning.

Rule 1 differs radically from the corresponding rule of play in classical game theory, in which players simultaneously choose strategies in a matrix game that determines an outcome. Instead of starting with strategy choices, I assume that players are already in some state at the start of play and receive payoffs from this

---

[17] Strategies may allow for sequential choices, as I showed in Section 3, but the classicial theory does not make endogenous who moves first, as TOM does, but instead specifies a fixed order of play (simultaneous or sequential)

[18] I do not use "strategy" in the usual sense to mean a complete plan of responses by the players to all possible contingencies allowed by rules 2 - 4, because this would make the normal form unduly complicated to analyze. Rather, *strategies* refer to the choices made by players that define a state, and *moves and countermoves* to their subsequent strategy switches from an initial state to a final state in an extensive-form game, as allowed by rules 2 - 4. For another approach to combining the normal and extensive forms, see [**17, 18**].

[19] An emendation in the rules of TOM that allows for backtracking would be appropriate in games of incomplete information, wherein players make mistakes that they wish to rectify. For more on possible rule changes under TOM, see [**6**] and [**26**], who does allow players to backtrack.

state if they stay. Based on these payoffs, they decide, individually, whether or not to change this state in order to try to do better.[20]

To be sure, some decisions are made collectively by players, in which case it would be reasonable to say that they choose strategies from scratch, either simultaneously or by coordinating their choices. But if, say, two countries are coordinating their choices, as when they agree to sign a treaty, the most important strategic question is what individualistic calculations led them to this point. The formality of jointly signing the treaty is the culmination of their negotiations, which covers up the move-countermove process that preceded it. This is precisely what TOM is designed to uncover.

In summary, play of a game starts in a state, at which players accrue payoffs only if they remain in that state so that it becomes the outcome of the game. If they do not remain, they still know what payoffs they would have accrued had they stayed; hence, they can make a rational calculation of the advantages of staying versus moving. They move precisely because they calculate that they can do better by switching states, anticipating a better outcome if and when the move-countermove process finally comes to rest.

Rules 1-4 say nothing about what causes a game to end, but only when: termination occurs when a "player whose turn it is to move next chooses not to switch its strategy" (rule 4). But when is it rational not to continue moving, or not to move in the first place from the initial state? To answer this question, I posit a rule of rational termination (first proposed in [**3**], pp. 106-107), which has been called "inertia" by ([**16**], p. 94). It prohibits a player from moving from an initial state unless doing so leads to a better (not just the same) final state:

5. A player will not move from an initial state if this move
   - (i) leads to a less preferred final state (i.e., outcome); or
   - (ii) returns play to the initial state (i.e., makes the initial state the outcome).

I will discuss shortly how rational players, starting from some initial state, determine, by using backward induction, what the outcome will be.

Condition (i) of rule 5, which precludes moves that result in an inferior state, needs no defense. But condition (ii), which precludes moves to the same state because of cycling back to the initial state, is worth some elaboration. It says that if it is rational, after $P1$ moves, for play of the game to cycle back to the initial state, $P1$ will not move in the first place. After all, what is the point of initiating the move-countermove process if play simply returns to "square one," given that the players receive no payoffs along the way (i.e., before an outcome is reached)?

Not only is there no gain from cycling, but in fact there may be a loss because of so-called transaction costs that players suffer by virtue of making moves that, ultimately, do not change the situation. Therefore, it seems sensible to assume that $P1$ will not trigger a move-countermove process if it only returns the players to the initial state, making it the outcome. I call rule 5 a *rationality rule*, because it provides the basis for players to determine whether they can do better by moving from a state or remaining in it. Still another rationality rule is needed to ensure

[20]Alternatively, players may be thought of as choosing strategies initially, after which they perform a thought experiment of where moves will carry them once a state is selected. The concept of an "anticipation game," developed later, advances this idea, which might be considered dynamic thinking about the static play of a matrix game. Generally, however, I assume that "moves" describe actions, not just thoughts, though I readily admit the possibility of the thought interpretation.

that both players take into account each other's calculations before deciding to move from the initial state. I call this rule the *two-sidedness rule*:

6. Each player takes into account the consequences of the other player's rational choices, as well as his or her own, in deciding whether to move from the initial state or subsequently, based on backward induction. If it is rational for one player to move and the other player not to move from the initial state, then the player who moves takes *precedence*: his or her move overrides the player who stays, so the outcome is that induced by the player who moves.

Because players are assumed to have complete information, they can look ahead and anticipate the consequences of their moves. I next show how, using backward induction, they do this. In the process, I link order-of-moves effects, due to unstable equilibria (Section 3), to the results given by TOM, which predicts the moves that players will make from these equilbria.

5 Unstable Nash equilibria and TOM

I begin by analyzing where the players will end up when the initial state is the unique dominant-strategy Nash equilibrium of (b_1, b_2) in the 2×2 Freedom Game (Figure 1(a)).[21]

Theorem 1 *In the 2×2 Freedom Game, (d_1, d_2) is the outcome to which the players will move, according to TOM, from the unique dominant-strategy Nash equilibrium of (b_1, b_2).*

Proof There are two cases to consider:

Case 1. *R Acts First*

If R acts first, the counterclockwise progression of moves from (b_1, b_2) back to (b_1, b_2) – with the player (R or C) who makes the next move shown below each state in the alternating sequence – is as follows (see Figure 1(a)):

	State 1		State 2		State 3		State 4		State 1
	R		C		R		C		
R starts	(b_1, b_2)	$\rightarrow$	(c_1, c_2)	$\rightarrow$	$(\underline{d}_1, \underline{d}_2)$	$\rightarrow\mid$	(a_1, a_2)	$\rightarrow$	(b_1, b_2)
Survivor:	(d_1, d_2)		(d_1, d_2)		(d_1, d_2)		(a_1, a_2)		

The *survivor* is determined by working backward, after a putative cycle has been completed.

Assume the players' alternating moves have taken them counterclockwise from (b_1, b_2) to (c_1, c_2) to (d_1, d_2) to (a_1, a_2), at which point C must decide whether to stop at (a_1, a_2) in state 4 or complete the cycle and return to (b_1, b_2). Given inequality (2), C prefers (b_1, b_2) to (a_1, a_2), so (b_1, b_2) is listed as the survivor below (a_1, a_2): because C would move the process back to (b_1, b_2) should it reach (a_1, a_2), the players know that if the move-countermove process reaches state 4, the outcome will be (b_1, b_2).

[21] Where the players end up depends on the endstate, or *anchor*, from which the backward induction proceeds, which I assume here – for reasons given in Section 4 – is after one complete cycle. This assumption defines a finite extensive-form game, or game tree, to which I apply backward induction.

Knowing this, would R at state 3 move to (a_1, a_2)? Given inequality (3), R prefers (d_1, d_2) to the survivor at state 3 – namely, (b_1, b_2) – so the answer is no. Hence, (d_1, d_2) becomes the survivor when R must choose between stopping at (d_1, d_2) and moving to (a_1, a_2) at state 4 – which, as I have just shown, would become (b_1, b_2) once (a_1, a_2) is reached.

At state 2, C prefers moving to (d_1, d_2) than stopping at (c_1, c_2) by inequality (4). Consequently, (d_1, d_2) again is the survivor if the process reaches (c_1, c_2) at state 2. Similarly, at state 1 R prefers the previous survivor, (d_1, d_2), to (b_1, b_2) from inequality (3), so (d_1, d_2) is the survivor at this state as well.

The fact that (d_1, d_2) is the survivor at state 1 means that it is rational for R initially to move to (c_1, c_2), and C subsequently to move to (d_1, d_2). At (d_1, d_2) the process will stop, making (d_1, d_2) the rational choice if R has the opportunity to move first from state 1. That is, after working *backward* from C's choice of completing the cycle or not at state 4, the players can reverse the process and, looking forward, determine that it is rational for R to move from (b_1, b_2) to (c_1, c_2), and C to move from (c_1, c_2) to (d_1, d_2), at which point R will stop the move-countermove process at (d_1, d_2).

Notice that R does better at (d_1, d_2) than at (b_1, b_2), where it could have terminated play at the outset, and C does better at (d_1, d_2) than at (c_1, c_2), where it could have terminated play, given that R moves first. I indicate that (d_1, d_2) is the consequence of backward induction by underscoring this state in the progression; it is the state at which *stoppage* of the process occurs, which I indicate by the vertical line blocking the arrow emanating from (d_1, d_2).

Case 2. *C Acts First* If C acts first, the progression of moves from (b_1, b_2) back to (b_1, b_2) is clockwise. But this time backward induction seems to leave open which state will be the final survivor. In particular, given that R chooses to complete the cycle at state 4 because $b_1 > c_1$ of inequality (1) – making (b_1, b_2) the survivor at state 4 – C's inequality (5) provides no information about which choice C will make at state 3 – stay at (d_1, d_2), or move to (c_1, c_2) and thence to (b_1, b_2):

	State 1		State 2		State 3		State 4		State 1
	C		R		C		R		
C starts:	(b_1, b_2)	?	(a_1, a_2)	?	(d_1, d_1)	?	(c_1, c_2)	$\rightarrow$	(b_1, b_2)
Survivor:							(b_1, b_2)		

This indeterminacy disappears, however, when the consequences of two possibilities at state 3 – either (b_1, b_2) survives or (d_1, d_2) survives, depending on whether $b_2 > d_2$ or $d_2 > b_2$ – are analyzed:

- (b_1, b_2) survives. Then R will choose (a_1, a_2) at state 2 because of R's inequality (5), and C will choose (b_1, b_2) at state 1 because of inequality (2), so (b_1, b_2) will be the survivor at state 1.
- (d_1, d_2) survives. Then R will choose (a_1, a_2) at state 2 because $a_1 > d_1$ of inequality (1), and C will choose (b_1, b_2) at state 1 because of inequality (2), so (b_1, b_2) will be the survivor at state 1.

In either case, (b_1, b_2) will be the outcome when C acts first, which is to say that C will not move from this state. But because I showed in case 1 that R will move from this state and induce (d_1, d_2), (d_1, d_2) will be the outcome because of the precedence of moving over staying (rule 6 in Section 4). □

Thus, the unique Nash equilibrium of (b_1, b_2) in the Figure 1(a) game is not only unstable (Section 3) but also one from which the players would depart and, according to TOM, move to (d_1, d_2). I next consider whether there are any additional 2×2 strict ordinal games, apart from the six given in Figure 2, that have unique dominant-strategy Nash equilibria from which the players would depart.

Theorem 2 *The six games in Figure 2 are the only 2×2 strict ordinal games in which the players will move from a unique dominant-strategy Nash equilibrium. Consequently, these games are the same as those in which order of play, due to instability, matters (analyzed in Section 3).*

Proof To show that there are no other games among the 78 in which players would move from a dominant-strategy Nash equilibrium, one must show that if either of the necessary conditions for instability given in Section 2 fails – inequality (3) or inequality (4) – the backward induction used in the proof of Theorem 1 will lead to the players' staying at (b_1, b_2) rather than moving to (d_1, d_2). Consider, first, case 1 when R acts first and the possible failure of each instability condition:

- Inequality (3) is reversed, so $b_1 > d_1$. Then the final survivor at state 1 will be (b_1, b_2) rather than (d_1, d_2).
- Inequality (4) is reversed, so $c_2 > d_2$. Then the next-to-final survivor at state 2 will be (c_1, c_2), which will be displaced as final survivor by (b_1, b_2) at state 1 since $b_1 > c_1$ by inequality (1).

Hence, R will not move from (b_1, b_2) if either inequality (3) or inequality (4) is reversed. As for C, it is easy to show that the reversal of either inequality does not change the backward-induction analysis in case 2, so C, as before, will not move from (b_1, b_2) with the reversal of either inequality. Therefore, only the satisfaction of both inequalities induces the players to move from (b_1, b_2), which are exactly the two conditions that produce the order-of-play effects (i.e., make the dominant-strategy Nash equilibrium unstable). □

Taken together, Theorems 1 and 2 demonstrate that order of play matters in 2×2 games with unique dominant-strategy Nash equilibria if and only if the players would, according to TOM, move from these equilibria to another state. The state that they would move to is the unique Nash equilibrium in the 2×4 game in which R (i.e., the player with the dominant strategy in the 2×2 game) moves first and chooses his dominated strategy.

For the six games in Figure 2, I indicate that the upper-right (b_1, b_2) will go into the lower-left (d_1, d_2) by putting (d_1, d_2) in brackets, just below (b_1, b_2), in each payoff matrix. In game 27, for example, [3,4] is shown below (2,3) in the upper-right cell.

The outcome into which a state goes is called the nonmyopic equilibrium (NME) from that state. NMEs may be viewed as the consequence of both players' looking ahead and making rational calculations of where the move-countermove process will transport them, based on the rules of TOM, from each of the four possible initial states.

Backward-induction analysis from each state in game 27 shows that each state will go into (3,4). Thus, wherever play starts, the players can anticipate that they will end up at (3,4), making it the unique NME in game 27. This is also true of (3,4) in game 28, but not in game 48, the third class I game. Starting at (4,2) in game 48, the players will not depart from this state, making (4,2) as well as (3,4) an NME in this game.

Like game 48, all games in class II contain at least two NMEs. But some of these NMEs are *indeterminate*, because there is a conflict over who will move first. In game 50, for example, if (1,1) is the initial state, [2,4]/[4,3] indicates that when R moves first from (1,1), (2,4) will be outcome, whereas when C moves first, (4,3) will be the outcome.[22]

Because R prefers (4,3) whereas C prefers (2,4), each player will try to hold out longer in order to induce the other player to move first. Who wins in this struggle will depend on which player has "order power" – that is, who can determine the order of moves, starting at (1,1) ([**6**], ch. 5). Every 2×2 game contains at least one NME, because from each initial state there is an outcome (perhaps indeterminate) of the move-countermove process. If this outcome is both determinate and the same from every initial state, then it is the only NME; otherwise, there is more than one NME.

In game 56 in Figure 2, there are three different NMEs, which is the maximum number that can occur in a 2×2 strict ordinal game; the minimum, as already noted, is one. All except two of the 78 2×2 games (game 56 and Chicken) have either one or two NMEs.

The four bracketed states of each game in Figure 2 define what I call the *anticipation matrix*, with each state in this matrix an NME. Insofar as players choose strategies as if they were playing a game based on this matrix, one can determine which NMEs are Nash equilibria in this game and therefore likely to be chosen.[23]

To summarize, where players start in the original games in Figure 2 – including the unique dominant-strategy Nash equilibria – may not be where they end up, according to TOM. Thus, an original game may mask a good deal of instability when the players can move and countermove from states.[24] And it is this instability that induces players to make choices that, while disallowed by the rules of standard game theory, are allowed by rules 2-4 of TOM.

[22]Actually, the result of backward induction by R from (1,1) in game 50 is (3,2) rather than (4,3). But, as I argue in ([**6**], p. 115, ftn. 20), the players would have a common interest in implementing the Pareto-superior (4,3) to (3,2) when there is clockwise movement from (1,1). However, the implementation of (4,3) would require a binding commitment on the part of R not to move on from (4,3) to (2,4), which is not assumed possible in noncooperative game theory. I conclude: "I do not see an airtight case being made for either (3,2) or (4,3) as *the* NME from (1,1) when Column moves first, which nicely illustrates the nuances that TOM surfaces that the rules of standard game theory keep well submerged." Incidentally, game 50 is the only game of the 78 2×2 strict ordinal games in which this kind of ambiguity about NMEs arises.

[23]Because the NMEs in games 27 and 28 are all the same, the strategies of the players in their anticipation games are indistinguishable, making all four states Nash equilibria. In the four other games, s_1 is R's (weakly) dominant strategy; C's best response leads to a dominant-strategy Nash equilibrium in each anticipation game (and sometimes others). Only in game 49 is the Nash equilibrium in the anticipation game, [2,4], also the Nash equilibrium, (2,4), in the original game.

[24]Even the choice of a state associated with a dominant-strategy Nash equilibrium in the anticipation games offers no assurance that players will stay at this state. Indeed, except for one of the two states associated with the [4,3] Nash equilibria in the anticipation game of game 50, the players will mvoe from *every* such state in the six anticipation games to some different NME.

6 Cyclic games and moving power

In this section, I postulate two new rationality rules (rules 5′ and 6′ instead of rules 5 and 6, given in Section 4); rule 5′ permits players to cycle in a matrix, and rule 6′ enables one player, if he or she possesses "moving power," to terminate the cycling.

Recall from Section 4 that the rules of play (i.e., rules 1-4) say nothing about what causes a game to end, only when. Rule 5, which forbids moves that do not lead to a better outcome and thus precludes cycling back to the initial state, provides one answer. But this ban on cycling may not be realistic, as many protracted conflicts (e.g., the Arab-Israeli conflict) – in which the protagonists have revisited the past again and again – make unmistakably clear.

To try to capture the cyclic aspect of certain conflicts, and give players the ability to make choices in which they repeat themselves (why they may want to do so will be considered shortly), I next define a class of games in which cycling is possible by precluding a class of games in which it is not. Rule 5 provides a sufficient condition for cycling not to occur:

5′ If, at any state in the move-countermove process, a player whose turn it is to move next receives its best payoff (i.e., 4), it will not move from this state.

Rule 5′, in fact, precludes cycling in 42 of the 78 2×2 games, 21 of which contain a mutually best (4,4) state. Of the remaining 36 games, they fall into three classes: strongly cyclic (9 games), moderately cyclic (18 games), and weakly cyclic (9 games).

The six 2×2 games subsumed by the Freedom Game are all moderately cyclic. Before illustrating this kind of cyclicity, consider the circumstances under which players, who know not only their own payoffs but also have complete information about the payoffs of their opponents, would have an incentive to cycle to try to outlast an opponent.

By "outlasting" an opponent, I mean that one (stronger) player can force the other (weaker) player to stop the move-countermove process at a state where the weaker player has the next move. Forcing stoppage at such a state, however, may not always lead to an outcome favorable to the stronger player. In fact, a stoppage-forcing strategy is unproductive in four of the six 2×2 games that the Freedom Game embraces, leading to a worse outcome than if the stronger player had stopped the cycling rather than forcing the weaker player to do so.

Rule 5′ specified what players would not do – namely, move from a best (4) state when it was their turn to move. By precluding cycling in 42 of the 78 games, however, this rule did not say anything about where cycling would stop in the remaining 36 cyclic games. In these games, a new rationality rule is needed to say at what, if any, state play will terminate:

6′ In a cyclic game, one player (say, $P1$) will move from an initial state, even if play returns to this state and repeatedly cycles, if $P1$
   (i) has "moving power"; and
   (ii) can induce a better outcome with this power.

$P1$ has moving power if he or she can induce player 2 ($P2$) eventually to stop, in the process of cycling, at one of the two states at which $P2$ has the next move. The state at which $P2$ stops is that which $P2$ prefers ([**6**], ch. 4). Define moving power to be *effective* if the outcome that each player can induce with this power is better for it than the outcome that the other player can induce. Given this definition, we have

Theorem 3 *Moving power is effective in two of the six games in Figure 2 (games 48 and 56).*

I will illustrate this result for game 48 (the reader can verify it for game 56) and later illustrate why moving power is not effective in game 49 (the reader can verify this for games 27, 28, and 50). The arrows shown below (ignore for now the distinction between the single and double arrows) illustrate the cyclicity of games 48 in a counterclockwise direction: starting at the upper right state, C benefits moving from $(4, 2)$ to $(2, 34)$; R does not benefit moving from $(2, 3)$ to $(1, 1)$ – that is why "i" (for impediment) is next to this arrow – but departs from a 2, not a 4, state; C benefits moving from $(1, 1)$ to $(3, 4)$; and R benefits moving from $(3, 4)$ to $(4, 2)$. Because no player, when it is his or her turn to move, ever departs from his or heer best (i.e., 4) state, and because game 48 has one impediment, this game is moderately cyclic (games with no impediments are stronlgy cyclic, and games with two impediments are weakly cyclic).[25]

R can induce (4,2) in game 48

$(2, 3)$	$\longleftarrow$	$\underline{(4, 2)}$
$i \Downarrow$		$\Uparrow$
$(1, 1)$	$\longrightarrow$	$(3, 4)$

C can induce (3,4) in game 48

$(2, 3)$	$\Longleftarrow$	$(4, 2)$
$i \downarrow$		$\uparrow$
$(1, 1)$	$\Longrightarrow$	$\underline{(3, 4)}$

To show what outcome R can induce if he has moving power, which might be thought of as greater endurance, let his moves (vertical, as illustrated on the left side of the above diagram) be represented by double arrows. C, whose (horizontal) moves are represented by single arrows, must stop in the cycling at either (1,1) or (4,2), from where her single arrows emanate that indicate she has the next move. Since she would prefer to stop at (4,2) than (1,1), R can induce his best outcome if he has moving power. On the other hand, if C has moving power (right side of the above diagram), she can force R to stop at either (2,3) or (3,4), from where his single arrows emanate that indicate he has the next move. Since he would prefer to stop at (3,4) rather than (2,3), C can induce her best outcome if she has moving power. Thus, the possession of moving power benefits the player who possesses it-compared with the other player's possession of it – so it is effective in game 48.

This is not the case in game 49. Applying the foregoing reasoning to this game, we see that R can induce only $(1, 2)$ – C prefers this to $(4, 1)$, the other state where she moves from – but C can induce $(3, 3)$ – R prefers this to $(2, 4)$, the other state he moves from. But since R also prefers $(3, 3)$ to $(1, 2)$, moving power is not effective: R cannot induce a better outcome when he has it than when C has it. Rather, it is *irrelevant* in game 49, because it would be in R's interest to stop at $(3, 3)$, even

[25]It is shown in ([**6**], Theorem 4.1, pp. 90-91) that the 782×2 strict ordinal games can be cyclic in at most one direction (either clockwise or counterclockwise). Furthermore, none of the 12 symmetric games (e.g., Chicken and Prisoners' Dilemma) is cyclic, because their noncyclicity in one direction implies it in the other direction.

if he has moving power, rather than to force C to stop at $1, 2$).[26] In Figure 2, the outcomes each player can induce with moving power are indicated by supersripts r (for row) and c (for column), with an outcome starred if moving power is irrelevant and *both* players would find it rational for play to stop at this outcome.

R can induce (1,2) in game 49

$(2,4)$	$\longleftarrow$	$(4,1)$
$i \Downarrow$		$\Uparrow$
$\underline{(1,2)}$	$\longrightarrow$	$(3,3)$

C can induce (3,3) in game 49

$(2,4)$	$\Longleftarrow$	$(4,1)$
$i \downarrow$		$\uparrow$
$(1,2)$	$\Longrightarrow$	$\underline{(3,3)}$

Moving power may be thought of as a constraint on free choice if the player who possesses it (i) can force the other player to stop and (ii) has good reason to do so. As we have seen in four of the six games that the Freedom Game subsumes, however, the possessor has no reason to exercise this kind of power – prefering to stop himself or herself in the move-countermove process – which renders its possession irrelevant. Hence, even if there is a power asymmetry in such games, it does not translate into an effective constraint. On the other hand, in the two games in which moving power is effective, the possession of this power by one player will prevent the other player from obtaining a preferred outcome.

One lesson in all this is that power asymmetries sometimes may stifle free choice, but such a consequence is far from inevitable. In four of the games under consideration, moving power is irrelevant. Nonetheless, other kinds of power based on TOM are effective in two games in which moving power is not ("threat power" in games 49 and 50; "order power" is effective in game 56, along with moving power). Thus, a second lesson is that free choice is also sensitive to what kinds of power can be exercised. When these different kinds are taken into account, only two of the six games (27 and 28) are unaffected by one player's possession of power [**6**].

Although one player's possession of power in these two games does not help it vis-à-vis the other, both players benefit from the exercise of either moving or (deterrent) threat power. This is because either R or C, with this power, can induce (3,4), which is Pareto-superior to the Nash equilibrium of (2,3) in game 27 and the Nash equilibrium of (2,2) in game 28. Hence, the exercise of power in these games has a salutary effect, reinforcing the choice of the unique NME (albeit under different rules).

[26]More generally, moving power is *irrelevant* when the outcome induced by one player is better for both. There is still a third category of movin gpower, which I call *ineffective* – each player prefers the outcome that the other player can induce with it to that which he or she can induce – but it does not apply to any of the six games in Figure 2 (it is applicable only to 4 of the $78 2 \times 2$ strict ordinal games).

7 Conclusions

In this paper I showed that in both standard game theory and TOM, the incentive to move from a unique dominant-strategy Nash equilibrium in the generic Freedom Game, which subsumes six specific 2×2 strict ordinal games, has a rational-choice justification. However, the justification in the standard theory is awkward. It requires comparing the Nash equilibrium in the 2×2 game with the Nash equilibrium that R can induce if he moves first in a 2×4 game. If the equilibrium in the 2×4 game is better for R, then R would have an incentive to move first, making the equilibrium in the 2×2 game unstable.

TOM, by contrast, does not presume different game forms or a first-moving player. Rather, play starts at an outcome, or state, with the order of moves endogenous. Players move and countermove according to rules of play 1-4; under the first set of rationality rules (5 and 6), they look ahead and are able to anticipate ending up at NMEs from each state. It turns out that the six 2×2 games in which players would move from the dominant-strategy Nash equilibrium to a different state (i.e., an NME) are exactly the 2×2 games with unstable Nash equilibria. What TOM offers that the standard theory does not is a dynamic and parsimonious rationale for the players' making their moves and countermoves, which were used to model two empirical cases in [**7**].

Specifically, I showed that the surprises of Pearl Harbor (modeled by game 56) and the Sadat peace initiative (modeled by game 27) stem from thinking that the player with the dominant strategy (Japan and Egypt, respectively) would have no incentive to depart from this strategy. But the outcome induced by the other player's (the United States' and Israel's, respectively) best response to these dominant strategy choices – that is, the unique Nash equilibrium in these games – has no equilibrium status, according to TOM, when play starts in this state. The migration of the players from the Nash equilibrium in these cases to a different NME, I argued, is therefore not surprising. The fact that there was so much surprise at the time led me to postulate that, even if the games played were games of complete information, there was incomplete information about either the rules of play or the rationality rules, at least on the part of the surprised player.

Because different rules may generate radically different equilibrium results, it is important to try to ascertain those that are most plausible in a situation. Rules 1-6, in my opinion, are eminently plausible in many situations. Players often do think more than one step ahead and calculate, on this basis, a series of moves and countermoves that they anticipate they and an opponent will make.

A theory should not deny players the opportunity to exercise fully their freedom of choice – and a more far-sighted rationality, based on this freedom to move and countermove, according to rules 1-4. This may entail not just the kind of nonmyopic thinking of rules 5 and 6, which precludes cycling, but also the possibility of cycling, as governed by rules $5'$ and $6'$, that occurred in the superpower rivalry and continues in the Arab-Israeli conflict. It also occurs in 16 so-called catch-22 and king-of-the-mountain games, which have been used to model medieval with trials [**9**] and mobilization decisions in international crises [**8**].

In the case of cycling, I showed that moving power is effective in two of the six games subsumed by the Freedom Game (48 and 56, which happen to be catch-22 games), making it rational for the player who possesses it to try to outlast the other player by restricting his or her freedom of choice. In the other four games, moving power is irrelevant, making it in the interest of both players to stop at the same outcome.

In the latter games – as in the first two games – the outcome induced by moving power is never the Nash equilibrium. In two of these games (27 and 28) it is Pareto-superior to it, and in the other two (49 and 50) it benefits one player at the expense of the other. Thus, even when moving power is irrelevant in cyclic games, one would expect it to be exercised by at least one player to try to undermine the Nash equilibrium.

To conclude, the rules of standard game theory fail, on occasion, to capture the more long-range choices that players can and do make in the Freedom Game, especially departures from its unique Nash equilibrium. The first set of new rules expands player choices by assuming that players have the ability to think ahead in plotting their moves, but it prohibits cycling. The second set allows for cycling, affording the players still more choices, but these may be limited by the exercise of moving power. Whichever set is most appropriate to model a given situation, each affords players a freedom to make moves that is constrained only by their (nonmyopic) rationality or the possibly greater moving power of an opponent.

References

[1] Amir, Rabah [1995], *Endogenous Timing in Two-Player Games: A Counterexample*, Games and Economic Behavior **9**, no. 2 (May), 234-237.

[2] Bagwell, Kyle [1995], *Commitment and Observability in Games*, Games and Economic Behavior **8**, no. 2 (February), 271-280.

[3] Brams, Steven J. [1983], *Superior Beings: If They Exist, How Would We Know? Game-Theoretic Implications of Omniscience, Omnipotence, Immortality, and Incomprehensibility*, Springer-Verlag, New York.

[4] Brams, Steven J. [1985], *Superpower Games: Applying Game Theory to Superpower Conflict*, Yale University Press, New Haven, CT.

[5] Brams, Steven J. [1993], *Theory of Moves*, American Scientist **81**, no. 6 (December), 562-570.

[6] Brams, Steven J. [1994], *Theory of Moves*, Cambridge University Press, Cambridge, UK.

[7] Brams, Steven J. [1997], *The Rationality of Surprise: Unstable Nash Equilibria and the Theory of Moves*, in Decisionmaking on War and Peace: The Cognitive-Rational Debate (N. Geva and A. Mintz, eds.), Lynne Rienner, Boulder, CO, 103-129.

[8] Brams, Steven J. [1998], *To Mobilize or Not to Mobilize: Catch-22 and King-of-the-Mountain Games in International Crises*, New York University, preprint.

[9] Brams, Steven J. and Jones, Christopher B. [1997], *Catch-22 and King-of-the-Mountain Games: Cycling, Frustration, and Power*, New York University, preprint.

[10] Brams, Steven J. and Kilgour, D. Marc [1988], *Game Theory and National Security*, Basil Blackwell, New York.

[11] Brams, Steven J. and Mattli, Walter [1993], *Theory of Moves: Overview and Examples*, Conflict Management and Peace Science **12**, no. 2 (Spring), 1-39.

[12] Casti, John L. [1994], *Complexification: Explaining a Paradoxical World Through the Science of Surprise*, HarperCollins, New York.

[13] Hamilton, Jonathan H. and Slutsky, Stephen M. [1990], *Endogenous Timing in Duopoly Games: Stackelberg or Cournot Equilibria*, Games and Economic Behavior **2**, no. 1 (March), 29-46.

[14] Hamilton, Jonathan H. and Slutsky, Stephen M. [1993], *Endogenizing the Order of Moves in Matrix Games*, Theory and Decision **34**, no. 1 (January), 47-62.

[15] Hirshleifer, Jack [1985], *Protocol, Payoff, and Equilibrium: Game Theory and Social Modelling*, Working Paper #366, Department of Economics, University of California, Los Angeles (March).

[16] Kilgour, D. Marc and Zagare, Frank C. [1987], *Holding Power in Sequential Games*, International Interactions **13**, no. 2, 91-114.

[17] Mailath, George J., Samuelson, Larry and Swinkels, Jereon [1993], *Extensive Form Reasoning in Normal Form Games*, Econometrica **61**, no. 2 (March), 273-302.

[18] Mailath, George J., Samuelson, Larry and Swinkels, Jereon [1994], *Normal Form Structures in Extensive Form Games*, Journal of Economic Theory **64**, no. 2 (December), 325-371.

[19] Maoz, Zeev and Felsenthal, Dan S. [1987], *Self-Binding Commitments, the Inducement of Trust, Social Choice, and the Theory of International Cooperation*, International Studies Quarterly **31**, no. 2 (June), 177-200.

[20] Powell, Robert [1989], *Nuclear Deterrence Theory: The Search for Credibility*, Cambridge University Press, Cambridge, U.K.

[21] Rapoport, Anatol and Guyer, Melvin J. [1966], *A Taxonomy of* 2×2 *Games*, General Systems: Yearbook of the Society for General Systems Reserach **11**, 203-214.

[22] Rosenthal, Robert W. [1991], *A Note on Robustness of Equilibria with Respect to Commitment Opporunities*, Games and Economic Behavior **3**, no. 2 (May), 237-243.

[23] van Damme, Eric [1991], *Stability and Perfection of Nash Equilibria*, 2nd ed., Springer-Verlag, Heidelberg, Germany.

[24] van Damme, Eric and Hurkens, Sjaak [1996], *Commitment Robust Equilibria and Endogenous Timing*, Games and Economic Behavior **15**, no. 2 (August), 290-311.

[25] von Stackelberg, Heinrich [1934], *Marktform und Gleichgewicht*, J. Springer, Berlin. Trans. by Alan Peacock as *The Theory of Market Economy*, William Hodge, London (1952).

[26] Willson, Stephen J. [1998], *Long-Term Behavior in the Theory of Moves*, Theory and Decision, forthcoming.

[27] Zagare, Frank C. [1987], *The Dynamics of Deterrence*, University of Chicago Press, Chicago.

Cooperative Game Theory

Fields Institute Communications
Volume **23**, 1999

Pairwise-Bargained Consistency and Game Theory: The Case of a Two-Sided Firm

Theo S. H. Driessen
Department of Applied Mathematics
University of Twente
P. O. Box 217, 7500 AE Enschede
The Netherlands
T.S.H.Driessen@math.utwente.nl

Abstract. The paper solves the allocation problem for a two-sided firm which is composed of two divisions, namely a management division and a production division. The corresponding monetary net profit function is supposed to satisfy some kind of non-increasing marginal returns to scale. The allocation problem is solved by using two approaches. The first approach is fully based on the standard two-person bargaining principle, applied to appropriately chosen pairwise-bargained claims, and its associated notion of pairwise-bargained consistency in that no employee should gain or lose from pairwise-bargaining on the total amount already allocated. It is established that there exists a unique (nonnegative) pairwise-bargained consistent allocation. According to this unique solution, each worker receives a half of his incremental net profit if this amount falls short of a common upper bound, awarded to each manager, where the common upper bound is so determined that the resulting allocation satiates the overall net profit. The second approach to the allocation problem is fully based on game-theoretic notions like the core and the prekernel of the associated transferable utility game. It is established that the set of all pairwise-bargained consistent core-allocations can be re-interpreted as the intersection of the prekernel and the core. Consequently, the unique (nonnegative) pairwise-bargained consistent allocation turns out to coincide with the nucleolus of the game. In summary, both approaches, although they are taken from two different fields, support each other in that they provide the same solution to the allocation problem the two-sided firm is dealing with.

1 An allocation problem in a two-sided economic model

Let a firm be composed of two divisions, namely a management division as well as a production division. Both divisions consist of workers, on the understanding that the members of the management division, called managers, are to be considered

1991 *Mathematics Subject Classification.* Primary 90D12; Secondary 90D40, 90D80.

as representatives of all the workers. The production process of a certain commodity is carried out only if all the committee members of the management division agree unanimously to pass the proposed project to produce the commodity. That is, there is no output whenever the managers do not achieve the unanimity agreement. In case the managers establish the unanimity agreement, then the monetary net profit, obtained by selling the commodity at an appropriate market, depends on the choice of the employees, including all managers, who are chosen to carry out the production process. It is supposed that the net profit function satisfies a special kind of "non-increasing marginal returns to scale" in that the incremental net profit of every group of workers is not less than the sum of the incremental net profits of each of those workers.

Given that the production process is considered to be most profitable whenever all employees take part in it, the allocation problem arises how to share the overall net profit among all employees. The latter allocation problem is the core topic of this paper and it will be solved by using two approaches, one taken from two-person bargaining theory and another taken from cooperative game theory. We aim to show that both approaches, although they are taken from two different fields, support each other in that they provide the same solution to the allocation problem the two-sided firm is dealing with. According to the solution in question, each worker receives half of his incremental net profit if this amount falls short of a common upper bound, awarded to each manager, where the common upper bound is so determined that the resulting allocation satiates the overall net profit.

The organization of this paper is as follows. In Section 2 we apply the two-person bargaining theory to the allocation problem the two-sided firm is dealing with, and we end up with the determination of the relevant allocation rule. Sections 3 and 4 are devoted to a game-theoretic analysis of the allocation problem on the basis of two fundamental solution concepts from cooperative game theory, namely the core and the prekernel. In the framework of the game-theoretic model, which replaces the two-sided economic model, we contribute new results on some notions related to the core and the prekernel. On the one hand, these new game-theoretic results enable us to reinterpret the two-person bargaining theory developed in Section 2 and on the other hand, to provide two different game-theoretic characterizations of the relevant allocation rule. In Section 5 we discuss a slightly adapted two-person bargaining theory as well as an alternative game-theoretic model. In Section 6 we establish that it is not advantageous for the managers to act as a syndicate (that is, the union of managers is treated as a single entity). Concluding remarks are given in Section 7.

2 A pairwise bargaining approach to the allocation problem

Let us cast the allocation problem presented in Section 1 in a mathematical form. To begin, the finite *set N of all the employees* of the two-sided firm is the disjoint union of the sets M of committee members of the management division and W of members of the production division. It is always supposed that the set W of *workers* is nonempty, whereas the set M of *managers* is not necessarily nonempty (in the degenerated case that there are no managers we say the workers run the one-sided firm by themselves). Thus, we have $N = M \cup W$ where $M \cap W = \emptyset$, $W \neq \emptyset$. Further, the *monetary net profit function* is denoted by $f : \mathcal{P}(N) \to \mathbb{R}$, so, for every subset S of N (notation: $S \subseteq N$ or $S \in \mathcal{P}(N)$), the net profit $f(S)$ represents the monetary net value of the output due to the production activities by the members of S. As usual, the incremental net profit of each employee is assumed to be nonnegative, that is $\Delta_k^f := f(N) - f(N\backslash\{k\}) \geq 0$ for all $k \in N$. In accordance with the two-sided economic model described in Section 1, the net profit function $f : \mathcal{P}(N) \to \mathbb{R}$ is supposed to satisfy the following two conditions:

$$f(S) = 0 \quad \text{for all } S \in \mathcal{P}(N),\ S \neq \emptyset, \text{ with } M \not\subseteq S \tag{2.1}$$

(no output if the managers in S cannot force the unanimity agreement);

$$f(S) \geq 0 \quad \text{and} \quad f(N) - f(S) \geq \sum_{k \in N\backslash S} [f(N) - f(N\backslash\{k\})] \tag{2.2}$$

for all $S \in \mathcal{P}(N)$, $S \neq \emptyset$, with $M \subseteq S$ (the so-called *pseudo-concavity* of the function f).

Notice that, in the nondegenerated case there are indeed managers, each employee (except one if there is a unique manager) has no individual net profit (i.e., $f(\{i\}) = 0$ for all $i \in N$ provided that $M \neq \emptyset$ and employee i is not a unique manager).

The statement that the production process is considered to be most profitable whenever all employees take part in it, refers to the achievement of the overall net profit the amount of $f(N)$. The allocation problem the two-sided firm is dealing with, is that of dividing the overall net profit among all employees of the two-sided firm. The solution part to this allocation problem prescribes so-called *allocations* in the form of vectors $x = (x_k)_{k\in N} \in \mathbb{R}^N$ of real numbers satisfying $\sum_{k\in N} x_k = f(N)$, where $x_k \in \mathbb{R}$ represents the monetary amount allocated to employee k, $k \in N$. In order to select allocations of a special kind, we will come up with a first approach taken from two-person bargaining theory.

For the moment, fix an allocation $x = (x_k)_{k\in N} \in \mathbb{R}^N$ and consider a pair of employees, say $i \in N$, $j \in N$, $i \neq j$. We would like to conclude whether both employees accept the proposed allocation, or not, on the basis of the following two-person bargaining principle. Confronted with their two-person situation involving the total amount $x_i + x_j$ (awarded to them by the proposed allocation x), employee i claims either his individual net profit $f(\{i\})$ or what is left of the total amount x_i+x_j after opponent j is conceded to get his incremental net profit Δ_j^f, whichever is more. Since employee j claims a similar part of the total amount x_i+x_j, the amount at issue is therefore the total amount minus the sum of the *pairwise-bargained claims* of both employees. The amount at issue is shared equally between the two employees and, in addition, each employee receives his pairwise-bargained claim.

According to this standard two-person bargaining principle, applied to pairwise-bargained claims of a special kind, one of the two employees may gain (and so, the other may lose) in comparison with their individual allocations according to x. If the latter situation happens to occur, we say the proposed allocation is pairwise-bargained inconsistent (with respect to the standard two-person bargaining principle) and as a matter of fact, we are interested in (the existence of) those allocations that are pairwise-bargained consistent for every pair of employees. In summary, the notion of pairwise-bargained consistency reflects the view that no employee should gain or lose from pairwise-bargaining on the total amount already allocated.

In the degenerated case that there are no managers, each worker claims what is left of the total amount after the other worker is conceded to get his incremental net profit (so pairwise-bargained claims in a one-sided model may be less than the worker's individual net profit).

Definition 2.1 *Let $N = M \cup W$ where $M \cap W = \emptyset$, $W \neq \emptyset$. Let $f : \mathcal{P}(N) \to \mathbb{R}$ be an arbitrary function with nonnegative incremental returns, that is $\Delta_k^f := f(N) - f(N \backslash \{k\}) \geq 0$ for all $k \in N$ (but not necessarily for conditions (2.1) and (2.2) to be satisfied).*

(i) *The* pairwise-bargained claim *of employee $i \in N$ against another employee $j \in N$, $i \neq j$, with respect to an allocation $x = (x_k)_{k \in N} \in \mathbb{R}^N$, is defined to be*

$$c_{ij}^f(x) = c_{ij}(x) := \begin{cases} \max\left[f(\{i\}),\ x_i + x_j - \Delta_j^f\right] & \text{if } M \neq \emptyset, \\ x_i + x_j - \Delta_j^f & \text{if } M = \emptyset. \end{cases} \tag{2.3}$$

(ii) *An allocation $x = (x_k)_{k \in N} \in \mathbb{R}^N$ is said to be* pairwise-bargained consistent *if for every pair $i \in N$, $j \in N$, $i \neq j$, of employees it holds*

$$x_i = c_{ij}(x) + \tfrac{1}{2} \cdot [x_i + x_j - c_{ij}(x) - c_{ji}(x)] \qquad \textit{and}$$

$$x_j = c_{ji}(x) + \tfrac{1}{2} \cdot [x_i + x_j - c_{ij}(x) - c_{ji}(x)] .$$

Remark 2.2 In the setting of Definition 2.1, we imposed no further constraints upon the underlying function f as well as the allocation $x = (x_k)_{k \in N} \in \mathbb{R}^N$ under consideration. Under the circumstances that the function f satisfies condition (2.1) and individual allocations are nonnegative (i.e., $x_k \geq 0$ for all $k \in N$), then the pairwise-bargained claim of any employee $i \in N$ against any manager $j \in M$, $i \neq j$, equals zero (that is, $j \in M$ implies $c_{ij}^f(x) = 0$, due to the facts that $f(\{i\}) = 0$ and $x_i + x_j \leq f(N) = \Delta_j^f$).

Remark 2.3 By Definition 2.1, it follows immediately that an allocation $x = (x_k)_{k \in N} \in \mathbb{R}^N$ is pairwise-bargained consistent if and only if the following system of equations holds:

$$x_i - c_{ij}^f(x) = x_j - c_{ji}^f(x) \qquad \text{for all} \quad i \in N, j \in N, i \neq j. \tag{2.4}$$

In words, (2.4) states that the surplus of the individual allocation over the pairwise-bargained claim is the same for each of the two employees involved in the two-person bargaining situation.

In the degenerated case that there are no managers (i.e., $M = \emptyset$), the system of Equations (2.4) simplifies to the following system of equations: $\Delta_j^f - x_j = \Delta_i^f - x_i$ for all $i \in N$, $j \in N$, $i \neq j$. Obviously, among all allocations, the latter system of equations has a unique solution x determined by $x_i = \Delta_i^f + \frac{1}{|N|} \cdot \left[f(N) - \sum_{k \in N} \Delta_k^f\right]$ for all $i \in N$. Here $|N|$ denotes the size (cardinality) of the set N.

Remark 2.4 Let us treat the case that there is a unique manager, say $M = \{j\}$, $W = N \backslash \{j\}$, and suppose the net profit function $f : \mathcal{P}(N) \to \mathbb{R}$ satisfies condition (2.1). Since $f(\{i\}) = 0$ for all employees i, $i \in W$, except for the manager j, we prefer to concentrate, instead of the given function f, on its zero-normalized version $g : \mathcal{P}(N) \to \mathbb{R}$ defined to be $g(S) := f(S) - f(\{j\})$ for all $S \in \mathcal{P}(N)$ with $j \in S$ and $g(S) := f(S)$ otherwise. Clearly, $g(\{i\}) = 0$ for all $i \in N$ and the corresponding incremental net profits are so that $\Delta_k^g = \Delta_k^f$ for all $k \in W$ and $\Delta_j^g = \Delta_j^f - f(\{j\})$. Now we assert that an allocation $x = (x_k)_{k \in N} \in \mathbb{R}^N$

satisfying $\sum_{k\in N} x_k = f(N)$ is pairwise-bargained consistent with respect to the function f if and only if the corresponding allocation $y = (y_k)_{k\in N} \in \mathbb{R}^N$, defined by $y_j := x_j - f(\{j\})$ and $y_k := x_k$ for all $k \in W$, is pairwise-bargained consistent with respect to the function g. The explanation of the latter assertion is based on a detailed, but straightforward study of the system of Equations (2.4) by noting that the pairwise-bargained claims of the form (2.3) satisfy $c^g_{jk}(y) = c^f_{jk}(x) - f(\{j\})$ for all $k \in W$ as well as $c^g_{ik}(y) = c^f_{ik}(x)$ for all $i \in N$, $k \in N$, $i \neq k$, $i \neq j$. In summary, the pairwise-bargained consistent allocations with respect to the function f and its zero-normalization g, respectively, differ only in the amount allocated to the unique manager j who receives a "bonus" the amount of $f(\{j\})$.

Remark 2.5 Let us determine the nonnegative solution(s) of the system of Equations (2.4) in the nondegenerated case that there are indeed managers, so suppose $M \neq \emptyset$. Further, we assume that the net profit function f satisfies condition (2.1), $f(N) \geq 0$, and, for convenience' sake but without loss of generality, $f(M) = 0$ whenever the set M of managers consists of a unique member (cf. Remark 2.4). Under these circumstances (cf. Remark 2.2), the system of Equations (2.4) reduces to the following system of equations:

$$x_i = x_j \qquad \text{for all } i \in M,\ j \in M; \tag{2.5}$$

$$x_i = \min\left[x_j,\ \Delta^f_i - x_i\right] \qquad \text{for all } i \in W,\ j \in M; \tag{2.6}$$

$$\min\left[x_i,\ \Delta^f_j - x_j\right] = \min\left[x_j,\ \Delta^f_i - x_i\right] \qquad \text{for all } i \in W,\ j \in W. \tag{2.7}$$

In view of Equation (2.5), managers receive identical amounts, so let us put $x_i = \lambda$ for all $i \in M$, where $\lambda \in \mathbb{R}$ will be chosen in an appropriate manner. Next, in turn, Equation (2.6) is fully equivalent to the equation $x_i = \min\left[\lambda,\ \frac{1}{2}\Delta^f_i\right]$, $i \in W$. That is, every worker receives half of his incremental net profit if this amount falls short of the common upper bound λ, awarded to each manager. Given that $x_i = \min\left[\lambda,\ \frac{1}{2}\Delta^f_i\right]$ for all $i \in W$, it is straightforward to verify that Equation (2.7) holds.

In summary, the system of Equations (2.4) has at most one nonnegative solution and, if it exists, the solution x is determined by $x_i = \lambda$ for all $i \in M$, and $x_i = \min\left[\lambda,\ \frac{1}{2}\Delta^f_i\right]$ for all $i \in W$, where $\lambda \in \mathbb{R}$ is fully determined by the efficiency constraint $\sum_{k\in N} x_k = f(N)$. The existence of a nonnegative solution, however, is guaranteed by the observation that the function $h : \mathbb{R}_+ \to \mathbb{R}$, defined to be $h(\lambda) := |M| \cdot \lambda + \sum_{i\in W} \min\left[\lambda,\ \frac{1}{2}\Delta^f_i\right]$ for all $\lambda \geq 0$, is continuous and strictly increasing so that $h(0) = 0 \leq f(N) \leq h(f(N))$.

We conclude that there exists a unique (nonnegative) pairwise-bargained consistent (PBC) allocation for the (nondegenerated) two-sided economic model. This unique PBC-allocation x is of the form $x_i = \lambda$ for all $i \in M$, and $x_i = \min\left[\lambda,\ \frac{1}{2}\Delta^f_i\right]$ for all $i \in W$, where $\lambda \in \mathbb{R}$ has to be chosen appropriately to fulfill the efficiency constraint $\sum_{k\in N} x_k = f(N)$. The reader who is familiar with the cost sharing problem (cf. [**7**]), may recognize the unique PBC-allocation as the *head tax* solution to the cost-sharing problem in which the cost c of the public good equals $f(N)$ and the agents' benefits b_i, $i \in N$, from consuming the public good equals $f(N)$ and $\frac{1}{2}\Delta^f_i$, respectively, for $i \in M$ and $i \in W$ (cf. [**7**], page 151).

Remark 2.6 Let us solve in a detailed manner the case that there is a unique manager, say $M = \{j\}$, $W = N \backslash \{j\}$, and assuming $f(M) = 0$. We claim that the unique pairwise-bargained consistent (PBC) allocation x is given by $x_i = \frac{1}{2}\Delta_i^f$ for all $i \in W$ and $x_j = f(N) - \sum_{k \in W} \frac{1}{2}\Delta_k^f$. (In words, according to the PBC-allocation, each employee, except for the manager, receives half of his incremental net profit). That is, we claim that the unique nonnegative solution λ of the equation $\lambda + \sum_{i \in W} \min\left[\lambda, \frac{1}{2}\Delta_i^f\right] = f(N)$ is determined by $\lambda = f(N) - \sum_{k \in W} \frac{1}{2}\Delta_k^f$. For that purpose, it suffices to show that $f(N) - \sum_{k \in W} \frac{1}{2}\Delta_k^f \geq \frac{1}{2}\Delta_i^f$ for all $i \in W$. From condition (2.2) applied to $S = M$, we deduce that $f(N) \geq f(M) + \sum_{k \in W} \Delta_k^f = \sum_{k \in W} \Delta_k^f$ and therefore, $f(N) - \sum_{k \in W} \frac{1}{2}\Delta_k^f \geq \sum_{k \in W} \frac{1}{2}\Delta_k^f \geq \frac{1}{2}\Delta_i^f$ for all $i \in W$, as was to be shown.

In the next section our major goal is to perform a game-theoretic analysis of the notions of pairwise-bargained claims and pairwise-bargained consistent allocations, given that the underlying net profit function f satisfies both conditions (2.1) and (2.2).

3 A first game-theoretic approach to the allocation problem: the case of the core

Generally speaking, a *TU-game (in coalitional form with transferable utility)* is a function $w : \mathcal{P}(N) \to \mathbb{R}$ that assigns a real number $w(S)$ to each subset S of a finite set N. The members of N are called *players* (notation: $i \in N$) and the subsets of the player set N are called *coalitions* (notation: $S \subseteq N$ or $S \in \mathcal{P}(N)$). For every coalition S, the associated real number $w(S)$ is called the *worth of* S and it represents the total amount of (monetary) profits that the members of S can achieve by cooperation, without the help of the other players.

In particular, the two-sided firm and the corresponding monetary net profit function introduced in Section 2 give rise to a TU-game $v : \mathcal{P}(N) \to \mathbb{R}$ of which the player set N consists of all the employees of the two-sided firm and the game agrees with the net profit function, that is $v(S) := f(S)$ for all $S \in \mathcal{P}(N)$. By convention, $v(\emptyset) = 0$. We refer to this specific game as the *two-sided TU-game* v. For convenience' sake, we say the two-sided TU-game v is *pseudo-concave* if its underlying net profit function f satisfies the two conditions (2.1) and (2.2). [1] [2]

With the two-sided pseudo-concave TU-game v at hand, our principal objective is twofold: firstly, to present a game-theoretic interpretation of the pairwise-bargained claim $c_{ij}^f(x)$, as given by (2.3), and secondly, to provide a game-theoretic characterization of pairwise-bargained consistent (PBC) allocations (with reference to the two-sided pseudo-concave TU-game v). For that purpose, we address our attention to one of the most appealing solution concepts for TU-games in general,

---

[1] The two-sided pseudo-concave TU-games with a degenerated player set, so $M = \emptyset$ and $N = W$, were already introduced in **[2]**, page 20, and were called 1-*convex games*. The interested reader should consult **[2]** or **[3]** for a detailed study of 1-convex games.

[2] The two-sided pseudo-concave TU-games with a nondegenerated player set, so $M \neq \emptyset$ and $W \neq \emptyset$, were already introduced in **[10]**, page 276, and were called *clan games* where the *clan* refers to the nonempty set M of managers. The interested reader should consult **[10]** for the purpose of two examples of clan games and **[8]** for the purpose of clan games of which the clan consists of a unique member called big boss.

called core. The *core of a TU-game* $w : \mathcal{P}(N) \to \mathbb{R}$ represents the set of all coalitionally stable allocations (i.e., allocations that are not rejected by any coalition) and is formally defined to be

$$CORE(w) := \{x \in \mathbb{R}^N \mid \sum_{k\in N} x_k = w(N) \text{ and } \sum_{k\in S} x_k \geq w(S) \text{ for all } S \subsetneq N, S \neq \emptyset\}. \tag{3.1}$$

The next auxiliary result states that the core of a two-sided pseudo-concave TU-game v with a nondegenerated player set N is fully characterized by the individual core-constraints $x_i \geq f(\{i\})$, $i \in N$, as well as the inequalities $x_i \leq \Delta_i^f$, $i \in W$. The latter inequalities agree with the core-constraints of the form (3.1) referring to coalitions excluding exactly one worker (since $x_i = w(N) - \sum_{k\in N\setminus\{i\}} x_k \leq w(N) - w(N\setminus\{i\})$ for all $i \in N$ and all $x \in CORE(w)$). The core-constraints of the form (3.1) referring to other coalitions turn out to be redundant because of the validity of the two conditions (2.1) and (2.2). In words, for an allocation to be coalitionally stable, it is necessary and sufficient that each individual allocation is nonnegative (bounded below by zero) and no worker receives more than his incremental net profit (bounded above by Δ_i^f, $i \in W$).

Lemma 3.1 *Let* $N = M \cup W$ *where* $M \cap W = \emptyset$, $W \neq \emptyset$, $M \neq \emptyset$, *and* $f : \mathcal{P}(N) \to \mathbb{R}$ *a monetary net profit function satisfying conditions (2.1) and (2.2). Let* $v : \mathcal{P}(N) \to \mathbb{R}$ *be the corresponding two-sided pseudo-concave TU-game. For every* $x = (x_k)_{k\in N} \in \mathbb{R}^N$ *satisfying* $\sum_{k\in N} x_k = f(N)$ *the following equivalences hold:*

(i) $x \in CORE(v)$.

(ii) $x_i \geq f(\{i\})$ for all $i \in N$ and $x_i \leq \Delta_i^f$ for all $i \in W$. [3]

(iii) $x_i \geq 0$ for all $i \in N$ and $x_i \leq \Delta_i^f$ for all $i \in W$.

Proof It remains to prove the implication $(iii) \Rightarrow (i)$. Suppose that $x_i \geq 0$ for all $i \in N$ and $x_i \leq \Delta_i^f$ for all $i \in W$. For every coalition $S \subsetneq N$, $S \neq \emptyset$, we distinguish two cases. If $M \not\subseteq S$, then it follows from $x_k \geq 0$ for all $k \in S$ and (2.1) respectively that $\sum_{k\in S} x_k \geq 0 = v(S)$ (notice that this type of a coalition S does occur because of $M \neq \emptyset$). If $M \subseteq S$, then it follows from $x_k \leq \Delta_k^f$ for all $k \in N\setminus S$ and (2.2) respectively that $\sum_{k\in S} x_k = v(N) - \sum_{k\in N\setminus S} x_k \geq v(N) - \sum_{k\in N\setminus S} \Delta_k^f \geq v(S)$. We conclude that $x \in CORE(v)$. So, (iii) implies (i). □

Remark 3.2 Let $v : \mathcal{P}(N) \to \mathbb{R}$ be a two-sided pseudo-concave TU-game with a degenerated player set, so $M = \emptyset$ and $N = W$. Then the following equivalence holds: $x \in CORE(v)$ if and only if $\sum_{k\in N} x_k = f(N)$ and $x_i \leq \Delta_i^f$ for all $i \in N$. [4] Its proof is similar to the proof above (but now there do not exist coalitions $S \subsetneq N$, $S \neq \emptyset$, satisfying $M \not\subseteq S$) and so, its proof is left to the reader. Notice that $CORE(v) \neq \emptyset$ if and only if $\sum_{k\in N} \Delta_k^f \geq f(N)$. In order to guarantee the nonemptiness of the core, we suppose throughout this paper that $\sum_{k\in N} \Delta_k^f \geq f(N)$ in case $M = \emptyset$.

---

[3]This characterization of the core for two-sided pseudo-concave TU-games with a nondegenerated player set, or so-called clan games, was already noted in [**10**], Theorem 2.1, page 279.

[4]This characterization of the core for two-sided pseudo-concave TU-games with a degenerated player set, or so-called 1-convex games, was already noted in [**2**], Theorem 2.4, page 21, or in [**3**], Proposition 5.2, page 73.

From the simplest characterization of the core mentioned in Lemma 3.1(iii), it is obvious that the core of a two-sided pseudo-concave TU-game with a nondegenerated player set is always nonempty (e.g., allocate to each worker the incremental net profit and allocate nothing to each manager, except one). In the remainder of this section we exploit more drastically the simple characterization of the core mentioned in Lemma 3.1(ii).

Given a proposed coalitionally stable allocation $x \in CORE(v)$ for a two-sided pseudo-concave TU-game v, let us describe the pairwise-bargained range within the core corresponding to a two-person situation.[5] Consider a pair of employees, say $i \in N$, $j \in N$, $i \neq j$. Confronted with their individual allocations x_i and x_j respectively (awarded to each employee by the proposed allocation x), employee i is willing to concede to employee j the largest amount that can be transferred from i to j with respect to the core-allocation x while fulfilling the principle of coalitional stability (that is, while remaining in the core of the game v). For convenience' sake, denote this largest amount that employee i concedes to employee j by $\delta^v_{ij}(x) \in \mathbb{R}$. Concerning the two-person situation involving employees i and j, the associated critical number $\delta^v_{ij}(x)$ represents the pairwise-bargained range of employee j over employee i within the core, starting at the core-allocation x, and as such, $x_i - \delta^v_{ij}(x)$ represents the *smallest* amount employee i can achieve according to a coalitionally stable allocation for the game v. In the context of a two-sided pseudo-concave TU-game v, the next auxiliary result states that the pairwise-bargained range of employee j over employee i within the core, starting at a core-allocation x, can easily be determined in terms of the employees' individual allocations x_i, x_j, and the incremental net profit Δ^f_j of employee j.

Proposition 3.3 *Let $N = M \cup W$ where $M \cap W = \emptyset$, $W \neq \emptyset$, and $f : \mathcal{P}(N) \to \mathbb{R}$ a monetary net profit function satisfying conditions (2.1) and (2.2). Let $v : \mathcal{P}(N) \to \mathbb{R}$ be the corresponding two-sided pseudo-concave TU-game. For every $x \in CORE(v)$ and every pair $i \in N$, $j \in N$, $i \neq j$, of employees it holds*

(i)

$$\delta^v_{ij}(x) = \begin{cases} \min\left[x_i - f(\{i\}),\ \Delta^f_j - x_j\right] & \text{if } M \neq \emptyset, \\ \Delta^f_j - x_j & \text{if } M = \emptyset. \end{cases} \tag{3.2}$$

(ii)

$$c^f_{ij}(x) = x_i - \delta^v_{ij}(x).$$

Proof Let $x \in CORE(v)$ and $i \in N$, $j \in N \backslash \{i\}$. By letting the i-th component x_i decrease and the j-th component x_j increase by the same nonnegative amount δ, the resulting allocation fulfils the core-constraints if and only if the following holds (cf. Remark 3.2 and Lemma 3.1(ii)):

- in case $M = \emptyset$: $x_j + \delta \leq \Delta^f_j$, so $\delta \leq \Delta^f_j - x_j$;
- in case $M \neq \emptyset$ and $j \in M$: $f(\{i\}) \leq x_i - \delta$, so $\delta \leq x_i - f(\{i\})$;
- in case $M \neq \emptyset$ and $j \in W$: $f(\{i\}) \leq x_i - \delta$ and $x_j + \delta \leq \Delta^f_j$, or equivalently, $\delta \leq \min\left[x_i - f(\{i\}), \Delta^f_j - x_j\right]$.

[5]For every TU-game with a nonempty core, the notion of the pairwise-bargained range within the core corresponding to a two-person situation was introduced in [**6**], page 314.

Furthermore, in case $j \in M$ we deduce from $x_k \geq 0$ for all $k \in N$ (cf. Lemma 3.1(iii)) that $\Delta_j^f - x_j = f(N) - x_j = \sum_{k \in N \setminus \{j\}} x_k \geq x_i \geq x_i - f(\{i\})$. From the previous reasonings and the definition of $\delta_{ij}^v(x)$, the statement mentioned in part (i) follows immediately. The statement mentioned in part (ii) is a direct consequence of part (i) and the definition of the pairwise-bargained claim $c_{ij}^f(x)$, as given by (2.3). This completes the proof. □

Theorem 3.4 *Let $N = M \cup W$ where $M \cap W = \emptyset$, $W \neq \emptyset$, and $f : \mathcal{P}(N) \to \mathbb{R}$ a monetary net profit function satisfying conditions (2.1) and (2.2). Let $v : \mathcal{P}(N) \to \mathbb{R}$ be the corresponding two-sided pseudo-concave TU-game. For every $x \in CORE(v)$, it holds that x is a pairwise-bargained consistent (PBC) allocation if and only if it satisfies the bisection property, i.e., $\delta_{ij}^v(x) = \delta_{ji}^v(x)$ for all $i \in N$, $j \in N$, $i \neq j$. (In words, the core-allocation x is pairwise-bargained consistent if and only if it is the midpoint of the symmetrically pairwise-bargained ranges within the core for* every *two-person situation involving two employees and the underlying core-allocation x).*

Proof For every $x \in CORE(v)$, we deduce the following equivalences:

$$
\begin{aligned}
& x \text{ is a PBC-allocation} \\
\Leftrightarrow\quad & x_i - c_{ij}^f(x) = x_j - c_{ji}^f(x) \quad \text{for all } i \in N, j \in N, i \neq j \text{ (cf. Remark 2.3)} \\
\Leftrightarrow\quad & \delta_{ij}^v(x) = \delta_{ji}^v(x) \quad\quad\quad\quad\;\; \text{for all } i \in N, j \in N, i \neq j \text{ (cf. Proposition 3.3(ii))}.
\end{aligned}
$$

□

In view of Proposition 3.3(ii), we arrived at the game-theoretic interpretation of the pairwise-bargained claim $c_{ij}^f(x)$ of employee i against employee j as the *smallest* core-allocation to employee i concerning their two-person situation with reference to the underlying core-allocation x. In other words, we conclude that the pairwise-bargained claim $c_{ij}^f(x)$, as given by (2.3), can be re-interpreted as some bargaining range between both employees i and j *outside* the core of the two-sided pseudo-concave TU-game v. See the Figure 1 depicted below.

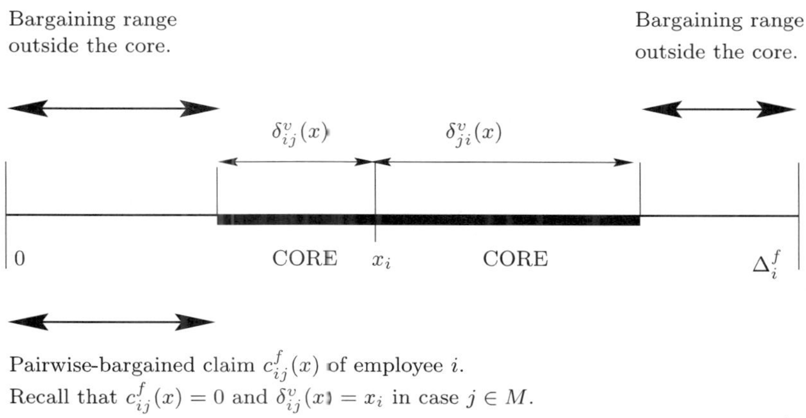

Figure 1

Remark 3.5 Theorem 3.4 provides a geometric characterization of pairwise-bargained consistent allocations within the core of a two-sided pseudo-concave TU-game v. This geometric characterization suggests that it is necessary to check all $\frac{1}{2} \cdot n \cdot (n-1)$ possibilities (i.e., the number of pairs of players out of n players) to conclude whether, or not, a core-allocation is the midpoint of any relevant pairwise-bargained range within the core. The Remarks 2.3 and 2.5, however, already illustrated that it suffices to check the bisection property corresponding to $n-1$ appropriately chosen pairs. In case $M = \emptyset$ (so $N = W$), it suffices to check $\delta_{ij}^v(x) = \delta_{ji}^v(x)$ for an arbitrarily fixed $j \in W$ and all $i \in N$, $i \neq j$ (cf. Remark 2.3). In case $M \neq \emptyset$, it suffices to check $\delta_{ij}^v(x) = \delta_{ji}^v(x)$ for an arbitrarily fixed $j \in M$ and all $i \in N$, $i \neq j$ (cf. Remark 2.5 and in particular, the system of Equations (2.5)-(2.7)).

Let us evaluate the case that there is a unique manager, say $M = \{j\}$, $W = N \backslash \{j\}$, and assuming $f(M) = 0$. By Remark 2.6, the unique pairwise-bargained consistent (PBC) allocation x is determined by $x_i = \frac{1}{2}\Delta_i^f$ for all $i \in W$ and $x_j = f(N) - \sum_{k \in W} \frac{1}{2}\Delta_k^f$. We verify that this allocation x does satisfy the bisection property, i.e., $\delta_{ij}^v(x) = \delta_{ji}^v(x)$ for all $i \in N$, $i \neq j$.

Fix $i \in W$. Firstly, starting at the core-allocation x, worker i can freely transfer all his money to manager j because there are no upper bounds for core-allocations to any manager (cf. Lemma 3.1(iii)). Hence, $\delta_{ij}^v(x) = \frac{1}{2}\Delta_i^f$. Secondly, starting at the core-allocation x, manager j can transfer all his money to worker i, provided that the total payoff to worker i does not exceed his incremental net profit Δ_i^f. Thus, $\delta_{ji}^v(x) = \min\left[f(N) - \sum_{k \in W} \frac{1}{2}\Delta_k^f, \frac{1}{2}\Delta_i^f\right] = \frac{1}{2}\Delta_i^f$. Here the latter equality was already explained in Remark 2.6. Finally we remark that the PBC-allocation x turns out to coincide with the center of gravity of the core of the corresponding two-sided pseudo-concave TU-game.

4 A second game-theoretic approach to the allocation problem: the case of the prekernel

So far the game-theoretic analysis of pairwise-bargained consistent (PBC) allocations dealt merely with the notion called core. In this section we extend the game-theoretic analysis by investigating another solution concept called prekernel. For the sake of completeness, we mention that Potters et al. (cf. [**10**]) already have studied two related solution concepts, called kernel and bargaining set, and concluded that the kernel of a nondegenerated two-sided pseudo-concave TU-game consists of exactly one allocation, namely the allocation rule discussed in our Remark 2.5 (and discovered by a pairwise bargaining approach).[6]

Our aim is to present, in the framework of the prekernel for two-sided pseudo-concave TU-games, one major result (not yet mentioned in [**10**]) and as a direct

[6]For two-sided pseudo-concave TU-games with a nondegenerated player set, called clan games in [**10**], it was shown that the so-called bargaining set coincides with the core (cf. [**10**], Theorem 2.1, page 279). Because the kernel is always contained in the bargaining set, it follows that the core of clan games includes the kernel. Generally speaking, it is known that the parts of the kernel and the prekernel inside the core do not differ (cf. [**6**], Theorem 3.2, page 314). Therefore, the kernel of clan games coincides with the part of the prekernel inside the core. The kernel, however, may differ from the prekernel since the pseudo-concavity condition (2.2) does not ensure the zero-normalized clan game to be monotonic. Because the prekernel is a simplified version of the kernel, we prefer to analyse the notion of prekernel by means of elementarily arithmetic tools.

consequence, we arrive at an alternative characterization of pairwise-bargained consistent core-allocations in terms of the prekernel.

We need to introduce some new notation involving an arbitrary TU-game. The *prekernel of a TU-game* $w : \mathcal{P}(N) \to \mathbb{R}$ represents the set of all allocations at which each pair of players is equally powerful and is formally defined to be

$$\mathcal{K}^*(w) := \{x \in \mathbb{R}^N \mid \sum_{k\in N} x_k = w(N) \text{ and } s^w_{ij}(x) = s^w_{ji}(x) \text{ for all } i \in N,\, j \in N,\, i \neq j\}, \tag{4.1}$$

where the *maximum surplus* $s^w_{ij}(x)$ of player i over another player j with respect to the allocation x in the TU-game w is defined to be

$$s^w_{ij}(x) := \max \left[w(S) - \sum_{k\in S} x_k \mid S \subseteq N,\ i \in S,\ j \notin S \right]. \tag{4.2}$$

In the context of two-sided pseudo-concave TU-games v, we are in a position to exploit the two properties (2.1)-(2.2) as well as the simple description of its core (cf. Lemma 3.1(iii) and Remark 3.2) in order to determine each maximum surplus (with respect to every coalitionally stable allocation $x \in CORE(v)$). We emphasize that the present approach is purely *arithmetic* in contradistinction to the *geometric* approach presented in the previous Section 3.

Proposition 4.1 *Let $N = M \cup W$ where $M \cap W = \emptyset$, $W \neq \emptyset$, and $f : \mathcal{P}(N) \to \mathbb{R}$ a monetary net profit function satisfying conditions (2.1) and (2.2). Let $v : \mathcal{P}(N) \to \mathbb{R}$ be the corresponding two-sided pseudo-concave TU-game. For every $x \in CORE(v)$ and every pair $i \in N$, $j \in N$, $i \neq j$, of employees it holds*

(i)

$$s^v_{ij}(x) = \begin{cases} -\min\left[x_i - f(\{i\}),\ \Delta^f_j - x_j\right] & \text{if } M \neq \emptyset, \\ x_j - \Delta^f_j & \text{if } M = \emptyset. \end{cases} \tag{4.3}$$

(ii)

$$\delta^v_{ij}(x) = -s^v_{ij}(x).$$

Proof Let $x \in CORE(v)$ and $i \in N$, $j \in N \setminus \{i\}$. Let $S \subseteq N$ be such that $i \in S$ and $j \notin S$. We distinguish two cases in order to determine the maximum in the expression at the right hand of (4.2) with reference to the two-sided pseudo-concave TU-game v.

Case one. Suppose $M = \emptyset$. Then it holds $M \subseteq S$, thus (2.2) applies. From (2.2), $j \notin S$, and $x_k \leq \Delta^f_k$ for all $k \in N \setminus S$, $k \neq j$, (cf. Remark 3.2) it follows that

$$v(S) - \sum_{k\in S} x_k = v(S) - v(N) + \sum_{k\in N\setminus S} x_k \leq \sum_{k\in N\setminus S} (x_k - \Delta^f_k) \leq x_j - \Delta^f_j \tag{4.4}$$

where the inequalities in (4.4) are equalities whenever $S = N\setminus\{j\}$. We conclude that the maximum in the expression at the right hand of (4.2) is attained at $S = N\setminus\{j\}$ and hence, $s^v_{ij}(x) = x_j - \Delta^f_j$ whenever $M = \emptyset$.

Case two. Suppose $M \neq \emptyset$. If $M \not\subseteq S$, then, by (2.1), $v(S) = 0$, and together with $i \in S$ and $x_k \geq 0$ for all $k \in S\setminus\{i\}$, (cf. Lemma 3.1(iii)) this implies that

$$v(S) - \sum_{k\in S} x_k = -\sum_{k\in S} x_k \leq -x_i \leq f(\{i\}) - x_i \qquad \text{whenever } M \not\subseteq S.$$

Therefore, coalitions S satisfying $M \not\subseteq S$ are redundant for the determination of the maximum in the expression at the right hand of (4.2) because their corresponding "excesses" are majorized by the "excess" associated with $S = \{i\}$. Concerning coalitions S satisfying $M \subseteq S$, it follows from (2.2), $j \notin S$, and $x_k \leq \Delta_k^f$ for all $k \in N \backslash S$, $k \neq j$, (cf. Lemma 3.1(iii)) that (4.4) holds once again, where the inequalities in (4.4) are equalities whenever $S = N \backslash \{j\}$. We distinguish two subcases.

Subcase one. Suppose $j \in M$. Since $j \notin S$, it holds $M \not\subseteq S$ (that is, there do not exist coalitions S satisfying $M \subseteq S$). We conclude that the maximum in the expression at the right hand of (4.2) is attained at $S = \{i\}$ and hence, $s_{ij}^v(x) = f(\{i\}) - x_i$. Recall that $\Delta_j^f - x_j = f(N) - x_j = \sum_{k \in N \backslash \{j\}} x_k \geq x_i \geq x_i - f(\{i\})$. This proves (4.3) in case $M \neq \emptyset$ and $j \in M$.

Subcase two. Suppose $j \in W$. Then it holds $M \subseteq N \backslash \{j\}$ and we conclude that the maximum in the expression at the right hand of (4.2) is attained at $S = \{i\}$ or $S = N \backslash \{j\}$, hence

$$s_{ij}^v(x) = \max\left[f(\{i\}) - x_i,\ x_j - \Delta_j^f\right] = -\min\left[x_i - f(\{i\}),\ \Delta_j^f - x_j\right].$$

This proves (4.3) in case $M \neq \emptyset$ and $j \in W$.

In summary, this completes the proof of the statement mentioned in part (i). The statement mentioned in part (ii) is a direct consequence of part (i) and Proposition 3.3(i). □

Corollary 4.2 *Let $N = M \cup W$ where $M \cap W = \emptyset$, $W \neq \emptyset$, and $f : \mathcal{P}(N) \to \mathbb{R}$ a monetary net profit function satisfying conditions (2.1) and (2.2).*

(i) *For every two-sided pseudo-concave TU-game, the set of all pairwise-bargained consistent core-allocations coincides with the intersection of the prekernel and the core.*

(ii) *Suppose $M \neq \emptyset$ and $f(M) = 0$ if $|M| = 1$. Then the unique nonnegative pairwise-bargained consistent allocation belongs to the core and coincides with the nucleolus of the two-sided pseudo-concave TU-game.*

Proof (i) Apply Theorem 3.4, Proposition 4.1(ii), and the definition (4.1) of the prekernel.
(ii) Generally speaking, it is known that, for games with a nonempty core, the parts of the kernel and the prekernel inside the core do not differ and moreover, the nucleolus belongs to the intersection of the kernel and the core. Under the given circumstances (like a nondegenerated player set), every two-sided pseudo-concave TU-game has a nonempty core consisting of nonnegative allocations (since $f(\{i\}) = 0$ for all $i \in N$). Hence, by part (i), the nucleolus of every two-sided pseudo-concave TU-game belongs to the set of all pairwise-bargained consistent core-allocations, which set, in turn, consists of at most one nonnegative pairwise-bargained consistent allocation (cf. Remark 2.5). From this we conclude that the unique nonnegative pairwise-bargained consistent allocation coincides with the nucleolus and thus, it belongs to the core as well. □

5 An alternative pairwise bargaining and game-theoretic approach: the dual case

In this section let $N = M \cup W$ where $M \cap W = \emptyset$, $W \neq \emptyset$, $M \neq \emptyset$, and $f : \mathcal{P}(N) \to \mathbb{R}$ a monetary net profit function satisfying conditions (2.1) and (2.2). In order to evaluate a proposed allocation $x = (x_k)_{k \in N} \in \mathbb{R}^N$, let us apply the *greedy bargaining principle* to every two-person situation. Consider a pair of employees, say $i \in N$, $j \in N$, $i \neq j$. Confronted with their two-person situation involving the total amount $x_i + x_j$ (awarded to them by the proposed allocation x), both employees are supposed to behave in a greedy manner in that each employee claims his incremental net profit or what is left of the total amount after the opponent is conceded to get his individual net profit, whichever is less. Formally, the *pairwise-bargained greedy claim* of employee i against another employee j, $i \neq j$, with respect to an allocation $x = (x_k)_{k \in N} \in \mathbb{R}^N$, is defined to be $d^f_{ij}(x) = d_{ij}(x) := \min\left[\Delta^f_i,\ x_i + x_j - f(\{j\})\right]$. Concerning the two-person situation with employees i and j, the amount at issue is therefore $x_i + x_j - d_{ij}(x) - d_{ji}(x)$; the latter amount is shared equally between the two employees and, in addition, each employee receives his pairwise-bargained greedy claim. We say the allocation x is *pairwise-bargained greedy-consistent* if the allocation can not be affected by any two-person bargaining procedure on the basis of the greedy bargaining principle, i.e., for every pair $i \in N$, $j \in N$, $i \neq j$, of employees it holds

$$x_i = d_{ij}(x) + \tfrac{1}{2} \cdot [x_i + x_j - d_{ij}(x) - d_{ji}(x)] \qquad \text{and}$$

$$x_j = d_{ji}(x) + \tfrac{1}{2} \cdot [x_i + x_j - d_{ij}(x) - d_{ji}(x)]$$

$$\text{or equivalently,} \qquad d^f_{ij}(x) - x_i = d^f_{ji}(x) - x_j \qquad \text{for all } i \in N,\ j \in N,\ i \neq j. \tag{5.1}$$

Let us compare the greedy two-person bargaining principle with the two-person bargaining principle discussed in Section 2. Both principles differ only in that the essential notion of, what we call from now on, *pairwise-bargained modest claim* $c^f_{ij}(x) = \max\left[f(\{i\}),\ x_i + x_j - \Delta^f_j\right]$ of employee i against employee j is replaced by the notion of *pairwise-bargained greedy claim* $d^f_{ij}(x) = \min\left[\Delta^f_i,\ x_i + x_j - f(\{j\})\right]$. Observe that, for allocations $x = (x_k)_{k \in N} \in \mathbb{R}^N$ that are bounded below by individual net profits and bounded above by incremental net profits (i.e., $f(\{k\}) \leq x_k \leq \Delta^f_k$ for all $k \in N$), it holds that $c^f_{ij}(x) \leq x_i \leq d^f_{ij}(x)$ for all $i \in N$, $j \in N$, $i \neq j$. By the latter observation, the prefix *modest* and *greedy*, respectively, is justified. From the definitions of the pairwise-bargained modest and greedy claims it is clear that these notions are related to each other so that they are complementary with respect to the total amount $x_i + x_j$, i.e., $c^f_{ij}(x) + d^f_{ji}(x) = x_i + x_j$ for all $i \in N$, $j \in N$, $i \neq j$. In words, the total amount is satiated whenever one employee claims in a modest manner and the opponent in a greedy manner. Consequently, an allocation is pairwise-bargained greedy-consistent if and only if it is pairwise-bargained modest-consistent (cf. the equivalence of (2.4) to (5.1)). We conclude that the greedy pairwise bargaining approach does not differ from the modest bargaining approach. Moreover, the equality $c^f_{ji}(x) = x_j - \delta^v_{ji}(x)$ (cf. Proposition 3.3(ii)) is fully equivalent to the equality $d^f_{ij}(x) = x_i + \delta^v_{ji}(x)$. That is, we arrived at the game-theoretic interpretation of the pairwise-bargained greedy claim $d^f_{ij}(x)$ of employee i against employee j as the *largest* core-allocation to employee i concerning their two-person situation with reference to the underlying core-allocation x. See the Figures 2 and 3 depicted below.

The game-theoretic treatment carried out in Sections 3 and 4 was fully based on the TU-game v which agrees with the underlying net profit function f in that $v(S) = f(S)$ for all $S \in \mathcal{P}(N)$. Surely, this is the most simple way to associate a

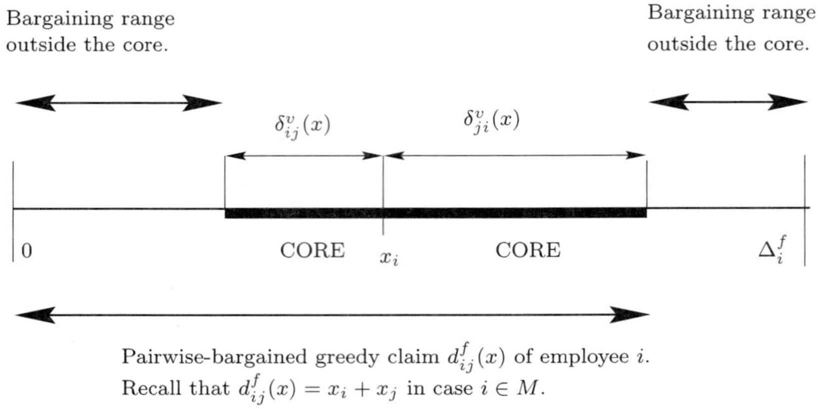

Figure 2

TU-game with the two-sided economic model presented in Section 1. Besides the two-sided pseudo-concave TU-game v, we want to put forward another TU-game that can be used to develop a similar kind of theory. The alternative TU-game reflects a very optimistic view in that the worth of every coalition represents an ideal achievement of profits. According to this optimistic view, the presence of at least one manager guarantees the achievement of the overall net profit, so the worth of every coalition with at least one manager equals the total amount of $f(N)$. Further, the worth of a single worker equals his incremental net profit and cooperation between a group of workers is supposed to be more profitable in that the worth of every coalition with no manager is not less than the sum of the incremental net profits of each of those workers.

Theorem 5.1 *Let $N = M \cup W$ where $M \cap W = \emptyset$, $W \neq \emptyset$, $M \neq \emptyset$, and $f : \mathcal{P}(N) \to \mathbb{R}$ a monetary net profit function satisfying conditions (2.1) and (2.2). Let $u : \mathcal{P}(N) \to \mathbb{R}$ be a TU-game satisfying*

$$u(S) = f(N) \qquad \text{for all } S \in \mathcal{P}(N) \text{ with } S \cap M \neq \emptyset \tag{5.2}$$

$$u(\{i\}) = \Delta^f_i \geq 0 \qquad \text{for all } i \in W \tag{5.3}$$

$$f(N) \geq u(S) \geq \sum_{k \in S} \Delta^f_k \text{ for all } S \in \mathcal{P}(N) \text{ with } S \subseteq W \text{ and } |S| \geq 2. \tag{5.4}$$

For every $y = (y_k)_{k\in N} \in \mathbb{R}^N$ satisfying $\sum_{k\in N} y_k = f(N)$ the following equivalences hold:

(i) $y \in CORE(u)$ i.e., $\sum_{k\in S} y_k \leq u(S)$ for all $S \subsetneq N$, $S \neq \emptyset$.

(ii) $y_i \geq 0$ for all $i \in N$ and $y_i \leq \Delta^f_i$ for all $i \in W$.

Proof The proof is similar to the proof of Lemma 3.1 and is left to the reader. □

6 A disadvantageous syndicate of managers

According to any pairwise-bargained consistent allocation, managers receive identical amounts (cf. (2.5)). The main goal of this section is to investigate the

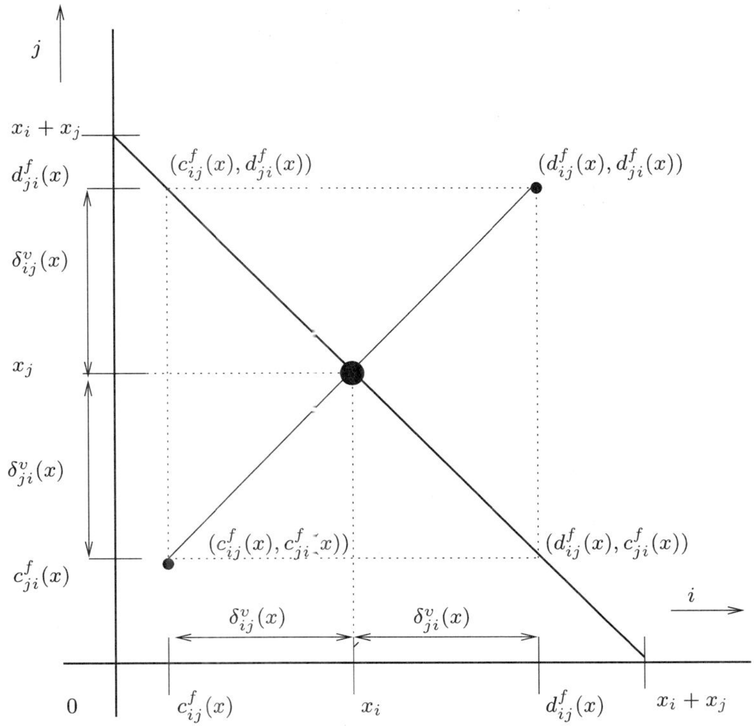

Sketch of pairwise-bargained modest claims $c^f_{ij}(x)$ and
pairwise-bargained greedy claims $d^f_{ij}(x)$ and
the pairwise bargaining ranges $\delta^v_{ij}(x)$ within the core.

Figure 3 This figure was suggested by William Thomson and Jim Schummer.

behaviour of pairwise-bargained consistent allocations when a syndicate is formed by the managers of the two-sided firm. By a *syndicate* of employees of the (two-sided) firm we mean a union of employees who have undertaken to operate as a single entity and hence, a syndicate is treated as a single agent. Particularly, we establish the disadvantageousness of the pairwise-bargained consistent allocations when a syndicate is formed by the managers of the two-sided firm. In other words, it is shown that the syndication of the managers does not increase the allocation to the syndicated managers if the employees are paid according to the pairwise-bargained consistent allocations.

Throughout this section we suppose that there is at least one manager and we are going to interpret the set M of managers as a single agent (denoted by 0) whose agreement on the production process of a certain commodity is indispensable to generate any output.

Definition 6.1 *Let $N = M \sqcup W$ where $M \cap W = \emptyset$, $W \neq \emptyset$, $M \neq \emptyset$, and $f : \mathcal{P}(N) \to \mathbb{R}$ a monetary net profit function satisfying conditions (2.1), (2.2), and $f(S) \geq f(M)$ for all $M \subseteq S \subseteq N$. Let $N^* = \{0\} \cup W$ and define the corresponding function $g : \mathcal{P}(N^*) \to \mathbb{R}$ by*

- $g(T) = 0$ for all $T \in \mathcal{P}(N^*)$, $T \neq \emptyset$, with $0 \notin T$
- $g(T) = f((T \backslash \{0\}) \cup M)$ for all $T \in \mathcal{P}(N^*)$ with $0 \in T$.

Lemma 6.2 *In the setting of Definition 6.1, let the agents' incremental returns with respect to the function* $g : \mathcal{P}(N^*) \to \mathbb{R}$ *be given by* $\Delta_k^g := g(N^*) - g(N^* \backslash \{k\})$ for all $k \in N^*$. *Then it holds*

(i) $\Delta_k^g = \Delta_k^f$ for all $k \in W$ and $\Delta_0^g = f(N)$.

(ii) $g(T) \geq 0$ and $g(N^*) - g(T) \geq \sum_{k \in N^* \backslash T} \Delta_k^g$ for all $T \in \mathcal{P}(N^*)$ with $0 \in T$

(that is, the function g *inherits the pseudo-concavity from the function* f*).*

The straightforward proof of Lemma 6.2 is left to the reader. In the framework of the pseudo-concave function g with a unique manager 0, the corresponding unique pairwise-bargained consistent (PBC) allocation $\bar{y} = (\bar{y}_k)_{k \in N^*} \in \mathbb{R}^{N^*}$ is determined by (cf. Remarks 2.4 and 2.6)

$$\bar{y}_i = \tfrac{1}{2}\Delta_i^g \quad \text{for all } i \in W \quad \text{and} \quad \bar{y}_0 = g(N^*) - \sum_{k \in W} \tfrac{1}{2}\Delta_k^g.$$

By Lemma 6.2(i), the incremental net profit of each worker is invariant under the transformation of the net profit function and therefore, the PBC-allocation with respect to the situation, in which the managers form a syndicate, is determined by

$$\bar{y}_i = \tfrac{1}{2}\Delta_i^f \quad \text{for all } i \in W \quad \text{and} \quad \bar{y}_0 = f(N) - \sum_{k \in W} \tfrac{1}{2}\Delta_k^f,$$

where $\bar{\lambda} = \bar{y}_0 - f(M)$ is the unique nonnegative solution λ of the equation

$$\lambda + f(M) + \sum_{k \in W} \min\left[\lambda,\ \tfrac{1}{2}\Delta_k^f\right] = f(N). \tag{6.1}$$

In comparison with the original situation, in which each manager behaves as one individual (but they are of the same type), the corresponding PBC-allocation $\bar{x} = (\bar{x}_k)_{k \in N} \in \mathbb{R}^N$ is of the form $\bar{x}_i = \bar{\rho}$ for all $i \in M$, and $\bar{x}_i = \min\left[\bar{\rho},\ \tfrac{1}{2}\Delta_i^f\right]$ for all $i \in W$, where $\bar{\rho} \in \mathbb{R}$ is the unique nonnegative solution ρ of the equation

$$|M| \cdot \rho + \sum_{k \in W} \min\left[\rho,\ \tfrac{1}{2}\Delta_k^f\right] = f(N). \tag{6.2}$$

Obviously, it holds that $\bar{y}_i = \tfrac{1}{2}\Delta_i^f \geq \min\left[\bar{\rho},\ \tfrac{1}{2}\Delta_i^f\right] = \bar{x}_i$ for all $i \in W$. Thus, the syndication of the managers does not decrease the allocation to any worker and hence, the syndicated managers do not profit from their syndication (i.e., $\bar{y}_0 \leq |M| \cdot \bar{\rho}$). We conclude that it is *not* advantageous for the managers to act as a syndicate if they are paid according to PBC-allocations. This disadvantageousness is due to the fact that the veto power of the new agent (representing the syndicated managers) is not so influential as the veto power of every individual manager in the original situation.

Note that $\bar{y}_i = \bar{x}_i$ for all $i \in W$ iff $\bar{\rho} \geq \tfrac{1}{2}\Delta_i^f$ for all $i \in W$ and $\bar{\rho} = \frac{1}{|M|} \cdot \left[f(N) - \sum_{k \in W} \tfrac{1}{2}\Delta_k^f\right]$. That is, both PBC-allocations with respect to the functions f and g respectively coincide if and only if $f(N) \geq \frac{|M|}{2} \cdot \max_{k \in W} \Delta_k^f + \sum_{k \in W} \tfrac{1}{2}\Delta_k^f$. As shown in Remark 2.6, the latter inequality holds whenever there is a unique manager. For example, if two of the three employees are managers, then the syndication of the two managers leads to a strict smaller PBC-allocation whenever the net profit function f satisfies $f(N) > 3 \cdot f(M) \geq 0$.

7 Concluding remarks

In the framework of two-sided pseudo-concave TU-games, the pairwise-bargained claim $c_{ij}(x)$ of employee i, as given by (2.3), may be re-interpreted as the smallest core-income to employee i concerning the two-person situation together with employee j (with respect to the underlying core-allocation x) or equivalently, the pairwise-bargained claim $c_{ij}(x)$ of employee i represents the pairwise-bargained range of employee i over employee j outside the core of the two-sided pseudo-concave game. Consequently, a core-allocation is pairwise-bargained consistent if and only if it is the midpoint of every pairwise-bargained range between two employees within the core of the two-sided pseudo-concave game. There exists a unique nonnegative pairwise-bargained consistent allocation (that belongs to the core of the two-sided pseudo-concave game).

The research into the pairwise-bargained consistency theory (including the essential notions of pairwise-bargained claims, consistency, core, prekernel, and pairwise-bargained ranges within and outside the core) took place in two stages. In the first stage (1994-1995), the development of this kind of theory (applied to different types of pairwise-bargained claims) has also been carried out in the context of the so-called bankruptcy games (cf. [**1**], [**5**]) as well as the bilateral assignment games (cf. [**11**], [**4**]). Together with the two-sided pseudo-concave games, each of these three classes of games has been treated separately because no overlap was noticed at the time of this pioneer work. In the second stage of the research (1996), the pairwise-bargained consistency theory has been further developed in that the relevant theory is applicable to an enlarged class of games of which the bankruptcy games and the two-sided pseudo-concave games form two particular subclasses. Roughly speaking, the games in that enlarged class are characterized by the fact that their core is nonempty and fully determined by the individual core-constraints as well as the core-constraints of the form (3.1) referring to coalitions excluding exactly one player (that is, core-constraints referring to other coalitions are supposed to be redundant). In other words, a game has to satisfy a slightly extended version of the pseudo-concavity condition (2.2) (cf. [**9**], Proposition 2.1, page 5). The bilateral assignment games, however, do not fit into this general framework (because the core of every bilateral assignment game is fully determined by the core-constraints referring to individuals and pairs of agents of different types). Moreover, for bilateral assignment games, it turns out that the set of all pairwise-bargained consistent core-allocations is not necessarily a singleton. Beyond the general class of games treated in [**9**] and the class of bilateral assignment games, the author is not aware of other classes of TU-games to which a similar pairwise-bargained consistency theory applies.

Acknowledgements

Thanks are given to William Thomson and Jim Schummer for helpful discussions and valuable comments.

References

[1] Aumann, R. J. and Maschler, M. [1985], *Game theoretic analysis of a bankruptcy problem from the Talmud*, J. Econom. Theory **36**, 195–213.

[2] Driessen, T. S. H. [1985], *Properties of* 1-*convex* n-*person games*, OR Spektrum **7**, 19–26.

[3] Driessen, T. S. H. [1988], *Cooperative Games, Solutions, and Applications*, Theory and Decision Library, Series C: Game Theory, Mathematical Programming and Operations Research **3** (S. H. Tijs, ed.), Kluwer Academic Publishers, Dordrecht, The Netherlands.

[4] Driessen, T. S. H. [1995], *On the core and kernel of the bilateral assignment market*, Memorandum No. 1284, Department of Applied Mathematics, University of Twente, Enschede, The Netherlands.

[5] Driessen, T. S. H. [1995], *An alternative game theoretic analysis of a bankruptcy problem from the Talmud: the case of the greedy bankruptcy game*, Memorandum No. 1286, Department of Applied Mathematics, University of Twente, Enschede, The Netherlands.

[6] Maschler, M., Peleg, B. and Shapley, L. S. [1979], *Geometric properties of the kernel, nucleolus, and related solution concepts*, Math. Oper. Res. **4**, 303–338.

[7] Moulin, H. [1988], *Axioms of cooperative decision making*, Econometric Society Monographs **15**, (J. M. Grandmont and C. F. Manski, eds.), Cambridge University Press, Cambridge, N. Y., U.S.A.

[8] Muto, S., Nakayama, M., Potters, J. and Tijs, S. [1988], *On big boss games*, Econom. Stud. Quart **39**, 303–321.

[9] Namekata, T. and Driessen, T. S. H. [1996], *Pairwise-bargained consistency: a relationship between the nucleolus and the* τ-*value through an associated bimatrix game*, Memorandum No. 1332, Department of Applied Mathematics, University of Twente, Enschede, The Netherlands.

[10] Potters, J., Poos, R., Tijs, S. and Muto, S. [1989], *Clan games*, Games Econom. Behav. **1**, 275–293.

[11] Rochford, S. C. [1984], *Symmetrically pairwise-bargained allocations in an assignment market*, J. Econom. Theory **34**, 262–281.

Fields Institute Communications
Volume **23**, 1999

A New Characterization of Totally Balanced Games

Juan Enrique Martínez-Legaz
Departament d'Economia and CODE
Universitat Autònoma de Barcelona
Edifici B 08193 Bellaterra (Barcelona) Spain
juanenrique.martinez@uab.es

Abstract. Totally balanced games are characterized in terms of indirect functions. The result is based on an earlier characterization due to Kalai and Zemel, according to which the class of totally balanced games is generated by that of additive games by means of the minimum operation.

1 Introduction

A totally balanced game was defined by Shapley and Shubik [**5**] as a game with the property that all subgames of the game have nonempty cores. Shapley and Shubik proved that totally balanced games are the same as market games. A market game is a game generated by an exchange economy whose traders have continuous, concave utility functions. Another characterization was provided by Kalai and Zemel in [**2**], who proved that totally balanced games coincide with games that can be expressed as the minimum of a finite collection of additive games. From this result they obtained that every totally balanced game is a flow game (i.e., a game associated with a directed network in such a way that the value of a coalition is defined as the maximum source-to-sink flow that can be sent by using only the edges which are owned by the members of the coalition). In fact, the converse also holds since using the max flow-min cut theorem of Ford and Fulkerson [**1**], Kalai and Zemel proved that every flow game is totally balanced.

The aim of this note is to give a new characterization of totally balanced games, in terms of their associated indirect functions. The concept of the indirect function of a cooperative game with transferable utility was introduced in [**3**]. It is a convex function defined on $\mathbb{R}^n$, n being the number of players, which provides an alternative definition of the game as it contains exactly the same information as the characteristic function. Indeed, a simple formula permits the recovery of the characteristic function from the indirect function.

1991 *Mathematics Subject Classification.* Primary 90D12; Secondary 26B25.

This work was partially supported by DGICYT (Spain), Project PB92-0615, and by the Comissionat per Universitats i Recerca de la Generalitat de Catalunya, Grant GRQ93-2044.

2 Preliminaries

An n-person cooperative game with transferable utility, or briefly, a TU game, is a pair $\Gamma = (N, v)$, where N is a finite set of players and $v : 2^N \longrightarrow \mathbb{R}$ is a function defined on the power set of N satisfying the condition $v(\emptyset) = 0$, called the characteristic function of the game. A fundamental solution concept for TU games is the core:

$$C(\Gamma) = \left\{ x \in \mathbb{R}^N \mid \sum_{i \in S} x_i \geq v(S) \quad \forall S \subset N, \sum_{i \in N} x_i = v(N) \right\}.$$

When this set is nonempty, Γ is called balanced. Given a coalition $S \in 2^N$, consider the subgame $\Gamma^S = (S, v^S)$, where v^S denotes the restriction of v to 2^S. If Γ^S is balanced for every $S \in 2^N$ then Γ is said to be totally balanced.

The indirect function of a TU game $\Gamma = (N, v)$, with $N = \{1, \ldots, n\}$, is $\pi : \mathbb{R}^n \longrightarrow \mathbb{R}$, defined by

$$\pi(w) = \max_{S \subset N} \left\{ v(S) - \sum_{i \in S} w_i \right\} \quad \forall w = (w_1, \ldots, w_n) \in \mathbb{R}^n.$$

The indirect function admits the following economic interpretation. Suppose an employer has to choose among the players those who will produce the maximum profit for him. If the subset S is selected then the total amount of output S will produce, using the resources owned by the employer, is $v(S)$. If the (possibly negative) salary demanded by player i is w_i then $\pi(w)$ represents the maximum net profit the employer can obtain under the vector of salaries w. Indirect functions are characterized in [**3**] by the following three properties:

(i) $\partial\pi(w) \bigcap \{0, -1\}^n \neq \emptyset \quad \forall w \in \mathbb{R}^n$

(ii) $\{0, -1\}^n \subset \bigcup_{w \in \mathbb{R}^n} \partial\pi(w)$

(iii) $\min_{w \in \mathbb{R}^n} \pi(w) = 0$;

the symbol ∂ stands for subdifferential in the sense of convex analysis (see [**4**]). Moreover, it is proved in [**3**] that the characteristic function of $\Gamma = (N, v)$ can be obtained from its associated indirect function according to the following formula:

$$v(S) = \min_{w \in \mathbb{R}^n} \left\{ \pi(w) + \sum_{i \in S} w_i \right\} \quad \forall S \subset N.$$

One can also find there some characterizations of standard properties of TU games, such as symmetry, superadditivity, decomposability, convexity and monotonicity, in terms of indirect functions. In the next section, we will add to this list a characterization of totally balanced games.

3 The characterization theorem

Our characterization of totally balanced games will be based on the fact, proved by Kalai and Zemel [**2**], that these games can be expressed as minima of finite collections of additive games:

Theorem 3.1 *Let* $\Gamma = (N, v)$ *be a TU game and let* $\pi : \mathbb{R}^n \longrightarrow \mathbb{R}$ *be its indirect function. The following statements are equivalent:*

1°. Γ *is totally balanced.*

2°. $\pi(w) = \max_{S \subset N} \min_{\bar{w} \in \pi^{-1}(0)} \sum_{i \in S} (\bar{w}_i - w_i) \qquad \forall w \in \mathbb{R}^n.$

3°. $\{0, -1\}^n \subset \bigcup_{w \in \pi^{-1}(0)} \partial\pi(w).$

Proof $1^\circ \Rightarrow 2^\circ$. Since Γ is the minimum game of a finite collection of additive games, for every $S \subset N$ one has

$$v(S) = \min\left\{\sum_{i \in S} \bar{w}_i \,|\, \sum_{i \in S'} \bar{w}_i \geq v(S') \quad \forall S' \subset N\right\}.$$

But condition $\sum_{i \in S'} \bar{w}_i \geq v(S')$ $\forall S' \subset N$ can be equivalently written as $\pi(\bar{w}) = \max_{S' \subset N} \{v(S') - \sum_{i \in S'} \bar{w}_i\} \leq 0$; in fact, the last inequality holds with the equal sign since the maximum is attained at $S' = \emptyset$. Therefore, one has

$$v(S) = \min_{\bar{w} \in \pi^{-1}(0)} \sum_{i \in S} \bar{w}_i \qquad \forall S \subset N,$$

which, together with the definition of π, gives 2°.

$2^\circ \Rightarrow 3^\circ$. Let -1_S be any point in $\{0, -1\}^n$, with $S \in 2^N$ (1_S denotes the characteristic vector of coalition S). According to (ii), there exists $\tilde{w} \in \mathbb{R}^n$ such that $-1_S \in \partial\pi(\tilde{w})$. By 2°, we have $\pi(\tilde{w}) \geq \min_{\bar{w} \in \pi^{-1}(0)} \sum_{i \in S} (\bar{w}_i - \tilde{w}_i)$, whence there exists $\bar{w} \in \pi^{-1}(0)$ such that $\pi(\tilde{w}) \geq \sum_{i \in S} (\bar{w}_i - \tilde{w}_i) =< \tilde{w} - \bar{w}, -1_S >$. Then, for any $w \in \mathbb{R}^n$ one has

$$\pi(w) \geq \pi(\tilde{w}) + < w - \tilde{w}, -1_S > \geq < \tilde{w} - \bar{w}, -1_S > + < w - \tilde{w}, -1_S >$$

$$= < w - \bar{w}, -1_S > = \pi(\bar{w}) + < w - \bar{w}, -1_S >,$$

i.e., $-1_S \in \partial\pi(\bar{w})$. This proves 2°.

$3^\circ \Rightarrow 1^\circ$. For each coalition $S \in 2^N$ we pick a vector $w^S \in \pi^{-1}(0)$ satisfying $-1_S \in \partial\pi(w^S)$. Given any coalition S and any vector $w \in \mathbb{R}^n$, one has

$$\pi(w) \geq \pi(w^S) + < w - w^S, -1_S >$$

or equivalently,

$$\pi(w) + \sum_{i\in S} w_i \geq \pi(w^S) + \sum_{i\in S} w_i^S.$$

Therefore, as $\pi(w^S) = 0$, one has $v(S) = v_S(S)$, v_S being the additive game defined by $v_S(\{i\}) = w_i^S$. On the other hand, for any $S' \in 2^N$ one has

$$v(S) \leq \pi(w^{S'}) + \sum_{i\in S} w_i^{S'} = v_{S'}(S).$$

It follows that $v = \min_{S'\subset N} v_{S'}$, so that Γ is the minimum game of a finite collection of additive games. This completes the proof. □

Combining this result with the characterization of indirect functions described in the preceding section, one gets:

Corollary 3.2 *A function $\pi : \mathbb{R}^n \longrightarrow \mathbb{R}$ is the indirect function of a (unique) totally balanced n-person TU game if and only if it satisfies (i), (iii) and*

$$\text{(ii')} \quad \{0,-1\}^n \subset \bigcup_{w\in\pi^{-1}(0)} \partial\pi(w).$$

Let us observe that balanced games (i.e., games with a nonempty core) also admit a simple characterization in terms of indirect functions, since the core of $\Gamma = (N, v)$ can be represented in terms of π (its indirect function), in the following way [**3**]:

$$C(\Gamma) = \{x \in \mathbb{R}^n | (-t,\ldots,-t) \in \partial\pi(x) \qquad \forall t \in [0,1]\}.$$

From this and the fact that the indirect function π^S of the subgame Γ^S induced by the coalition S is given by

$$\pi^S(w_S) = \min_{z_{N\setminus S}} \pi(w_S, z_{N\setminus S}) = \lim_{z_{N\setminus S}\to+\infty} \pi(w_S, z_{N\setminus S})$$

[cf. Proposition 12 and the subsequent comments in [**3**]], one can derive a (somewhat complicated) alternative characterization of totally balanced games in terms of indirect functions.

Another consequence of the Kalai and Zemel [**2**] characterization of totally balanced games underlying our results is that, for an arbitrary TU game $\Gamma = (N, v)$, there exists a smallest majorant $\bar{v}$ of v such that the game $\bar{\Gamma} = (N, \bar{v})$ is totally balanced. Indeed, one can define $\bar{v}$ as the pointwise infimum of the collection of additive majorants of v, namely,

$$\bar{v}(S) = \inf\left\{\sum_{i\in S} w_i | \sum_{i\in S'} w_i \geq v(S') \quad \forall S' \in 2^N\right\} \qquad \forall S \in 2^N.$$

Being a linear programming problem with a nonempty feasible set and a bounded from below (by $v(S)$) objective function, the minimization problem defined by the right-hand side of this equality has an optimal solution $w^S = (w_1^S, \ldots, w_n^S)$. One can easily prove that $\bar{v} = \min_{S\subset N} v_S$, with v_S denoting the additive game given by $v_S(\{i\}) = w_i^S$. Hence, $\bar{v}$ is totally balanced. By the linear programming duality theorem, an equivalent definition of $\bar{v}$ is

$$\bar{v}(S) = \max\left\{\sum_{S'\subset N}\gamma_{S'}v(S') \mid \sum_{\substack{S'\subset N\\ S'\ni i}}\gamma_{S'} = 1_S(i)\,\forall i\in N, \gamma_{S'}\geq 0\;\forall S'\in 2^N\right\}$$

$$= \max\left\{\sum_{S'\subset S}\gamma_{S'}v(S') \mid \sum_{\substack{S'\subset S\\ S'\ni i}}\gamma_{S'} = 1\;\forall i\in S, \gamma_{S'}\geq 0\;\forall S'\in 2^N\right\},$$

which coincides with the definition of the cover of a game given by Shapley and Shubik [**5**]. Although these authors proved that $\bar{v} \geq v$ with equality if and only if the game is totally balanced, they did not observe that $\bar{v}$ is just the smallest majorant of v that defines a totally balanced game.

The equivalence between totally balanced games and market games was proved by Shapley and Shubik [**5**] by showing that the cover of a game $\Gamma = (N, v)$ is generated by the direct market induced by Γ, defined as follows. The traders of the direct market are the players of Γ, the commodity space is $\mathbb{R}^n_+$, the initial endowments are the unit vectors (the interpretation of which is that each trader possesses one unit of "himself") and all traders share the same utility function $u : \mathbb{R}^n_+ \longrightarrow \mathbb{R}$, given by

$$u(x) = \max\left\{\sum_{S\subset N}\gamma_S\, v(S) \mid \gamma_S \geq 0 \quad \forall S\subset N, \sum_{S\ni i}\gamma_S = x_i \qquad \forall i = 1,\ldots,n\right\}.$$

This utility function can be explained by supposing that one tries to distribute in an optimal way the existing x_i "units of trader i" among all possible coalitions. Using the duality theorem of linear programming, the utility function can be written as

$$u(x) = \min\left\{\sum_{i=1}^{n} x_i w_i \mid \sum_{i\in S} w_i \geq v(S) \quad \forall S\subset N\right\}.$$

But, as observed in the proof of the theorem, the inequality system appearing in the latter expression can be simply written as $\pi(w) = 0$. Thus, we obtain the following expression for u,

$$u(x) = \min_{w\in\pi^{-1}(0)} < w, x >,$$

which gives the utility function of the direct market in terms of the indirect function of the game. In the converse direction, the indirect function of the market game generated by a direct market can be obtained from the utility function by means of

$$\pi(w) = \max_{S\subset N}\left\{u(1_S) - \sum_{i\in S} w_i\right\},$$

since one has $v(S) = u(1_S) \quad \forall S\subset N$ (see [**5**]).

To conclude, let us briefly consider the case of monotone games. Recall that $\Gamma = (N, v)$ is said to be monotone if v is nondecreasing with respect to inclusion, i.e., if one has $v(S) \leq v(T)$ whenever $S \subset T$. As proved in [**3**], for monotone games one only needs to know the restriction of the indirect function to the nonnegative orthant in order to recover the characteristic function. This restriction, which will be denoted by $\pi^+ : \mathbb{R}^n_+ \longrightarrow \mathbb{R}$, is called the restricted indirect function of Γ. The formula expressing v in terms of π^+ is simpler than the one relating v to π:

$$v(S) = \min_{w_{N\setminus S}} \pi^+(0_S, w_{N\setminus S}) \qquad \forall S \subset N.$$

Also, two properties suffice to characterize restricted indirect functions, namely, (i) and (iii) of Section 1 with π and $\mathbb{R}^n$ replaced by π^+ and $\mathbb{R}^n_+$, respectively. One can easily check that the proofs of our results in this paper can be adapted to demonstrate that they remain valid in the case of monotone games, with those replacements.

Acknowledgements

I wish to thank Salvador Barberà, Theo Driessen and Myrna Wooders for helpful comments and suggestions.

References

[1] Ford, L. R. and Fulkerson, D. R. [1962], *Flows in Networks*, Princeton University Press, Princeton.

[2] Kalai, E. and Zemel, E. [1982], *Totally Balanced Games and Games of Flow*, Math. Oper. Res. **7**, 476-478.

[3] Martínez-Legaz, J.-E. [1996], *Dual Representation of Cooperative Games Based on Fenchel-Moreau Conjugation*, Optimization **36**, 291–319.

[4] Rockafellar, R. T. [1970] *Convex Analysis*, Princeton University Press, Princeton.

[5] Shapley, L. S. and Shubik, M. [1969], *On Market Games*, J. Econom. Theory **1**, 9-25.

Fields Institute Communications
Volume **23**, 1999

The Worth of a Cooperative Enterprise to Each Member

Michael Maschler
Department of Mathematics
The Hebrew University
Jerusalem, Israel
maschler@vms.huji.ac.il

1 Introduction

It is well known that game theory – especially cooperative game theory – is rich in "solution concepts"; namely, recommendations to choose particular payoff-vectors as outcomes of games in a given class. This state of affairs makes it necessary to choose a particular solution, whenever one encounters a real-life conflict. Thus, it is highly desirable to know when and why a certain solution should be preferred to others. Putting it more mildly, it is desirable to know, what aspects does each solution reflect in any given real case. Contributions in these areas increase the applicability of game theory. They also shed light on the nature of the various solutions.

The present paper is a contribution of this type. We address ourselves to the following "real conflict": Consider a cooperative enterprise whose members have decided to sell it out and distribute its worth among themselves. How should they share the revenue? One method, often employed, is to share it in proportion to the investments of the members. Variants of this method appear in the form of shares, which are distributed by many companies. But, there is another method, based on a different principle, which is often employed by an "enterprise" consisting of husband and wife, who want to decide on a divorce settlement. According to this method, wife takes back every item that she brought ot the marriage, husband does the same and the rest is divided equally among the couple[1]. Can this procedure be generalized to more than two members?

In Section 2 we describe the various real situations we have in mind. We then construct abstract games to represent them.

Section 3 suggests a procedure that generalizes the divorce procedure. *A priori*, this procedure raises some queries and suggests conjectures. In Section 4 we resolve the queries and prove the conjectures. It turns out that our procedure leads to a

1991 *Mathematics Subject Classification.* Primary 90D10; Secondary 90D40, 90D80 .

Reprinted with permission from *Games, Ecomonics Dynamics and Time Series Analysis* M. Deistler, F. Fürst and G. Schwödiauer, eds , Physica Verlag, Wien-Würzburg, 1982, pp. 103–110.

[1]I owe this observation to Heinz Schleicher, who drew my attention to the fact that this procedure is often used. In fact, there are countries in which this is the legal procedure, unless marriage arrangements specify otherwise.

celebrated solution of the corresponding abstract game[2]. Section 5 is devoted to various comments, concerning, among other things, the insight gained on the nature of the solution concept.

2 Cooperative enterprises and their representation as games

To generalize a "marriage enterprise", we consider a set $N = \{1, 2, \ldots, n\}$ of *members* of the enterprise, whom we allow to *own* assets of the enterprise. As long as the enterprise is functioning, these assets serve its needs and all profits are distributed to the members in a way that need not concern us[3]. However, the assets are still legally owned by the members, who are entitled to claim their worths when the enterprise is sold out.

Of course if there are debts on such assets, such as mortgages and other loans, the owners are responsible to pay them off.

We also allow for the possibility that certain assets are owned by sets consisting of several members. Thus, if one identifies a member i with the set $\{i\}$, one can associate with each subset S of N, called "coalition", the set $a(S)$ of all the assets owned by S. We can also speak about $b(S)$, which is the set of all assets owned by subsets of S:

$$b(S) = \cup\{a(R) : R \subseteq S\}. \tag{2.1}$$

Example 2.1 A cooperative enterprise of three members 1, 2 and 3 operates a garage, a gasoline station, a store and a restaurant. Historically, it developed as follows: 1, who owned the garage, formed a partnership with 2, who owned the gasoline station. They worked together for a while, and since business was prosperous, they both bought and opened the store, hiring someone to serve the customers. Later, their neighbour 3, who owned the restaurant, joined the partnership and they all hoped that this enlarged enterprise will attract additional customers. Unfortunately, difficulties originating from differences in personalities made it impossible to continue and all agreed to sell everything and terminate the partnership. Of course, 1 still owned the garage, etc. Thus[4]

$$\begin{array}{lll} a(\{1\}) = \{\text{garage}\}, & a(\{2\}) = \{\text{gasoline station}\}, & \\ a(\{3\}) = \{\text{restaurant}\}, & a(\{1,2\}) = \{\text{store}\}, & a(\{1,3\}) = \emptyset, \\ a(\{2,3\} = \emptyset, & a(\{1,2,3\}) = \emptyset \quad \textit{and} & a(\emptyset) = \emptyset. \end{array}$$

The function $b(\cdot)$ can now be computed, using (2.1).

Our next step will be to represent situations such as the above as characteristic-function games $(N; \mathrm{v})$. To do so, we shall define:

$$\mathrm{v}(S) = \textit{the present net worth of the set of assets } b(S). \tag{2.2}$$

The worth is measured in a fixed monetary unit. We base the definition of $\mathrm{v}(S)$ on $b(S)$, instead of on $a(S)$, in order to reflect the idea that a coalition S can claim all the benefits when $b(S)$ is sold. It may well be, for example, that $b(S)$ can get a better price if it is sold as one package instead of as separate items. Of course, when $\mathrm{v}(S)$ is claimed by S, one still has to decide on its distribution among the members of S. Such distribution should take into account the various claims $\mathrm{v}(R), R \subseteq S$.

---

[2]To keep an element of surprise, we do not disclose at present the identity of the solution. Everyone who wants to miss the fun may turn directly to Section 4.

[3]We do not rule out the possibility that this distribution is also influenced by arguments used in this paper.

[4]For simplicity we ignore items that were bought, say, for the garage, after the partnership started; these should be included in a $(\{1, 2\})$ or in $a(\{1, 2, 3\})$.

The characteristic function for this example could be:

$$\begin{array}{lll} v(\{1\}) = 30, & v(\{2\}) = 12, & v(\{3\}) = 6, \\ v(\{1,2\}) = 36, & v(\{1,3\}) = 36, & v(\{2,3\}) = 48, \\ v(\{1,2,3\}) = 81, & v(\emptyset) = 0, & \end{array}$$

where the worths are measured in a certain monetary unit. There is a high mortgage on the store, which accounts for the fact that $v(\{1,2\}) < v(\{1\}) + v(\{2\})$. The gasoline station and the restaurant can get a higher price if sold as one package instead of as separate items; therefore, $v(\{2,3\}) > v(\{2\}) + v(\{3\})$, etc.

3 Sharing the worth of the enterprise

To find a procedure for sharing the worth of the enterprise in a way which resembles the divorce settlement, let us proceed naively, working with the figures in Example 2.1.

Let us assume that $\{3\}$ claims the worth of his restaurant, then 3 gets paid 6 units, and this amount should be deducted from all coalitions containing $\{3\}$. Indeed, their worth is based on a package of items that contains $(\{3\})$, but they have to return $v(\{3\})$ to 3, whenever he claims it. We thus arrive at a new game $(N; v^1)$, where

$$\begin{array}{llll} v^1(\{1\}) = 30, & v^1(\{2\}) = 12 & v^1(\{3\}) = 0, & v^1(\{1,2\}) = 36, \\ v^1(\{1,3\}) = 30, & v^1(\{2,3\}) = 42, & v^1(\{1,2,3\}) = 75, & v^1(\emptyset) = 0. \end{array}$$

Suppose now that $\{1,2\}$ claims its worth of 36 units, then 1 and 2 get 18 each, and we arrive at a game (N, v^2) by deducting 36 from each coalition containing $\{1,2\}$, leaving the rest unchanged. Later, $\{1\}$ will claim the worth of his assets. He will get paid 30 units and a debt of 30 will be imposed on all coalitions containing 1, including $\{1,2\}$. Similar debts will be imposed when 2 claims his worth. These debts will be shared by $\{1,2\}$ in equal parts, so that in fact, $\{1,2\}$ will be sharing equally only $v(\{1,2\}) - v(\{1\}) - v(\{2\})$. Similar reasonings hold for all coalitions. Table 1 exhibits a possible sequence of events. It leads to a final payoff-vector: $(30, 27, 24)$.

1	2	3	{1}	{2}	{3}	{1, 2}	{1, 3}	{2, 3}	{1, 2, 3}	
0	0	0	30	12	6	36	36	48	81	$:v^0$
0	0	6	30	12	0	36	30	42	75	$:v^1$
18	18	0	30	12	0	0	30	42	39	$:v^2$
30	0	0	0	12	0	-30	0	42	9	$:v^3$
0	12	0	0	0	0	-42	0	30	-3	$:v^4$
0	15	15	0	0	0	-42	0	0	-33	$:v^5$
-11	-11	-11	0	0	0	-42	0	0	0	$:v^6$
-21	-21	0	0	0	0	0	0	0	42	$:v^7$
14	14	14	0	0	0	0	0	0	0	$:v^8$
30	27	24								

Table 1 A possible sequence for arriving at a share

The above procedure certainly generalizes the divorce settlement, but it raises several doubts: Does every sequence lead to the same payoff-vector? Must it be a finite process? Shall one always obtain a share of $v(N)$? These questions will be answered in Section 4. Here we would like to examine a particular sequence;

namely, a shortest sequence obtained when 1-person coalitions claim their worths first, then – 2-person coalitions and finally – the grand coalition. Table 2 exhibits such a sequence.

1	2	3	{1}	{2}	{3}	{1,2}	{1,3}	{2, 3}	{1,2,3}	
0	0	0	30	12	6	36	36	48	81	$:\mathrm{v}^0$
30	0	0	0	12	6	6	6	48	51	$:\mathrm{v}^1$
0	12	0	0	0	6	-6	6	36	39	$:\mathrm{v}^2$
0	0	6	0	0	0	-6	0	30	33	$:\mathrm{v}^3$
-3	-3	0	0	0	0	0	0	30	39	$:\mathrm{v}^4$
0	15	15	0	0	0	0	0	0	9	$:\mathrm{v}^5$
3	3	3	0	0	0	0	0	0	0	$:\mathrm{v}^6$
30	27	24								

Table 2 A shortest sequence for arriving at a share.

The fact that we have arrived at the same final payoff-vector suggests that the procedure may have some merits. Let us therefore formulate it precisely:

Procedure 3.1 *Starting with the original game* $(N;\mathrm{v})$, *define a sequence of games* $(N;\mathrm{v}) \equiv (N;\mathrm{v}^0), (N,\mathrm{v}^1), (N;\mathrm{v}^2), \ldots$ *and payoff-vectors as follows: At game* $(N;\mathrm{v}^k)$, *a coalition* S_k *for which* $\mathrm{v}^k(S_k) \neq 0$, *claims its worth* $\mathrm{v}^k(S_k)$, *which is shared by the members of* S_k *in equal parts. The game* $(N;\mathrm{v}^{k+1})$ *is given by*

$$\mathrm{v}^{k+1}(R) = \begin{cases} \mathrm{v}^k(R) - \mathrm{v}^k(S_k), & \textit{whenever } R \supseteq S_k, \\ \mathrm{v}^k(R), & \textit{otherwise.} \end{cases} \tag{3.1}$$

The process terminates when all coalitions arrive at a zero worth. The final payoff is the amount of money that was accumulated at each step.

4 Justifying the procedure

The next two theorems will show that Procedure 3.1 is sound.

Theorem 4.1 *Procedure 3.1 must always terminate in a finite number of steps.*

Proof When a coalition claims its worth, its new worth becomes zero. It can regain a non-zero worth only after a proper subcoalition has claimed its own worth. Suppose that a process is endless, then at least one coalition claims its worth infinitely often, so that there exists a coalition S of minimal size, that claims its worth infinitely often. This can only occur, if a proper subcoalition claims its own worth infinitely often, contrary to the minimality of the size of S. The contradiction shows that each sequence of Procedure 3.1 must be finite. □

Theorem 4.2 *Regardless of the sequence chosen for Procedure 3.1, the final share is the Shapley value of the game* $(N;\mathrm{v})$.

Proof By (3.1) $\mathrm{v}^k = \mathrm{v}^{k+1} + w^k$, where

$$w^k(R) = \begin{cases} \mathrm{v}^k(S_k), & \text{whenever } N \supseteq R \supseteq S_k, \\ 0 & \text{otherwise.} \end{cases} \tag{4.1}$$

Here, $(N;\mathrm{v}^k), k = 0,1,\ldots,p$, is a particular sequence. Thus, $\mathrm{v} = \mathrm{v}^1 + w^0 = \mathrm{v}^2 + w^1 + w^0 = \cdots = \mathrm{v}^p + w^{p-1} + w^{p-2} + \cdots + w^0$. Now, $\mathrm{v}^p = 0$, therefore, $\mathrm{v} = w^0 + w^1 + \cdots + w^{p-1}$. In the game $(N;w^k)$, the players of S_k are symmetric and the others are zero-players; therefore, its Shapley value specifies $\mathrm{v}(S_k)/|S_k|$

to each member of S_k and 0 to the others. This is precisely the way $\mathrm{v}(S_k)$ was shared in the procedure. The proof concludes by using the additivity property of the Shapley value (see [**3**]). □

Remark 4.3 The shortest sequences obtained if 1-person coalitions claim their worths first, then – two-person coalitions, etc. In such sequences no coalition claims its worth more than once. For a game "in a general position" it takes $2^n - 1$ steps. This short sequence was treated by Harsanyi in [**2**], and he already observed that it leads to the Shapley value. Thus, Theorem 4.2 is a generalization of Harsanyi's result. The proof is new. Harsanyi used the procedure in order to describe a bargaining process that leads participants of an arbitrary cooperative game to the Shapley value. We apply the procedure for the particular case of liquidating a cooperative enterprise, where, we believe, it is particularly attractive; at any rate, it is not a game in the usual sense (see Section 5).

Remark 4.4 [5] Largest sequences are obtained if a coalition is allowed to claim its non-zero worth only if each coalition containing it properly has a zero worth. Such sequences may be quite long even for small values of n. Let s_n be the number of steps needed by such a sequence for a game "in a general position". It can easily be shown that s_n satisfies the recursion formula:

$$s_n = \binom{n}{0}(s_0+1) + \binom{n}{1}(s_1+1) + \ldots + \binom{n}{n-1}(s_{n-1}+1),\ s_0 = 0. \tag{4.2}$$

This recursive formula can be used to compute s_n for a small n; however, for theoretical purposes it is worthwhile to note that

$$s_n = \sum_{k=1}^{\infty} \frac{k^n}{2^k} - 1. \tag{4.3}$$

(The proof is by varifying (4.2).) Using the Stirling numbers of the second kind, $S(n,k)$, we can express s_n as a finite sum:

$$s_n = 1!S(n,1) + 2!S(n,2) + \ldots + n!S(n,n). \tag{4.4}$$

Estimating (4.3) by the integral $\int_0^\infty x^n/2^x dx$, we can obtain a rough estimate for s_n; namely,

$$s_n = \frac{n!}{(ln\, 2)^{n+1}}\left(1 + 0\left(\frac{1}{\sqrt{n}}\right)\right), \tag{4.5}$$

which yields a good idea about the speed of the growth of s_n, as n grows.

5 Afterthought

It is well known that many solution concepts for side-payment cooperative games $(N;\mathrm{v})$ can be employed, even if v is measured in monetary units instead of utilities. No requirement of transferable utility is needed; one merely requires that utility of money is a concave function for each player or, alternatively, that the rules of the game prohibit performing lotteries. Among such solutions one can name the core, the von-Neumann Morgenstern stable sets and the bargaining set. Indeed, one can employ them directly on "money games", because all they need

[5]The problems described in this remark were raised and solved by Micha A. Perles. We suppy here enough information so that the reader can prove the results for himself, but we omit the details in order not to depart significantly from the main line of this paper.

to be justified is the obviously true assumption, that each player prefers to have more money than less.[6] The Shapley value is an exception: Indeed, if one uses the Shapley value to find the expectation from playing a game, one must measure v in von-Neumann Morgenstern utilities, and one must assume that these utilities are transferable. This paper shows that the Shapley value can also be employed for finite "money games" – although of a certain kind.

If one agrees that Procedure 3.1 is sound, it is natural for him to ask whether the Shapley value could be recommended without the procedure – merely on the merit of the axioms that define it in [**3**]. Clearly, one wants to share $\mathrm{v}(N)$. The symmetry axiom imposes an obvious fairness requirement. Null players are players without assets – clearly they cannot claim anything, if the property of the enterprise is sold. It remains to check the additivity axiom: Imagine that the same members participate in two enterprises (N, v) and $(N; w)$. Suppose also that there are no complementarity relations between the assets of the two enterprises so that $b^{\mathrm{v}}(S) \cup b^{w}(S)$ is worth exactly $\mathrm{v}(S) + w(S)$ for each $S, S \subseteq N$. Under such circumstances it seems only reasonable that $\mathrm{v}(N)$ and $w(N)$ be shared separately,[7]

Mathematically, $(N; \mathrm{v})$ is an abstract game. When applied to real cases, it may have various interpretations. For example, $\mathrm{v}(S)$ can be interpreted as the worth of S, *if it is formed.* Under such an interpretation, one may argue that $(N; \mathrm{v}+w)$ does not represent "properly" the aggregate of the games $(N; \mathrm{v})$ and $(N; w)$, because S may be "more likely" to form in one of these games and less likely – in the other. No such objection can be raised in the case of the cooperative enterprise. Here, *all coalitions already exist.* They were formed before and while the organization of the enterprise was completed. These coalitions are responsible for their assets (including debts) and they all must liquidate. Under such circumstances, the additivity axiom is particularly attractive. Having observed this, perhaps we should not be surprised to find out that the Shapley value merged.

References

[1] Aumann, R. J. [1967], *A survey of cooperative games without side payments.* Essays in Mathematical Economics in Honor of Oskar Morgenstern (M. Shubik, ed.), Princeton, 3 - 27.

[2] Harsanyi, J. C. [1959], *A bargaining model for the cooperative n-person game.* Contributions to the Theory of Games, IV, (A. W. Tucker and R. D. Luce, eds.), Annals of Mathematics Studies **40**, Princeton, 325-355.

[3] Shapley, L. S. [1953], *A value for n-person games.* Contributions to the Theory of Games, II (H. W. Kuhn and A. W. Tucker, eds.), Annals of Mathematics Studies **28**, Princeton, 307-317.

[6] Thus an objection (a domination relation) is a sound concept for "money games", because it yields each player of a coalition actually more money. (See [**1**], Section 8, where this issue is dealt more thoroughly and in relation to the theory of cooperative games without side payments.)

[7] In fact, the method of sharing in proportion to the investments (see Section 1) can be criticized, *because it is not additive.* Try the games: $N = \{1,2\}, \mathrm{v}(\{1\}) = 20, \mathrm{v}(\{2\}) = 30, \mathrm{v}(\{1,2\}) = 100, w(\{1\}) = 10, w(\{2\}) = 40, w(\{1,2\}) = 1000, \mathrm{v}(\emptyset) = w(\emptyset) = 0$.

Fields Institute Communications
Volume **23**, 1999

Bargaining Between Heterogeneous Organizations

Gordon H. McCormick
Center for Terrorism and Irregular Warfare
Naval Postgraduate School
Monterey, CA 93943 USA
gmccormick@nps.navy.mil

Guillermo Owen
Department of Mathematics
Naval Postgraduate School
Monterey, CA 93943 USA
gowen@math.nps.navy.mil

Abstract. We consider a bargaining situation between two organizations which wish to divide some 'pie.' While members of the two organizations generally agree as to the value of the pie, not all members are eager for an agreement; rather, they assign different levels of utility to the possibility of continued conflict.

We represent this situation as an n-person game among all the members of the organizations, and analyze the game by determining its core. We consider both a no-side-payments model and a model in which some (restricted) side payments are allowed.

1 Introduction

We consider a conflict bargaining situation between two organizations, A and B, for a share of some "pie". The principal complication is that neither A nor B is a unitary actor. Each is modeled as a heterogeneous group in which there is a division of opinion over the merits of negotiation. Some percentage of the membership of each organization support continuing negotiations with the other side. The balance of each membership, however, assign a positive payoff to continuing conflict. Members of the latter group on each side will only agree to participate in the bargaining process if their side is able to obtain a large enough share of the pie.

2 The models of heterogeneous bargaining

We assume that there are two possible results for this game: either the two organizations come to an agreement, or they do not. However, while the disagreement outcome is unique (continuing conflict, represented by C), there are many possible agreement outcomes.

1991 *Mathematics Subject Classification.* Primary 90D12; Secondary 90D80.

As usual, we will represent the possible agreements by a set K in the positive quadrant of the plane: K is assumed to be non-empty, compact, and convex. For normalization, we will also assume that it consists of all points (s, t) in this quadrant, satisfying the inequalities $0 \leq t \leq \psi(s)$, where ψ is a monotonically decreasing concave function. The points on the curve $t = \psi(s)$ are the Pareto-optimal points of K. (In the simplest of cases, this could be the straight line $s + t =$ constant ($=$ size of the pie), but we allow the possibility of a nonhomogeneous pie, which means that the convex compact set K can be bounded by a curve rather than a line.)

Thus, we will characterize agreements by pairs (s, t) in K (usually, but not necessarily, on the Pareto-optimal boundary of K), where A receives a portion s of the pie, while B receives t. This s is treated as a public good for the members of A, while t is a public good for the members of B.

Our assumption, then, is that at an agreement, all members of A receive the same amount, as do all members of B. However, the two sets A and B are heterogeneous, some members being more avid for a peaceful settlement than others. We will represent this heterogeneity by assigning different utilities to the several players for the conflict outcome C.

Formally, we assume that each of the two groups is distributed along a $[0, 1]$ spectrum, according to two distribution functions, F_A and F_B. We assume that each of these two functions F_i is a monotone non-decreasing function of the real variable x, continuous on the right, satisfying

$$\begin{array}{lll} 0 \leq F_i(x) \leq 1 & \text{for all} & x \\ F_i(x) = 0 & \text{for all} & x < 0 \\ F_i(x) = 1 & \text{for all} & x > 1 \end{array}$$

where we interpret $F_i(x)$ to be the portion of the membership of organization i located between 0 and x (inclusive). In each case, the "doves" (i.e., those in favor of a rapprochement) lie close to 0, while the "hawks" (i.e., those who favor continuing conflict), lie close to 1 on the spectrum. Specifically, a player located at a position x $(0 \leq x \leq 1)$ on the spectrum receives utility x from outcome C.

For a given a εA, we will use $x(a)$ to denote $a's$ position; similarily, we use $y(b)$ for b's position, $b \varepsilon B$.

An agreement can only be reached if a sufficiently large set of players (a **winning coalition**) agrees to it. In the simplest case, a winning coalition is any set of players which contains at least a fraction α of organization A, and at least a fraction of β of organization B, where both α and β lie between 0 and 1. Similarly, a **blocking coalition** is any coalition which is not winning, but whose complement is not winning. In the simplest case, a blocking coalition would be any non-winning coalition with either more than $1 - \alpha$ of the members of A, or more than $1 - \beta$ of the members of B. We specify the game by assuming that a winning coalition can enforce any agreement, whereas a blocking coalition can always guarantee continued conflict.

A. The case without side payments. Model I.

In our first model, each individual player's utility in the event of an agreement depends only on the share of the pie that is obtained by his organization. Thus, if a division (s, t) is reached, the resulting utility will be

$$u_a(s, t) = s \quad \text{for each} \quad a \varepsilon A \tag{2.1}$$

$$u_b(s, t) = t \quad \text{for each} \quad b \varepsilon B. \tag{2.2}$$

On the other hand, the utility gained from conflict depends only on each person's position on the "moderate-radical" axis. Thus, in case of conflict (C)

$$u_a(C) = x(a) \tag{2.3}$$

$$u_b(C) = y(b). \tag{2.4}$$

Note that an individual's utility, in this case, does not depend on his or her organizational affiliation. Thus while a member of A at position 0.9 and a member of B at position 0.9 are bitter enemies, they have a common interest: both are likely to be better off with outcome C, continued conflict, than with any negotiated settlement that does not give their respective sides an overwhelming share of the pie.

In the simplest case, as we have discussed above (for winning coalitions), let x_0 be $F_A^{-1}(\alpha)$, defined as

$$F_A^{-1}(\alpha) = \min \{s \mid F_A(s) \geq \alpha\} \tag{2.5}$$

and similarly, let y_0 be $F_B^{-1}(\beta)$, where

$$F_B^{-1}(\beta) = \min \{t \mid F_B(t) \geq \beta\}. \tag{2.6}$$

We will analyze this particular bargaining situation by looking for the game-theoretic **core**[1].

It is clear that the easiest winning coalition to satisfy will be one that contains all members of A located between 0 and x_0, together with all members of B between 0 and y_0. Now the former of these sets will all prefer (s, t) to C if $s \geq x_0$, while the latter (s, t) to C if $t \geq y_0$. Thus, we conclude that there will be an agreement in the core of the game if

$$y_0 \leq \psi(x_0). \tag{2.7}$$

In this case, a division (s, t) will be in the core if it is Pareto-optimal and satisfies

$$x_0 \leq s, \quad y_0 \leq t. \tag{2.8}$$

On the other hand, if $s < x_0$, then all members of A who are situated between s and 1 will prefer conflict to the division (s, t). By definition (2.5) of x_0, there are more than $1 - \alpha$ of the members of A in this situation, and they thus form a blocking coalition, which can enforce C. Similarily, if $t < y_0$, then a blocking coalition consisting of more than $1 - \beta$ of the members of B will prefer C to (s, t). We conclude that conditions (2.7) and (2.8) are both necessary and sufficient for a core agreement. □

Suppose, next, that

$$y_0 \geq \psi(x_0). \tag{2.9}$$

We find that, in this case, C (the conflict outcome) is in the core of the game. In fact, suppose C were dominated by some agreement, (s, t). From (2.9), it is clear that, for any $(s,\ t)$, we must have either $s \leq x_0$, or $t \leq y_o$. Suppose, then, that $s \leq x_0$.

In this case, only those members of A located between 0 and s (exclusive of s) will prefer $(s,\ t)$ to C, and by (2.5), there are less than α of these. Thus, we see that there is no winning coalition preferring $(s,\ t)$ to C, and we conclude that $(s,\ t)$ cannot dominate C.

Similarily, if $t \leq y_0$, then less than β of the members of B will prefer $(s,\ t)$ to C, and again we conclude that $(s,\ t)$ does not dominate C.

Next, note that no Pareto-optimal agreement $(s,\ t)$ can be dominated by another agreement $(s',\ t')$. The reason for this is that, since agreements can only be enforced by coalitions containing some members of each organization, this would require $s' > s$ and $t' > t$. But this is impossible if $(s,\ t)$ is Pareto-optimal.

Finally, any non-Pareto-optimal agreement $(s,\ t)$ is dominated by some other agreement. The reason for this is that, if $(s,\ t)$ is not Pareto-optimal, then $t < \psi(s)$. Now, ψ is a continuous function and therefore there will be some $s'' > s$ such that $t'' = \psi(s'') > t$. Then all players prefer $(s'',\ t'')$ to $(s,\ t)$, and thus $(s'',\ t'')$ dominates $(s,\ t)$.

We express all these results in the form of a theorem:

Theorem 2.1 *In the game above, there will be an agreement, $(s,\ t)$, in the core, if and only if (2.7) holds (See Figure 1). In this case, the core agreements will be the Pareto-optimal $(s,\ t)$ satisfying (2.8). The conflict point, C, will be in the core of the game if and only if (2.9) holds (See Figure 2).*

It should be noted that in the bounding case, when $(x_0,\ y_0)$ is Pareto-optimal in K, the core of the game will consist of exactly two outcomes: the single agreement $(x_0,\ y_0)$ and the conflict point C. □

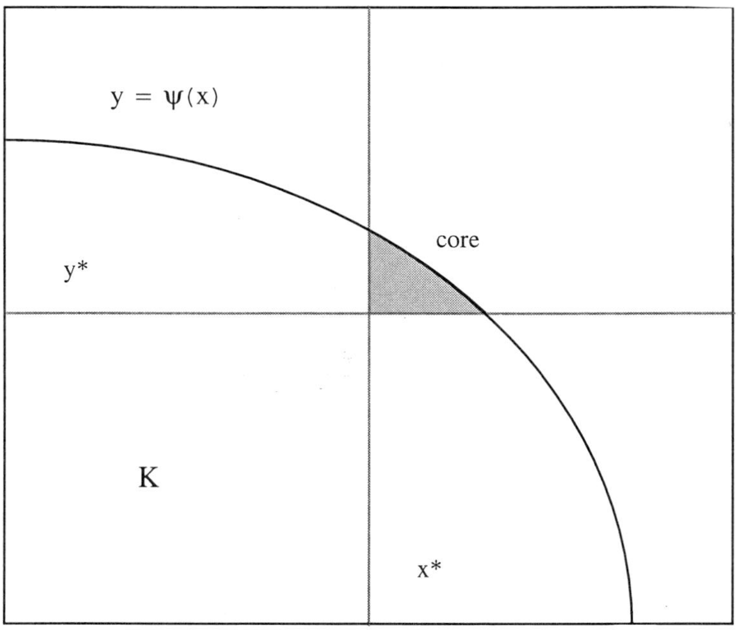

Figure 1 The core is the (upper-right) boundary of the darker area

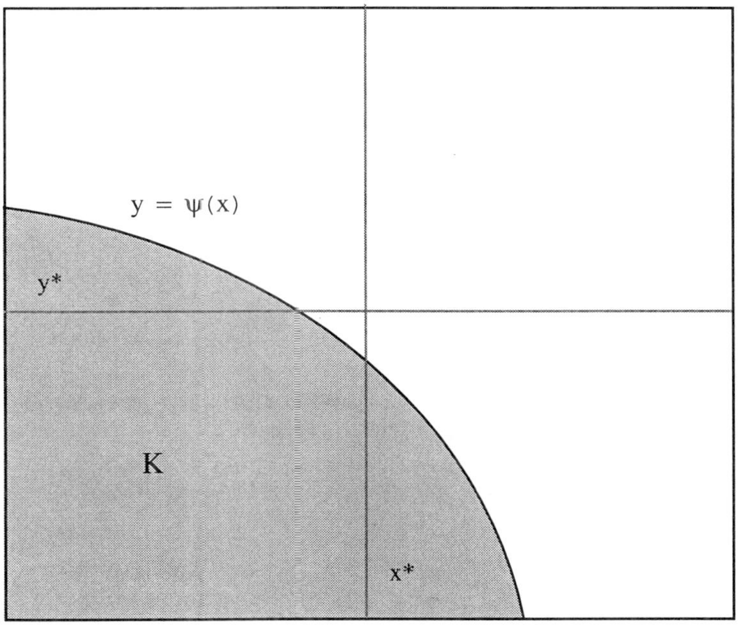

Figure 2 The core consists of the conflict outcome, C, only.

Model II

In a somewhat more complicated model, the two sets A and B are not homogenoeus, i.e., it is not sufficient to obtain some fraction α or β of the membership to form a winning coalition. We assume, rather, the existence of certain collections, denoted W_A and W_B of subsets of A and B respectively. We will call these the **controlling** coalitions for A and for B (respectively); a winning coalition will now be the union of any controlling coalition for A, with any controlling coalition for B. (By way of example, in the case of a state, a controlling coalition might be any coalition that contains at least some defined percentage of the ruling elite, the armed forces, and the general population. In the case of a terrorist organization or other sub-state actor, it might be any coalition that contains at least some defined percentage of the membership of each of the group's regional commands. Controlling coalitions are clearly not limited by these examples. They can vary widely.)

Though this model looks considerably more complicated than Model I, the underlying idea is similar: we can define x_0 as the minimum of all those s such that the members of A between 0 and s can form a controlling coalition for side A:

$$x_0 = \min \{s \quad \{a \mid 0 \leq a \leq s\} \,\varepsilon\, W_A\} \tag{2.10}$$

and, similarly, y_0 as the minimum of all those r such that the members of B between 0 and r are a controlling coalition for B.

Of course it may be difficult, in this case, to compute x_0 and y_0 - - where the form of the controlling collections is complicated, this computation may require a large number of mathematical operations. Assuming that an analytical result in fact can be computed, however, the situation becomes quite simple: the results of Theorem 1 above hold as stated.

B. The case with side payments[2]

Model I

In case B, we assume that, apart from the public goods (s and t respectively) obtained from a division of the pie, there is also some private good available to the **leadership** of the two organizations, if an agreement is reached. Each bargaining coalition is in a position to use this private good for side payments to persuade recalcitrant members of their respective organizations to sign on to the peace process. In the simplest case, this private good can be pecuniary: coalitions pay off their members. In other cases the side payments may consist of cabinet positions, choice of policies, etc. In any case, we will assume that A has a total amount Q, and B a total amount R, of this good, to distribute to its membership. Thus, an outcome, in the case of agreement, will be defined not only as the division, t, but will also include a pair of vectors (allocations), $\mathbf{q} = (q(a))_{a \,\varepsilon\, A}$ and $\mathbf{r} = (r(b))_{b \,\varepsilon\, B}$, satisfying

$$q(a) \geq 0 \quad \text{for all} \quad a \,\varepsilon\, A \tag{2.11}$$

$$\sum_A q(a) = Q \tag{2.12}$$

$$r(b) \geq 0 \quad \text{for all} \quad b \,\varepsilon\, B \tag{2.13}$$

$$\sum_B r(b) = R. \tag{2.14}$$

In the event outcome $T = (s,\, t,\, \mathbf{q},\, \mathbf{r})$ is reached, then the payoff to a member a and b, of A and B, respectively, is

$$u_a\,(T) = s + q\,(a) \quad \text{if} \quad a \,\varepsilon\, A \tag{2.15}$$

$$u_b\,(T) = t + r\,(b) \quad \text{if} \quad b \,\varepsilon\, B^3 \tag{2.16}$$

In the event of conflict, the payoff is simply as given, previously, by Equations (2.3) and (2.4).

To formalize the game, we will assume that the leadership of each organization controls the side payments. We also assume that this leadership is not interested in receiving any side payments; they merely wish to bring about an agreement. The effect of this is that an outcome which includes side payments can only be effected if the leadership agrees, i.e., a coalition which does not include the leadership cannot enforce side payments.

Because of the role played by the side payments, the core conditions are now more complicated than before. In this case, we will say that the agreement T is **implementable** if there exist side payment vectors $\mathbf{q}(a)$ and $\mathbf{r}(b)$ such that the outcome $T = (s,\, t,\, \mathbf{q},\, \mathbf{r})$ is in the core.

The point is that it may be possible to "buy off" some members of A or B by suitable selective incentives. A member of A at position x will be willing to collaborate in agreement $(s,\, t)$, even if $s < x$, so long as he receives some side payment $q(a)$ such that $x < s + q(a)$.

Each bargaining coalitions's optimal move in such a situation will be to offer side payments to those who, though not in favor of the bargaining process, are "close enough" to s or t that they might be affordably induced to change sides. In fact, to achieve agreement (s, t), there is no need to offer additional incentives to those coalition members (in A) located between 0 and s, since these already prefer the agreement. Similarly, the price of "purchasing" the support of those located at some x much larger than s (much closer to 1) will be unnecessarily high if the required support can be found closer to s. Thus, the available side payments will be offered to each side's marginal opponents, defined as those located between s and x_0 and between t and y_0 respectively (where x_0 and y_0 are defined as in (2.5) and (2.6)), as the collaboration of these individuals will be had at the smallest possible price.

Assuming agreement (s, t) has been reached, then, the minimal amount of side payments necessary to group A is given by

$$z(A, s) \;=\; \sum_{a:\, s \le x(a) \le x_0} (x(a) - s) \tag{2.17}$$

while the minimal necessary for B is

$$z(B, t) \;=\; \sum_{b:\, t \le y(b) \le y_0} (x(b) - t) \tag{2.18}$$

It is possible of course, that the $z\,(A, s)$ and $z\,(B, t)$ needed to actually reach an agreement may not be available for side payments (i.e., they might be larger than Q or R respectively). In that case, agreement (s, t) cannot be implemented because neither side will be able to obtain the necessary minimum level of support.

To make this clearer, we will define two new quantities, x^* and y^*, as follows. Note, first, that $z\,(A, s)$ is a strictly decreasing function of s for $s < x_0$; moreover, $z\,(A, x_0) = 0$. Thus there will be a unique x^* (depending on Q) such that

$$z\,(A, x^*) \;=\; Q. \tag{2.19}$$

In a similar way, there will be a unique number y^* such that

$$z\,(B, y^*) \;=\; R. \tag{2.20}$$

We then obtain the following theorem:

Theorem 2.2 *The agreement (s, t) is implementable if it is Pareto-optimal and if, moreover, $s \ge x^*, \quad t \ge y^*$.*

In such a case, any vector $\mathbf{q}$ satisfying

$$q\,(a) \;\ge\; x\,(a) \;-\; s \qquad \text{if} \quad s \;\le\; x\,(a) \;\le\; x_0 \tag{2.21}$$

$$q\,(a) \;\ge\; 0 \qquad \text{otherwise} \tag{2.22}$$

$$\sum_{a\,\varepsilon\,A} q\,(a) \;\le\; Q \tag{2.23}$$

and any vector $\mathbf{r}$ satisfying

$$r\,(b) \;\ge\; y\,(b) \;-\; t \qquad \text{if} \quad t \;\le\; y\,(b) \;\le\; y_0 \tag{2.24}$$

$$r\,(b) \;\ge\; 0 \qquad \text{otherwise} \tag{2.25}$$

$$\sum_{b \varepsilon B} r(b) \leq R \tag{2.26}$$

will give us an outcome $T = (s, t, \mathbf{q}, \mathbf{r})$ which is in the core. Moreover, if Equations (2.21) and (2.24) hold as strict inequalities, this outcome dominates C and thus C is not in the core.

Proof First of all, a non-Pareto-optimal agreement (s, t) can never be implementable. For, let $T = (s, t, \mathbf{q}, \mathbf{r})$ for some side payments $\mathbf{q}$ and $\mathbf{r}$. Since (s, t) is not Pareto-optimal, there is, as in Theorem 2.1, some Pareto-optimal (s', t') with $(s' > s, t' > t)$. Then clearly all players prefer $(s', t', \mathbf{q}, \mathbf{r})$ to $(s, t, \mathbf{q}, \mathbf{r})$, and it follows that T is dominated, so (s, t) is not implementable.

Next, if (s, t) is a Pareto-optimal agreement, then $(s, t, \mathbf{q}, \mathbf{r})$ cannot be dominated by any agreement outcome. The reason is that the leadership controls all side payments. Now, we have assumed that the leadership gains no direct benefits from side payments, and hence the leadership of A will only prefer a different agreement outcome, $T' = (s', t', \mathbf{q}', \mathbf{r}')$ if $s' > s$. Similarly, the leadership of B will only prefer T' to T, if $t' > t$. Since (s, t) is Pareto-optimal, it is not possible that $s' > s$ and $t' > t$ simultaneously. Thus, if T' dominates T, the effective coalition cannot include the leadership groups. But in the case we would have $\mathbf{q}' = 0$, $\mathbf{r}' = 0$ (since non-leadership groups cannot be effective for positive side payments) and it is then not possible that $s' + q'(a) > s + q(a)$ for some $a \varepsilon A$ and also $t' + r'(b) > t + r(b)$ for some $b \varepsilon B$.

Next, suppose that $s \geq x^*$ and $t \geq y^*$. We can then choose side payments $\mathbf{q}$ and $\mathbf{r}$ satisfying (2.21) - (2.26), and we see that, for all $a \ \varepsilon \ A$ with

$$\begin{aligned} 0 \leq x(a) \quad &\leq x_0, \\ s + q(a) \quad &\geq x(a). \end{aligned}$$

Similarily, for all $b \ \varepsilon \ B$ with $0 \leq y(b) \leq y_0$,

$$t + r(b) \geq y(b).$$

It follows that C is preferred to the outcome $T = (s, t, \mathbf{q}, \mathbf{r})$ only by those players located strictly to the right of x_0 and y_0 respectively. But these do not form a blocking coalition, and it follows that C cannot dominate T. If we assume moreover that (s, t) is Pareto-optimal, then T cannot be dominated by any agreement outcome, and we conclude that T is in the core. □

Suppose, on the other hand, that for agreement (s, t), either $s < x^*$ or $t < y^*$. For specificity, assume $s < x^*$, i.e. $z(A, s) > Q$. It follows that no side payment vector $\mathbf{q}$ satisfying (2.22 - 2.23) can satisfy (2.21) as well, i.e., at least one member $a \ \varepsilon \ A$ with $x(a) \leq x_0$ will receive total utility $s + q(a) < x(a)$. Thus at least one of these players will prefer C to any agreement outcome based on (s, t). Moreover, it is impossible to replace such players with other members of A, since these would require even higher side payments. It follows that C is preferred to T by more than $1 - \alpha$ of the members of A, and so C dominates T. Thus (s, t) is not implementable.

We conclude that the implementable agreements are precisely those Pareto-optimal ones for which (2.19) - (2.20) both hold.

Theorem 2.3 *A necessary and sufficient condition for the conflict outcome C to be in the core of the game is that $y^* \geq \psi(x^*)$.*

Proof (Necessity) Suppose, first that $y^* < \psi(x^*)$. Then, as in Theorem 2.1 above, there is an agreement (s, t) with $s > x^*$ and $t > y^*$. We can then find side payments $\mathbf{q}$ and $\mathbf{r}$ that satisfy conditions (2.21) - (2.26), with (2.21) and (2.24) as strict inequalities for each a and b in that sum. This will mean that $T = (s, t, \mathbf{q}, \mathbf{r})$ is preferred to C by all a with $x(a) \leq x_0$ and by all b with $y(b) \leq y_0$ respectively. Since these form a winning coalition, T dominates C, and C is not in the core.

Conversely, suppose that $y^* \geq \psi(x^*)$. Then, for any agreement (s, t), either $s \leq x^*$ or $t \leq y^*$.

We shall show a contradiction by assuming that C is not in the core. This means C is dominated by some outcome $T = (s, t, \mathbf{q}, \mathbf{r})$. For specificity, let us assume $s \leq x^*$.

Since T dominates C, there must be at least α members of A all of whom prefer T to C. For each such a, we must have

$$s + q(a) > x(a) \tag{2.27}$$

or

$$q(a) > x(a) - s. \tag{2.28}$$

Now, (2.28) holds trivially for each a with $x(a) < s$. Assume there are v such. Then $q(a)$ must be sufficiently large to satisfy (2.28) for another $\alpha - v$ members. Summing over all these, we must then have

$$Q \geq \sum q(a) > \sum (x(a) - s)$$

but this means that $z(A, s) < Q$, since in the definition (2.17) of $z(A, s)$, those a were chosen for which the corresponding sum was minimal. This in turn means that $s < x^*$, which contradicts our above assumption.

Similarly, if $t \leq y^*$, we will find it is not possible to obtain a sufficient number, β, of members of B to prefer T over C. In either case, we reach a contradiction. □

Model II

In this case, as in the no-side-payments case, winning coalitions are not simply defined as coalitions that contain at least α of the members of A and at least β of the members of B, but rather, by two collections, W_A and W_B of subsets of A and B respectively: the controlling coalitions for A and for B. A winning coalition, as above, consists of one coalition from W_A and one from W_B.

In contrast to Model 1, the optimal strategy for side payments in cases where each side must put together a controlling coalition to achieve an agreement is not always clear. In our initial case, the optimal strategy consisted of buying off all opponents of the bargaining process lying between t and x_0. In this case, a successful controlling coalition can often assume a large number of different forms. All things being equal, the optimal form will be the least costly. This is likely to be situation-specific. Finding this coalition may itself be a difficult mathematical problem. Assuming an optimal side payments policy can be determined, the minimal necessary payoffs to achieve an agreement will again be defined by $z(A, s)$ and $z(B, t)$. The only question will be whether these are $\leq Q$ and R, respectively. These then allow us to obtain quantities x^* and y^*, as above. Theorems 2.2 and 2.3, above, will then hold, with only slight modifications in the proof.

Endnotes

1. In game theory, an outcome X is said to **dominate** an outcome Y if there exists a coalition S such that (a) all members of S receive greater utility from X than from Y, and (b) the coalition S is in fact strong enough to enforce X. The **core** of a game is the set of all **undominated outcomes**. The core is generally considered to represent the strongest type of stability for n-person cooperative games. Note that many games have empty cores; on the other hand, when the core is non-empty, it frequently contains more than one outcome.

2. Our use of the term side payment does not correspond fully to the standard game-theoretic idea of side payments. In fact, in the game-theorectic literature, a game with side payments is one in which all types of side payments are allowed from any player to any other, so that utility can be transferred from one to another at a one-for-one rate. In the current article, only a very restricted type of side payment is allowed.

3. In positioning this type of payoff, we are assuming a rather strong interpersonal comparison of utility: the relationship between money and the public good is rather strong here, in the sense that equal amounts of money increase each player's utility in the same scale as increases in the share of the pie. This seems to suggest all members of the organization are equally venal. In reality, different players may have different utilities for money, so that a person only slightly to the right of s may require a larger payment than someone far to the right. Indeed some players may simply refuse to be bought. In such a case we would have to redefine the payoff (2.15) as $s + k(a)\,q(a)$, where $k(a)$ is a coefficient, depending on a and measuring a's love of money. A similar change would be made for equation (2.16). The problem of course is that in this case the optimal side payments (those which would put together a winning coalition with minimal expenditure) would be considerably more difficult to calculate. Such a situation would have to be treated more or less as Model II is treated.

References

[1] McCormick, Gordon H. and Owen, Guillermo [1997], *Violence, Factionalism, and State-Terrorist Bargaining*, Technical Report, Department of Mathematics, Naval Postgraduate School.

[2] McCormick, Gordon H. and Owen, Guillermo [1997], *Unraveling of a Peace Accord*, Technical Report, Department of Mathematics, Naval Postgraduate School.

Fields Institute Communications
Volume **23**, 1999

Coalition Structure Values of Mixed Games

Richard P. McLean
Department of Economics
Rutgers University
New Brunswick, NJ 08903 USA
rpmclean@rci.rutgers.edu

Abstract. A coalition structure is a partition of the player set that arises when players organize themselves into groups for the purpose of bargaining. The player set of a mixed game consists of a set of individually negligible players (the "ocean") and a finite set of individually nonnegligible players (the "atoms"). In this paper, we study coalition structure values for mixed games. These values resolve the bargaining problems at the intercoaltional and intracoalitional levels and our results extend previous work in the finite and nonatomic cases.

1 Introduction

The value theory of nonatomic games, developed in [**2**], has proved to be a very powerful tool in the analysis of economic and political problems where large numbers of individually negligible players are present. Mixed games, i.e., games with a set of individually negligible players (the "ocean") as well as some nonnegligible individual players (the "atoms") provide an important generalization of the nonatomic case and are particularly well suited to the study of voting problems and associated "political-economic games". Building on earlier studies of [**15**] and [**10**], [**4**] studied Aumann-Shapley values of mixed games from the axiomatic viewpoint while [**3**], [**11**] and [**13**] studied them from the asymptotic viewpoint.

In this paper, we study mixed games with coalition structures. A coalition structure is a partition of the player set that arises when players organize themselves into groups (such as unions, cartels, or syndicates) for the purpose of voting or bargaining. Given such a partition, bargaining occurs at two levels. At the "intercoalitional level", the coalitions comprising the coalition structure negotiate for coalitional payoffs. After these coalitional payoffs are determined, each coalition then resolves an "intracoalitional" bargaining problem to determine how its coalitional payoff will be divided among its members. A **coalition structure value** (CS value for short) is a solution concept for TU games that resolves the bargaining problem at the "intercoalitional" level and the "intracoalitional" level. For finite games, the CS value was defined in [**14**] where an axiomatic characterization is

1991 *Mathematics Subject Classification.* Primary 90D12; Secondary 90D13.

also provided. An alternative axiomatic characterization (and an application to endogenous coalition formation in finite TU games) can be found in [**5**] and [**6**].

In a previous paper [**8**], we extended the notion of CS value to nonatomic games from the axiomatic and asymptotic viewpoints. In this paper, we extend these results to the case of mixed games with coalition structures and obtain a synthesis of the work described above. The paper is outlined as follows. In Section 2, we provide notation and definitions. In Section 3, we provide a survey of results for coalition structure values of finite and nonatomic games. In Section 4, these results are extended to the case of mixed games from the axiomatic viewpoint and in Section 5 we present the asymptotic approach.

2 Notation and definitions

2.1 Nonatomic games. In this section, our notation is taken from [**2**] and a familiarity with the definitions and concepts in that book will be assumed. Let $(I,\mathcal{C})$ be a measurable space isomorphic to $([0,1],\mathcal{B})$ where $\mathcal{B}$ denotes the Borel $\sigma-$ algebra of subsets of [0,1]. The members of $\mathcal{C}$ are called *coalitions*. A *game* is a set function $v : \mathcal{C} \to \Re$ satisfying $v(\emptyset) = 0$. A game is *monotonic* if $v(S) \leq v(T)$ whenever $S \subseteq T$ and a game is said to be of *bounded variation* if it is the difference of two monotonic games. The Banach algebra of all games of bounded variation is denoted BV. If Q is a subspace of BV, then Q^+ denotes the set of monotonic games in Q. If $Z \subseteq BV$ is a set of games, then the *subspace of BV spanned by* Z is defined as the (variation norm) closure of the set of all linear combinations of members of Z.

The vector subspace of BV consisting of finitely additive, bounded signed measures is denoted FA. Let NA, NA^+, and NA^1 denote, respectively, the spaces of nonatomic measures, nonnegative nonatomic measures and nonatomic probability measures on $(I,\mathcal{C})$. A coalition C is a *carrier* of v if $v(S) = v(S \cap C)$ for every $S \in \mathcal{C}$. Let FIN denote the subspace of BV consisting of games with finite carrier and let FC denote the subspace of BV consisting of measures with finite carrier. Finally, define $FL = NA + FC$, i.e., FL is the subspace of BV consisting of measures that can be written as the sum of a nonatomic measure and a measure with finite carrier. The spaces FL^+ and FL^1 are defined analagously. If $\omega = \mu + \nu \in FL$ with $\mu \in NA$ and $\nu \in FC$, then we will engage in a mild abuse of standard nomenclature and refer to the elements of the (finite) carrier of ν as "atoms".

The space of all real valued functions f of bounded variation on $[0,1]$ such that $f(0) = 0$ will be denoted bv. Let bv' denote the subspace of bv consisting of functions that are continuous at 0 and 1 and let c' denote the subspace of bv consisting of functions that are continuous. For any subset $W \subseteq FL^1$, let $bv'W$ (resp. $c'W$) be the subspace of BV spanned by games of the form $f \circ \omega$ where $f \in bv'$ ($f \in c'$) and $\omega \in W$. The elements of $bv'FL$ are called mixed games. The subspace spanned by powers of NA^+ measures is denoted pNA.

2.2 Coalition structures. A *coalition structure* $\Pi = \{B_1, .., B_m\}$ is a partition of I satisfying (i) $B_k \in \mathcal{C}$ for each k and (ii) the measurable space $(B_k, \mathcal{C}_k)$ is isomorphic to $([0,1],\mathcal{B})$ where $\mathcal{C}_k := \{S \cap B_k | S \in \mathcal{C}\}$. In this paper, we will deal only with finite coalition structures. If $\Pi = \{B_1, .., B_m\}$ is a coalition structure, let $M = \{1, .., m\}$ denote the index set of Π. A game on M is a set function $w : 2^M \to \Re$ satisfying $w(\emptyset) = 0$ and the $2^{m-1}-$ dimensional vector space of games on M is denoted G_M. The collection of $m!$ orderings of the elements of M will be denoted Ω. If $R = (i_1, ..., i_m)$ is an element of Ω and $k = i_t$, let

$X_k(R) := \{i_1, .., i_{t-1}\}$ be the set of players that precede k in the ordering R. (If $k = i_0$, then $X_k(R) = \emptyset$). If $w \in G_M$, then the marginal contribution of k in the order R is defined as $w(X_k(R) \cup k) - w(X_k(R))$. The *Shapley value* on G_M is the linear operator $\varphi : G_M \to \Re^M$ defined as $\varphi_k(w) = \sum_{R\in\Omega} \frac{1}{m!}[w(X_k(R) \cup k) - w(X_k(R))]$ for each $k \in M$.

Given a game $v \in BV$, each coalition structure $\Pi = \{B_1, .., B_m\}$ gives rise to a naturally defined "*game between coalitions*" $v_\Pi \in G_M$ where $v_\Pi(K) = v(\cup_{k\in K} B_k)$ for each $K \subseteq M$. Each ordering of the coalitions in Π corresponds to an ordering in Ω and vice versa. If $R \in \Omega$, let $X_k^\Pi(R) = \cup_{\ell \in X_k(R)} B_\ell$. Hence, $v(X_k^\Pi(R) \cup B_k) - v(X_k^\Pi(R)) = v_\Pi(X_k(R) \cup k) - v_\Pi(X_k(R))$ for each $v \in BV$. Coalitions B_k and B_ℓ are *substitutes* in v_Π if $v_\Pi(K \cup k) = v_\Pi(K \cup \ell)$ whenever $K \subseteq M \backslash \{k, \ell\}$. Coalition B_k is a *null player* in v_Π if $v_\Pi(K \cup k) = v_\Pi(K)$ whenever $K \subseteq M \backslash \{k\}$.

If $\omega \in FL^1$, let $bv'_{\Pi,\omega}$ denote the subspace of bv defined as

$$bv'_{\Pi,\omega} = \{f \in bv | f \text{ is continuous at } \omega(\cup_{k\in K} B_k) \text{ for each } K \subseteq M.$$

From the definition, it follows that $bv'_{\Pi,\omega} \subseteq bv'$ (since $K = \emptyset$ and $K = M$ are subsets of M) and that $bv'_{\{I\},\omega} = bv'$ for all $\omega \in FL^1$. Finally, let $bv'_\Pi FL$ denote the space spanned by all functions of the form $f \circ \omega$ where $\omega \in FL^1$ and $f \in bv'_{\Pi,\omega}$ (again, $bv'_{\{I\},\omega} FL = bv' FL$). Let BV_k denote the space of set functions of bounded variation on $(B_k, \mathcal{C}_k)$, let NA_k^1 denote the space of nonatomic probability measures on $(B_k, \mathcal{C}_k)$ and let $bv'_k NA$ denote the subspace of games on $(B_k, \mathcal{C}_k)$ spanned by all games of the form $v = f \circ \mu$ where $f \in bv'$ and $\mu \in NA_k^1$.

If $\omega \in FL^1$, then the $\Pi-$ *support of* ω is defined as

$$\text{supp}_\Pi(\omega) = \{k \in M | \omega(B_k) > 0\}.$$

The group of bimeasurable isomorphisms of I onto itself is denoted $\mathcal{G}$. If $\Theta \in \mathcal{G}$ and $v \in BV$, define the game $\Theta_* v$ as $(\Theta_* v)(S) = v(\Theta S)$. If $\Pi = \{B_1, .., B_m\}$ is a coalition structure, define $\mathcal{G}_\Pi = \{\Theta \in \mathcal{G} \mid \Theta B_k = B_k \text{ for each } k \in M\}$. A subspace $Q \subseteq BV$ is $\Pi-$ *symmetric* if $\Theta_* v \in Q$ whenever $v \in Q$ and $\Theta \in \mathcal{G}_\Pi$. An $\{I\}$-symmetric subspace will be referred to simply as *symmetric*. Obviously, every symmetric subspace of BV is $\Pi-$ symmetric for all coalition structures Π.

3 Coalition structure values

We next present the axioms that will be used to define coalition structure values on certain subspaces of BV. In the statements of the axioms, $\Pi = \{B_1, ..., B_m\}$ is a coalition structure, Q is a Π-symmetric subspace of BV and $\psi : Q \to FA$ is a function.

A.1: (*Linearity*) ψ is a linear mapping of Q into FA.

A.2: (*Efficiency*) For each $v \in Q$. $(\psi v)(I) = v(I)$.

A.3: (*Monotonicity*) ψv is monotonic whenever v is monotonic.

A.4: (*Intracoalitional Symmetry*) For all $v \in Q, \psi(\Theta_* v) = \Theta_*(\psi v)$ whenever $\Theta \in \mathcal{G}_\Pi$.

A.5: (*Intercoalitional Symmetry*) If B_k and B_ℓ are substitutes in v_Π, then $(\psi v)(B_k) = (\psi v)(B_\ell)$.

A.6: (*Null Coalitions*) If B_k is a null player in v_Π , then $(\psi v)(B_k) = 0$.

The linearity and monotonicity axioms A.1 and A.3 are precisely the same as those used by Aumann and Shapley in the value theory of nonatomic games (without coalition structures). We have chosen the carrier form of the efficiency axiom A.2 in order to insure that null coalitions (i.e., coalitions with zero worth) receive a payoff of zero.

Axiom A.4 says that the names of players belonging to the same coalition in Π do not matter. If $\Pi = \{I\}$ (so that $\mathcal{G}_\Pi = \mathcal{G}$), then A.4 is precisely the symmetry axiom of Aumann and Shapley. Axiom A.5 states that two coalitions that affect the game between coalitions in the same way will receive the same coalitional payoff under ψ while A.6 says that a coalition that has no effect on the game between coalitions receives a coalitional payoff of zero. If $\Pi = \{I\}$, then A.5 is trivially satisfied and A.6 is a consequence of A.2.

Given a coalition structure Π and a Π-symmetric subspace Q of BV, a **Π-CS value** is a function $\psi : Q \to FA$ satisfying axioms A.1-A.6. If $\Pi = \{I\}$, then a **Π-CS value** on Q is a value on Q in the sense of Aumann and Shapley.

3.1 Coalition structure values of finite games. Let $\Pi = \{B_1, .., B_m\}$ be a coalition structure and let $w \in FIN$ with carrier A. If $B_k \cap A \neq \emptyset$, let Ξ_k denote the set of 1:1 mappings of $B_k \cap A$ onto $J_{|B_k \cap A|}$ where $J_q = \{1, .., q\}$ for each positive integer q.

Theorem A Let $\Pi = \{B_1, .., B_m\}$ be a coalition structure. There exists a unique Π−CS value ψ^Π on FIN. If $w \in FIN$ and A is a carrier of w and if $a \in A \cap B_k$, then

$$\psi^\Pi(w)(\{a\}) = \sum_{R \in \Omega} \frac{1}{m!} \sum_{\pi \in \Xi_k} \frac{1}{|B_k \cap A|!} [w(X_k^\Pi(R) \cup \pi^{-1}(J_{\pi(a)-1}) \cup \{a\})$$
$$- w(X_k^\Pi(R) \cup \pi^{-1}(J_{\pi(a)-1}))].$$

This result is a version of the main theorem in [**14**] and [**6**]. To interpret this result, suppose that we are given an ordering of the atoms in the carrier A and for simplicity, suppose that each $A_k \equiv B_k \cap A \neq \emptyset$. Such an ordering is "consistent" if, whenever $a, a' \in B_k$, the players "between" a and a' in the ordering are also members of B_k. A consistent ordering of A induces an ordering of the elements of Π and an ordering of the members of each A_k in an obvious manner. The coalition structure value ψ^Π assigns to each atom in A his expected marginal contribution in a consistent ordering of the atoms in A where all consistent orderings are equally likely. This is equivalent to a multistep procedure in which an ordering $R \in \Omega$ of the members of Π is first chosen with probability $\frac{1}{m!}$, then an ordering (i.e., a permutation) $\pi \in \Xi_k$ of the members of A_k is chosen with probability $\frac{1}{|A_k|!}$. The set of players that precede a in the induced consistent ordering is equal to $X_k^\Pi(R) \cup \pi^{-1}(J_{\pi(a)-1})$ and the formula for ψ^Π is precisely the expected marginal contribution of a.

At the intercoalitional level, the coalitional payoff to each B_k in Π is determined by applying the Shapley value to the game between coalitions w_Π. At the intracoalitional level, the payoff to each individual in A_k is then determined by applying the Shapley value to a particular game on A_k that specifies what each $S \subseteq A_k$ can obtain if it secedes from A_k and bargains with the remaining players. There are several possible candidates for this game on A_k, depending on whether or not the members of $A_k \backslash S$ "stay" in the original game. We will use the original game proposed in [**14**] (for others, see [**6**]). In this scenario, each $S \subseteq A_k$ receives its Shapley value payoff as a player in the m-player game defined by restricting w

to the field generated by the set $\{A_1, .., A_{k-1}, S, A_{k+1}, ..A_m\}$. More formally, let $w \in FIN$ with carrier A and $S \subseteq A_k = A \cap B_k$, define

$$r_k^w(S) = \sum_{R \in \Omega} \frac{1}{m!}[w(X_k^\Pi(R) \cup S) - w(X_k^\Pi(R))].$$

Thus r_k^w is a game with player set A_k and $r_k^w(S)$ is precisely the Shapley value payoff in the m-player game with player set $\{B_1, .., B_{k-1}, S, B_{k+1}, .., B_m\}$ described above. As a consequence of the definitions, we have the following result.

Proposition 1 Let $\Pi = \{B_1, .., B_m\}$ be a coalition structure and let ψ^Π denote the unique $\Pi-$CS value on FIN. If $w \in FIN$ and A is a carrier of w and if $a \in A \cap B_k$, then

$$\psi^\Pi(w)(\{a\}) = \sum_{\pi \in \Xi_k} \frac{1}{|A_k|!}[r_k^w(\pi^{-1}(J_{\pi(a)-1}) \cup \{a\}) - r_k^w(\pi^{-1}(J_{\pi(a)-1}))].$$

Proposition 1 identifies a "consistency" property of the CS value: the CS value payoff to player $a \in A_k$ in a game w is exactly that player's (ordinary) Shapley value payoff in the $|A_k|-$ player "reduced" game r_k^w. For other notions of consistency and reduced game in the context of values, see, e.g., [**7**].

3.2 Coalition structure values of nonatomic games. In a previous paper [**8**], we extended the notion of coalition structure value for finite games presented above to the nonatomic case and we review some of the main results in that paper.

Theorem B Let $\Pi = \{B_1, .., B_m\}$ be a coalition structure. There exists a unique Π-CS value ψ on $bv'_\Pi NA$. If $v = f \circ \mu$ where $\mu \in NA^1$ and $f \in bv'_{\Pi,\mu}$, then

$$\psi(v)(S) = \sum_{k \in \, supp_\Pi(\mu)} \varphi_k(v_\Pi)\frac{\mu(S \cap B_k)}{\mu(B_k)}$$

where $\varphi_k(v_\Pi)$ denotes the Shapley value of player B_k in the m-player game v_Π.

Theorem B is the nonatomic analogue of Theorem A. The theorem states that the overall bargaining problem represented by $v = f \circ \mu$ is resolved by first resolving the intercoalitional problem and paying each B_k its Shapley value in the game between coalitions (i.e., $\psi(v)(B_k) = \varphi_k(v_\Pi)$). Then (loosely speaking) the intracoalitional problem is resolved by giving each player in B_k an individual payoff equal to $\varphi_k(v_\Pi)$ times his relative weight in B_k (more precisely, $\psi(v)(T) = \varphi_k(v_\Pi)\frac{\mu(T \cap B_k)}{\mu(B_k)}$ whenever $T \subseteq B_k$.). Games that can be expressed as the composition of a continuously differentiable function and a nonatomic vector measure play a fundamental role in the theory and applications of the Aumann-Shapley value and a great deal of research has focused on the the "diagonal formula" for the value on pNA. In [**8**], we extended the diagonal formula to the coalition structure case.

Theorem C Let $\Pi = \{B_1, .., B_m\}$ be a coalition structure. There exists a unique $\Pi-$CS value ψ on pNA. Furthermore, let $v \in pNA$ be such that $v = f \circ \mu$ where μ is a n-vector of nonatomic measures and $f : \Re^n \to \Re$ is continuously differentiable. Then for each $S \in \mathcal{C}$

$$\psi v(S) = \sum_{R \in \Omega} \frac{1}{m!} \sum_{k=1}^{m} \int_0^1 \nabla f(X_k(R) + t\mu(B_k)) \cdot \mu(S \cap B_k) dt.$$

If $\Pi = \{I\}$, then the formula in Theorem C reduces to the ordinary diagonal formula for nonatomic games without coalition structures. For an application of the $\Pi-$CS value on pNA to cost allocation problems, see [**9**].

The consistency of the CS value in the finite case can also be extended to the nonatomic case by adapting the scenario in which the members of $B_k \backslash S$ disappear from the bargaining process whenever the members of $S \subseteq B_k$ decide to secede from B_k and bargain as a separate unit. More formally, let $v \in BV$ and for each $S \in \mathcal{C}_k$, define

$$r_k^v(S) = \sum_{R \in \Omega} \frac{1}{m!}[v(X_k^\Pi(R) \cup S) - v(X_k^\Pi(R))].$$

Hence, r_k^v is a set function defined on $(B_k, \mathcal{C}_k)$ and in fact we have the following result from [**8**] for the $\Pi-$ CS value on $bv'_\Pi NA$ that may be considered the nonatomic version of Proposition 1 above.

Proposition 2 Let ψ_Π denote the unique $\Pi-$ CS value on $bv'_\Pi NA$. If $v \in bv'_\Pi NA$, then $r_k^v \in bv'_k NA$ and

$$(\psi^\Pi v)(S) = \sum_{k=1}^{m} (\psi^k r_k^v)(S \cap B_k)$$

where ψ^k denotes the unique Aumann-Shapley value on $bv'_k NA$. In particular, $(\psi^\Pi v)(S) = (\psi^k r_k)(S)$ for each $S \in \mathcal{C}_k$.

4 Coalition structure values of mixed games: axiomatic approach

4.1 Values of mixed games. Loosely speaking, a value assigns to each player in a game his expected marginal contribution to the players that precede him in a "random ordering" of the players in I. In games with finite carrier, random orderings are easy to describe and the definition of a value is relatively straightforward. In the nonatomic case, the situation is substantially more complicated since the formalization of the concept of a "random order" becomes more delicate. Further complications arise when we consider values of mixed games. Each ordering of the players in I will induce an ordering of the atoms of a mixed game v. Although there are only finitely many such orderings of the atoms in v, there are infinitely many ways in which a given ordering of the atoms can be "blended" into the ocean. As a result, there are infinitely many Aumann-Shapley values ($\{I\}$-CS values in our terminology) on $bv'FL$. In [**4**], Hart has studied a class of Aumann-Shapley values on $bv'FL$ in which each member of the class corresponds to a different "blending" rule and we will extend his analysis to the coalition structure case.

A CS value assigns to a player his expected marginal contribution to the players that precede him in a "consistent random ordering" of the players in I. In games with finite carrier, consistent orderings are also easy to construct: first, order the members of Π and then order the atoms in each element of Π. In the mixed case however, the same "blending" problem that gives rise to the nonuniqueness of the value in games without coalition structures introduces similar complexities in the CS case.

4.2 Coalition structure values of mixed games. Suppose that $f \in bv'$ and q is a positive integer. Let ξ be a measure on $J_q = \{1, .., q\}$ satisfying $0 \leq \xi(J_q) \leq 1$

and for each $s \in [0,1]$ and each $t \in [0,1]$, define

$$\Delta(s,i,t,\xi,\gamma,f,q) = f(s+t[\gamma-\xi(J_q)]+\xi(J_i)) - f(s+t[\gamma-\xi(J_q)]+\xi(J_{i-1})).$$

To interpret this expression, suppose that J_q is the set of atoms in coalition B_k. Let $\xi(J_q)$ be the measure of J_q and suppose that B_k has total measure γ. Thus, the measure of the "oceanic" part of B_k is equal to $\gamma - \xi(J_q)$. An ordering of the players in I is consistent if it is constructed by first ordering the members of Π and then ordering the players in each element of Π. We then interpret $\Delta(s,i,t,\xi,\gamma,f)$ as the marginal contribution of atom $i \in J_q$ in a consistent ordering of the players in I when the measure of the coalitions preceding B_k is s, the measure of the oceanic part of B_k preceding i is a fraction t of the total measure of this oceanic part (i.e., $t(\gamma-\xi(J_q))$ and the atoms in J_q are arranged in increasing order.

Now suppose $v \in bv'_\Pi FL$ with $v = f \circ \omega$ where $f \in bv'_{\Pi,\mu}$ and $\omega = \mu + \nu$ is a probability measure, $\mu \in NA^+$ and $\nu \in FC^+$. Let A be a finite carrier of ν. If $k \in \text{supp}_\Pi(\nu)$, let $A_k = A \cap B_k$, $|A_k| = n_k$ and let Ξ_k denote the collection of one-to-one mappings of A_k onto J_{n_k}. For each $k \in M$, let p_k be a probability measure on the Borel sets of $[0,1]$ and let $p = (p_1,..,p_m)$. The operator E_{p_k} will denote expectation computed with respect to p_k. If $k \in \text{supp}_\Pi(\nu)$, let $T_k^1,...,T_k^{n_k}$ be a collection of n_k random variables each with distribution p_k and let $0 < T_k^{(1)} < \cdots < T_k^{(n_k)} < 1$ denote the associated order statistics. Now define a set function $\psi_p^\Pi(v) \in FA$ as follows:

If $k \in \text{supp}_\Pi(\nu)$ and $a \in A_k$, then

$$\psi_p^\Pi(v)(a) = \frac{1}{m!}\sum_{R\in\Omega}\frac{1}{n_k!}E_{p_k}\left[\sum_{\pi\in\Xi_k}\Delta(\omega(X_k^\Pi(R)),\pi(a),T_k^{\pi(a)},\nu\circ\pi^{-1},\omega(B_k),f,n_k)\right].$$

If $k \notin \text{supp}_\Pi(\nu)$ and $a \in A_k$, then

$$\psi_p^\Pi(v)(a) = 0.$$

If $\text{supp}_\Pi(\mu) \neq \emptyset$ and $S \subseteq I\backslash A$, then

$$\psi_p^\Pi(v)(S) = \sum_{k\in\,\text{supp}_\Pi(\mu)} \alpha_k(p,\Pi,v)\frac{\mu(S\cap B_k)}{\mu(B_k)}$$

where $\alpha_k(p,\Pi,v) = \varphi_k(v_\Pi) - \sum_{a\in A_k}\psi_p^\Pi(v)(a)$ and $\varphi_k(v_\Pi)$ denotes the Shapley value payoff to player B_k in the game v_Π.

If $\text{supp}_\Pi(\mu) = \emptyset$ and $S \subseteq I\backslash A$, then

$$\psi_p^\Pi(v)(S) = 0.$$

The operator ψ_p^Π generalizes the construction of a value on $bv'FL$ in [**4**] and reduces to his definition if $\Pi = \{I\}$ (in which case $m = 1$). If $\omega \in FC^1$, then $\psi_p^\Pi(v)$ reduces to the formula of Theorem A for each p. If $\omega \in NA^1$, then $\psi_p^\Pi(v)$ reduces to the formula of Theorem B for each p. The main result on Π-CS values of mixed games can now be stated. A probability measure is *continuous* if its distribution function is continuous.

Theorem D Let $\Pi = \{B_1,..,B_m\}$ be a coalition structure and let $p = (p_1,\ldots,p_m)$ be a vector of continuous probability measures. Then there exists a

Π-CS value ψ on $bv'_\Pi FL$ such that, if $v = f \circ \omega, f \in bv'_{\Pi,\omega}$ and $\omega \in FL^+$ is a probability measure, then $\psi(v) = \psi_p^\Pi(v)$.

This theorem generalizes the aforementioned results of [**4**] ($\Pi = \{I\}$), [**8**] ($\omega \in NA$) and [**6**] ($\omega \in FC$). As in the nonatomic case, the theorem states that the overall bargaining problem represented by $v = f \circ (\mu + \nu)$ is resolved by first resolving the intercoalitional problem and paying each B_k its Shapley value in the game between coalitions (i.e., $\psi_p^\Pi(v)(B_k) = \psi_p^\Pi(v)(B_k \backslash A_k) + \psi_p^\Pi(v)(A_k) = \varphi_k(v_\Pi)$). Then (loosely speaking) the intracoalitional problem is resolved by giving each player $a \in A_k$ an individual payoff equal to $\psi_p^\Pi(v)(a)$ as defined above while each player in $B_k \backslash A_k$ receives $\varphi_k(v_\Pi) - \sum_{i=1}^{n_k} \psi_p^\Pi(v)(a)$ times his relative weight in $B_k \backslash A_k$ (more precisely, $\psi_p^\Pi(v)(T) = [\varphi_k(v_\Pi) - \sum_{a \in A_k} \psi_p^\Pi(v)(a)]\frac{\mu(T \cap B_k)}{\mu(B_k)}$ whenever $T \subseteq B_k \backslash A_k$).

The proof of Theorem D will follow from the next few lemmas.

Lemma 1 Let

$$v = f_1 \circ \omega_1 + \cdots f_r \circ \omega_r$$

where $\omega_q = \mu_q + \nu_q \in FL^1$, $\mu_q \in NA^+, \nu_q \in FC^+$ and $f_q \in bv'_{\Pi,\omega_q}$ for each $1 \le q \le r$. Then

$$\|\sum_{q=1}^{r} \psi_p^\Pi(f_q \circ \omega_q)\| \le \|\sum_{q=1}^{r} f_q \circ \omega_q\|.$$

Proof Choose $R \in \Omega$ and without loss of generality, let R be the natural ordering $(1, .., m)$. Furthermore, we may assume without loss of generality that the measures ν_q have the same carrier which we denote A and that $A_k = A \cap B_k \neq \emptyset$ for all k. Let $|A_k| = n_k$.

For each q and each $k \in \mathrm{supp}_\Pi(\omega_q)$, let $\hat{f}^R_{q,k} : [0, 1] \to \Re$ be defined as

$$\hat{f}^R_{q,k}(x) = f_q(\omega_q(X_k^\Pi(R) + \omega_q(B_k)x) - f_q(\omega_q(X_k^\Pi(R))$$

and let

$$\hat{\mu}_{q,k}(S) = \frac{\mu_q(S)}{\omega_q(B_k)} \text{ and } \hat{\nu}_{q,k}(S) = \frac{\nu_q(S)}{\omega_q(B_k)}$$

for each $S \in \mathcal{C}_k$ so that

$$\hat{\omega}_{q,k}(S) = \frac{\mu_q(S) + \nu_q(S)}{\omega_q(B_k)} = \hat{\mu}_{q,k}(S) + \hat{\nu}_{q,k}(S).$$

Hence, $\hat{f}^R_{q,k} \circ \hat{\omega}_{q,k} \in bv'_k FL$. Note that $\mathrm{supp}_\Pi(\omega_q) = \mathrm{supp}_\Pi(\mu_q) \cup \mathrm{supp}_\Pi(\nu_q)$. Hence, for each $k \in \mathrm{supp}_\Pi(\omega_q)$, each $\pi_k \in \Xi_k$ (= the set of 1:1 mappings of A_k onto $J_{n_k} = \{1, .., n_k\}$) and each n_k- tuple $\tau_k = (\tau_k^1, .., \tau_k^{n_k})$ with $0 < \tau_k^1 < \cdots < \tau_k^{n_k} < 1$, we can define a finitely additive measure $\hat{\rho}_{\pi_k,\tau_k}(\hat{f}^R_{q,k} \circ \hat{\omega}_{q,k})$ on $(B_k, \mathcal{C}_k)$ as follows:

If $k \in \mathrm{supp}_\Pi(\nu_q)$ and $a \in A_k$, then

$$\hat{\rho}_{\pi_k,\tau_k}(\hat{f}^R_{q,k} \circ \hat{\omega}_{q,k})(a) = \Delta(0, \pi_k(a), \tau_k^{\pi_k(a)}, \hat{\nu} \circ \pi_k^{-1}, \hat{\omega}_{q,k}(B_k), \hat{f}^R_{q,k}, n_k).$$

If $k \notin \mathrm{supp}_\Pi(\nu_q)$ and $a \in A_k$, then

$$\hat{\rho}_{\pi_k,\tau_k}(\hat{f}^R_{q,k} \circ \hat{\omega}_{q,k})(a) = 0.$$

If $k \in \mathrm{supp}_\Pi(\mu_q)$ and $S \subseteq B_k \backslash A_k$, then

$$\hat{\rho}_{\pi_k,\tau_k}(\hat{f}^R_{q,k} \circ \hat{\omega}_{q,k})(S) = \alpha_k(\hat{f}^R_{q,k} \circ \hat{\omega}_{q,k})\hat{\mu}_{q,k}(S),$$

where

$$\hat{\alpha}_k(\hat{f}^R_{q,k} \circ \hat{\omega}_{q,k}) = \frac{\hat{f}^R_{q,k}(1) - \sum_{a \in A_k} \hat{\rho}_{\pi_k,\tau_k}(\hat{f}^R_{q,k} \circ \hat{\omega}_{q,k})(a)}{\hat{\mu}_{q,k}(B_k)}.$$

If $k \notin \text{supp}_\Pi(\mu_q)$ and $S \subseteq B_k \backslash A_k$, then

$$\hat{\rho}_{\pi_k,\tau_k}(\hat{f}^R_{q,k} \circ \hat{\omega}_{q,k})(S) = 0.$$

Claim 1. For each $k \in M$,

$$\| \sum_{q:\omega_q(B_k)>0} E_{p_k} \left[\sum_{\pi_k \in \Xi_k} \frac{1}{n_k!} \hat{\rho}_{\pi_k,T_k}(\hat{f}^R_{q,k} \circ \hat{\omega}_{q,k}) \right] \|_k \leq \| \sum_{q:\omega_q(B_k)>0} \hat{f}^R_{q,k} \circ \hat{\omega}_{q,k} \|_k$$

where $|| \cdot ||_k$ denotes the variation norm on BV_k.

Proof of Claim 1. This is simply the proof of Theorem A in [**4**] (up to equation 4.3) applied to the game $\sum_{q:\omega_q(B_k)>0} \hat{f}^R_{q,k} \circ \hat{\omega}_{q,k} \in bv'_k FL$.

Claim 2.

$$\sum_{k \in M} \| \sum_{q:\omega_q(B_k)>0} \hat{f}^R_{q,k} \circ \hat{\omega}_{q,k} \|_k \leq \| \sum_{q=1}^{r} f_q \circ \omega_q \|.$$

Proof of Claim 2. Choose $\epsilon > 0$ and for each $k \in M$, let

$$w^R_k = \sum_{q:\omega_q(B_k)>0} \hat{f}^R_{q,k} \circ \hat{\omega}_{q,k}.$$

For each k, choose a chain $\emptyset = C^k_0 \subseteq C^k_1 \subseteq \cdots \subseteq C^k_{t_k} = B_k$ on B_k such that

$$\sum_{t=1}^{t_k} \left| w^R_k(C^k_t) - w^R_k(C^k_{t-1}) \right| \geq ||w^R_k||_k - \epsilon/k$$

Now construct a chain

$$\emptyset = S_0 \subseteq S^1_1 \subseteq \cdots \subseteq S^1_{t_1} \subseteq S^2_1 \subseteq \cdots \subseteq S^2_{t_2} \subseteq \cdots \subseteq S^m_1 \subseteq \cdots \subseteq S^m_{t_m} = I.$$

where $S_t^k = X_k^\Pi(R) \cup C_t^k$ for each $t \in \{1,..,t_k\}$. Hence,

$$\begin{aligned}
||\sum_{q=1}^{r} f_q \circ \omega_q|| &\geq \sum_{k \in M} \Bigg[\Big| \sum_{q=1}^{r} \{(f_q \circ \omega_q)(S_1^k) - (f_q \circ \omega_q)(X_k^\Pi(R))\} \Big| \\
&\qquad + \sum_{t=1}^{t_k - 1} \Big| \sum_{q=1}^{r} \{(f_q \circ \omega_q)(S_{t+1}^k) - (f_q \circ \omega_q)(S_t^k)\} \Big| \Bigg] \\
&= \sum_{k \in M} \sum_{t=1}^{t_k} \Big| \sum_{q=1}^{r} \{(f_q \circ \omega_q)(X_k^\Pi(R) \cup C_t^k) \\
&\qquad - (f_q \circ \omega_q)(X_k^\Pi(R) \cup C_{t-1}^k)\} \Big| \\
&= \sum_{k \in M} \sum_{t=1}^{t_k} \Big| \sum_{q:\omega_q(B_k)>0} \{(f_q \circ \omega_q)(X_k^\Pi(R) \cup C_t^k) \\
&\qquad - (f_q \circ \omega_q)(X_k^\Pi(R) \cup C_{t-1}^k)\} \Big| \\
&= \sum_{k \in M} \sum_{t=1}^{t_k} \Big| w_k^R(C_t^k) - w_k^R(C_{t-1}^k) \Big| \\
&\geq \sum_{k \in M} ||w_k^R||_k - \epsilon
\end{aligned}$$

Therefore,

$$||\sum_{q=1}^{r} f_q \circ \omega_q|| \geq \sum_{k \in M} || \sum_{q:\omega_q(B_k)>0} \hat{f}_{q,k}^R \circ \hat{\omega}_{q,k}||_k$$

and the proof of Claim 2 is complete.

Claim 3.

$$||\sum_{q=1}^{r} \psi_p^\Pi(f_q \circ \omega_q)|| \leq ||\sum_{q=1}^{r} f_q \circ \omega_q||$$

Proof of Claim 3. From the definitions above, it follows that

$$||\sum_{q=1}^{r}\psi_p^{\Pi}(f_q\circ\omega_q)||$$

$$=\sum_{k\in M}\sum_{a\in A_k}\Big|\sum_{q:\nu_q(B_k)>0}\Big[\sum_{R\in\Omega}\frac{1}{m!}E_{p_k}\Big[\sum_{\pi_k\in\Xi_k}\frac{1}{n_k!}\rho_{\pi_k,T_k}(\hat{f}^R_{q,k}\circ\hat{\omega}_{q,k})(a)\Big]\Big]\Big|$$

$$+||\sum_{k\in M}\sum_{q:\mu_q(B_k)>0}\sum_{R\in\Omega}\frac{1}{m!}\alpha_k(\hat{f}^R_{q,k}\circ\hat{\omega}_{q,k})\hat{\mu}_{q,k}||_k$$

$$\leq\frac{1}{m!}\sum_{R\in\Omega}\sum_{k\in M}\sum_{a\in A_k}\Big|E_{p_k}\sum_{\pi\in\Xi_k}\frac{1}{n_k!}\sum_{q:\nu_q(B_k)>0}\rho_{\pi_k,T_k}(\hat{f}^R_{q,k}\circ\hat{\omega}_{q,k})(a)\Big|$$

$$+\sum_{k\in M}\sum_{R\in\Omega}\frac{1}{m!}||\sum_{q:\mu_q(B_k)>0}\alpha_k(\hat{f}^R_{q,k}\circ\hat{\omega}_{q,k})\hat{\mu}_{q,k}||_k$$

Since

$$||\sum_{q:\omega_q(B_k)>0}\sum_{\pi_k\in\Xi_k}\frac{1}{n_k!}E_{p_k}[\rho_{\pi_k,T_k}(\hat{f}^R_{q,k}\circ\hat{\omega}_{q,k})]||_k$$

$$=\sum_{a\in A_k}\Big|\sum_{q:\nu_q(B_k)>0}\sum_{\pi_k\in\Xi_k}\frac{1}{n_k!}E_{p_k}[\rho_{\pi_k,T_k}(\hat{f}^R_{q,k}\circ\hat{\omega}_{q,k})(a)]\Big|$$

$$+||\sum_{q:\mu_q(B_k)>0}\alpha_{p_k}(\hat{f}^R_{q,k}\circ\hat{\omega}_{q,k})\hat{\mu}_{q,k}||_k$$

we combine the results above and apply Claim 2 to obtain

$$||\sum_{q=1}^{r}\psi_p^{\Pi}(f_q\circ\omega_q)||$$

$$\leq\frac{1}{m!}\sum_{R\in\Omega}\sum_{k\in M}\left(||\sum_{q:\omega_q(B_k)>0}\sum_{\pi_k\in\Xi_k}\frac{1}{n_k!}E_{p_k}[\rho_{\pi_k,T_k}(\hat{f}^R_{q,k}\circ\hat{\omega}_{q,k})]||_k\right)$$

$$\leq\frac{1}{m!}\sum_{R\in\Omega}\sum_{k\in M}||\sum_{q:\omega_q(B_k)>0}\hat{f}^R_{q,k}\circ\hat{\omega}_{q,k}||_k$$

$$\leq\tfrac{1}{m!}\sum_{R\in\Omega}||\sum_{q=1}^{r}f_q\circ\omega_q||$$

$$=||\sum_{q=1}^{r}f_q\circ\omega_q||$$

This completes the proof of Claim 3 and the Lemma. □

Proof of Theorem D The proof is an adaptation of the argument in [**4**] and we outline the details for the sake of completeness. Let Y be the subspace of $bv'_{\Pi}FL$ consisting of all games of the form

$$v = f_1 \circ \omega_1 + \cdots f_r \circ \omega_r$$

where $\omega_q \in FL^1$ and $f_q \in bv'_{\Pi,\omega_q}$ for each $1 \leq q \leq r$. Define an operator $\psi : Y \to FA$ as follows: if $v \in Y$ has the representation $v = f_1 \circ \omega_1 + \cdots f_r \circ \omega_r$, then

$$\psi(v) = \psi(f_1 \circ \omega_1 + \cdots f_r \circ \omega_r) = \sum_{q=1}^{r} \psi_p^{\Pi}(f_q \circ \omega_q).$$

It follows from Lemma 1 that ψ is well defined (i.e., $\psi(v)$ independent of the particular representation of v). Furthermore, $bv'_{\Pi}FL$ is spanned by Y so we conclude that ψ can be uniquely extended to a continuous linear operator (also denoted ψ) from $bv'_{\Pi}FL$ to FA such that $||\psi|| \leq 1$. To show that ψ is a $\Pi - CS$ value on $bv'_{\Pi}FL$, note that ψ is linear by definition so A.1 holds. The arguments that A.3 and A.4 hold are identical to those in the proof of Theorem A in [**4**]. If $\omega \in FL^1$ and $f \in bv'_{\Pi,\omega}$, then direct computation shows that $\psi(f \circ \omega)(B_k) = \varphi_k((f \circ \omega)_{\Pi})$. Therefore, the continuous linear functional $v \to \psi(v)(B_k) - \varphi_k(v_{\Pi})$ is identically 0 on Y for each k, hence identically 0 on $bv'_{\Pi}FL$ for each k. As a result, A.2, A.5 and A.6 are satisfied and the proof of Theorem D is complete. □

Remark When the measures p_k are not continuous, then a coalition structure extension of Theorem B in [**4**] can be obtained for the space $c'_{\Pi}FL$ as well as the spaces $lc'_{\Pi}FL$ and $rc'_{\Pi}FL$.

4.3 Consistency of CS values of mixed games. In this section, we show that the consistency property of the CS value in the finite and nonatomic cases can be extended to the mixed case as well.

Let $v \in BV$ and for each $S \in \mathcal{C}_k$, define

$$r_k^v(S) = \sum_{R \in \Omega} \frac{1}{m!}[v(X_k^{\Pi}(R) \cup S) - v(X_k^{\Pi}(R))].$$

Hence, r_k^v is a set function defined on $(B_k, \mathcal{C}_k)$ and in fact we have the following result whose proof is identical to that of the case of $bv'_{\Pi}NA$ found in [**8**].

Lemma 2 If $v \in bv'_{\Pi}FL$, then $r_k^v \in bv'_kFL$ for each k. Furthermore, the mapping $\Lambda_k : bv'_{\Pi}FL \to bv'_kFL$ where $\Lambda_k(v) = r_k^v$ is continuous (with respect to the variation norm topologies on $bv'_{\Pi}FL$ and bv'_kFL).

Our last result in this section extends Propositions 1 and 2 to the mixed case.

Proposition 3 Let $p = (p_1, .., p_m)$ be a vector of continuous probability measures with associated $\Pi-$ CS value ψ_p^{Π} on $bv'_{\Pi}FL$. Then

$$(\psi_p^{\Pi} v)(S) = \sum_{k \in M} (\varphi_{p_k}^k r_k^v)(S \cap B_k)$$

where $\varphi_{p_k}^k$ denotes the (ordinary) value on bv'_kFL associated with the probability measure p_k. In particular, $(\psi_p^{\Pi} v)(S) = (\varphi_{p_k}^k r_k)(S)$ for each $S \in \mathcal{C}_k$.

Proof Choose $v = f \circ \omega$ where $\omega \in FL^1$ and $f \in bv'_{\Pi,\omega}$ and define r_k^v for each $k \in M$ as above. If $k \in \text{supp}_\Pi(\omega)$, define a probability measure on B_k where $\hat{\omega}_k(S) = \frac{\omega(S)}{\omega(B_k)}$ for each $S \in \mathcal{C}_k$ and a function $g_k^R : [0,1] \to \Re$ where

$$g_k^R(x) = f(\omega(X_k^\Pi(R)) + \omega(B_k)x) - f(\omega(X_k^\Pi(R)).$$

Note that each $g_k^R \in bv'$ since $f \in bv'_{\Pi,\omega}$ and therefore, $g_k^R \circ \hat{\omega}_k \in bv'_k FL$. Applying Theorem D above, it follows that

$$\begin{aligned}(\psi_p^\Pi(f \circ \omega))(S) &= \sum_{k \in \text{supp}_\Pi(\omega)} \left[\sum_{R \in \Omega} \frac{1}{m!} \varphi_{p_k}^k (g_k^R \circ \hat{\omega}_k)(S \cap B_k)\right] \\ &= \sum_{k \in \text{supp}_\Pi(\omega)} \left(\varphi_{p_k}^k \left(\sum_{R \in \Omega} \frac{1}{m!} (g_k^R \circ \hat{\omega}_k)\right)\right)(S \cap B_k) \\ &= \sum_{k \in \text{supp}_\Pi(\omega)} (\varphi_{p_k}^k r_k^v)(S \cap B_k).\end{aligned}$$

To complete the proof, let $\{v_n\}$ be a sequence with limit v such that $v_n = \sum_i \alpha_i^n (f_i^n \circ \omega_i^n)$ where for each i and n, $\omega_i^n \in FL^1$, $f_i^n \in bv'_{\Pi,\omega_i^n}$, and α_i^n is a real number.

Let

$$r_k^{i,n}(S) = \sum_{R \in \Omega} \frac{1}{m!} [(f_i^n \circ \omega_i^n)(X_k^\Pi(R) \cup S) - (f_i^n \circ \omega_i^n)(X_k^\Pi(R))]$$

for each $S \subseteq B_k$ so that

$$r_k^{v_n} = \sum_i \alpha_i^n r_k^{i,n} = \sum_{i : \omega_i^n(B_k) > 0} \alpha_i^n r_k^{i,n}$$

Lemma 2 implies that each $r_k^{i,n} \in bv'_k FL$ (so $r_k^{v_n} \in bv'_k FL$ for all n) and from the preceding argument, we conclude that

$$\psi_p^\Pi (f_i^n \circ \omega_i^n)(S) = \sum_{k \in \text{supp}_\Pi(\omega_i^n)} (\varphi_{p_k}^k r_k^{i,n})(S \cap B_k).$$

Each $v_n \in bv'_\Pi FL$ so the linearity ψ_p^Π implies that

$$\begin{aligned}\psi_p^\Pi v_n(S) &= \sum_i \alpha_i^n \psi_p^\Pi (f_i^n \circ \omega_i^n)(S) \\ &= \sum_i \alpha_i^n \left[\sum_{k \in \text{supp}_\Pi(\omega_i^n)} (\varphi_{p_k}^k r_k^{i,n})(S \cap B_k)\right] \\ &= \sum_{k \in M} \left(\varphi_{p_k}^k \left[\sum_{i : \omega_i^n(B_k) > 0} \alpha_i^n r_k^{i,n}\right]\right)(S \cap B_k) \\ &= \sum_{k \in M} (\varphi_{p_k}^k r_k^{v_n})(S \cap B_k).\end{aligned}$$

Note that each $\varphi^k_{p_k}$ is continuous on $bv'_k FL$ and ψ^Π_p is continuous on $bv'_\Pi FL$ as a result of Theorem D above. Since $v_n \to v$, it follows from Lemma 2 that $r_k^{v_n} \to r_k^v$ and therefore,

$$\psi^\Pi_p v = \lim \psi^\Pi_p v_n = \lim \sum_{k\in M} \varphi^k_{p_k} r_k^{v_n} = \sum_{k\in M} \varphi^k_{p_k} r_k^v$$

completing the proof of the proposition. □

Remark The proof of existence of a $\Pi-$ CS value on $bv'_\Pi FL$ given in Section 4 generalizes the approach used in [**4**]. An alternative proof of existence utilizing Lemma 2 is possible. In this second approach, one would choose a vector of continuous probability measures $p = (p_1, .., p_m)$ and define an operator ψ^Π_p on $bv'_\Pi FL$ by

$$(\psi^\Pi_p v)(S) = \sum_{k\in M} (\varphi^k_{p_k} r_k^v)(S \cap B_k).$$

It can then be shown that such an operator satisfies A.1-A.6 and is the same as the value given in Theorem D.

5 CS values of mixed games: asymptotic approach

In this section, we present the asymptotic approach to CS values of mixed games. Choose a coalition S and let $\{\Xi_n\}_{n=1}^\infty$ be an *admissible* (i.e., decreasing and separating) sequence of measurable partitions of I with $\Xi_1 = \{S, I\backslash S\} \vee \Pi$. (The "join" of two partitions Π and Π' , denoted $\Pi \vee \Pi'$, is the coarsest common refinement of Π and Π'). Let $B_k^{(n)} = \{C \in \Xi_n | C \subseteq B_k\}$ and let $\Pi^{(n)} = \{B_1^{(n)}, .., B_m^{(n)}\}$. Hence, $\Pi^{(n)}$ is a coalition structure for the finite player game v_{Ξ_n}. Let $\psi^{\Pi^{(n)}}(v_{\Xi_n})$ denote the Owen CS value of the $|\Xi_n|-$ player game v_{Ξ_n} for the coalition structure $\Pi^{(n)}$ and define

$$\psi^{\Pi^{(n)}}(v_{\Xi_n})(S) = \sum_{\substack{C\in\Xi_n \\ :C\subseteq S}} \psi^{\Pi^{(n)}}(v_{\Xi_n})(C)$$

If the sequence $\{\psi^{\Pi^{(n)}}(v_{\Xi_n})(S)\}_{n=1}^\infty$ has a limit for each $S \in \mathcal{C}$ and each admissible sequence $\{\Xi_n\}$ and if the limit is independent of the sequence $\{\Xi_n\}$, then we will denote the limit $\psi^\Pi(v)(S)$. The resulting set function $\psi^\Pi(v)$ is called the $\Pi-$ *asymptotic value* of v. For the special coalition structure $\Pi = \{I\}$, our construction reduces to that of the original asymptotic value of Aumann and Shapley and we will call the $\{I\}$-asymptotic value of v simply the asymptotic value of v. As in the case of the asymptotic value, the $\Pi-$ asymptotic value of a game, when it exists, is unique and finitely additive.

The set of games $v \in BV$ possessing a $\Pi-$ asymptotic value will be denoted $ASYMP(\Pi)$. $ASYMP(\{I\})$ is the set of games possessing an asymptotic value in the original sense of AS and will be simply denoted $ASYMP$. The subspace of games in BV_k possessing an (ordinary) asymptotic value will be denoted $ASYMP_k$. In [**8**], we proved

Theorem E Let Π be a coalition structure. $ASYMP(\Pi)$ is a closed, $\Pi-$ symmetric subspace of BV containing $bv'_\Pi NA$. The operator ψ that assigns to each $v \in ASYMP(\Pi)$ its $\Pi-$ asymptotic value is a $\Pi-$ CS value on $ASYMP(\Pi)$.

In [**8**], we also proved the following consistency result.

Proposition 4 Let $\Pi = \{B_1, .., B_m\}$ be a coalition structure. Then $v \in ASYMP(\Pi)$ if and only if $r_k^v \in ASYMP_k$ for each $k \in M$. If ψ is the operator that assigns to each $v \in ASYMP(\Pi)$ its $\Pi-$ asymptotic value and if ψ^k is the operator that assigns to each $w \in ASYMP_k$ its (ordinary) asymptotic value, then

$$\psi(v)(S) = \sum_{k \in M} \psi^k(r_k^v)(S \cap B_k)$$

for all $v \in ASYMP(\Pi)$.

The extension of Theorem E to mixed games is the following.

Theorem F Let $\Pi = \{B_1, ..., B_m\}$ be a coalition structure. Then $bv'_\Pi FL \subseteq ASYMP(\Pi)$. If ψ is the operator that associates to each game in $bv'FL$ its asymptotic Π-CS value, then $\psi(v) = \psi_p^\Pi(v)$ where $p = (p_1, .., p_m)$ is an m-tuple of uniform probability measures.

Proof Let ψ denote the operator that associates to each game in $ASYMP(\Pi)$ its asymptotic Π-CS value and choose $v \in bv'_\Pi FL$. From Lemma 2, it follows that $r_k^v \in bv'_k FL$ for each k. Applying the Main Theorem in [**11**], we conclude that $r_k^v \in ASYMP_k$ for each $k \in M$ and therefore $v \in ASYMP(\Pi)$ (apply Proposition 4 above.) If S is a coalition, then the asymptotic value payoff on $bv'_k FL$ to the coalition $S \cap B_k$ is equal to

$$\varphi_{p_k}^k(r_k^v)(S \cap B_k)$$

where $\varphi_{p_k}^k$ is the value on $bv'_k FL$ corresponding to the uniform probability measure p_k. Therefore, we can apply Proposition 4 again and conclude that

$$\psi v(S) = \sum_{k \in M} \varphi_{p_k}^k(r_k^v)(S \cap B_k) = \psi_p^\Pi(v)(S)$$

where $p = (p_1, .., p_m)$ is the m-tuple of uniform probability measures. □

Remark These asymptotic results can be generalized to the space $bv'_\Pi \mathfrak{M}$ where $\mathfrak{M}$ is the space of measures (including those with countably many nonnull atoms). Then $bv'_\Pi M$ is the space spanned by all functions of the form $f \circ \omega$ where $\omega \in \mathfrak{M}^1$ and $f \in bv'_{\Pi,\omega}$ (again, $bv'_{\{I\},\omega}\mathfrak{M}$ is the same as $bv'\mathfrak{M}$ in [**13**]). Lemma 2 holds when $bv'_\Pi FL$ is replaced by $bv'_\Pi \mathfrak{M}$ i.e., $r_k^v \in bv'_k \mathfrak{M}$ whenever $v \in bv'_\Pi \mathfrak{M}$. Applying Theorem A in [**13**], it follows that $r_k^v \in ASYMP_k$ for each $k \in M$ and therefore $bv'_\Pi \mathfrak{M} \subseteq ASYMP(\Pi)$.

References

[1] Artstein, Z. [1971], *Values of Games with Denumerably Many Players*, Int. J. Game Theory **1**, 27-37.

[2] Aumann, R. J. and Shapley, L. S. [1974], *Values of Non Atomic Games,* Princeton University Press, Princeton, NJ.

[3] Fogelman, F. and Quinzii, M. [1980], *Asymptotic Values of Mixed Games*, Math. Opns Res. **5**, 86-93.

[4] Hart, S. [1973], *Values of Mixed Games*, Int. J. Game Theory **2**, 69-85.

[5] Hart, S. and Kurz, M. [1982], *Stable Coalitions Structures*, in Forming Coalitions (M. J. Holler, ed.), Springer Verlag, Berlin.

[6] Hart, S. and Kurz, M. [1983], *Endogenous Formation of Coalitions*, Econometrica **51**, 1047-1064.

[7] Hart, S. and Mas-Colell, A. [1989], *Potential, Value and Consistency*, Econometrica **57**, 589-614.

[8] McLean, Richard P. [1995], *Coalition Structure Values of Nonatomic Games*, mimeo, Department of Economics, Rutgers University

[9] McLean, Richard P. [1995], *Customer Specific Aumann-Shapley Pricing*, mimeo, Department of Economics, Rutgers University

[10] Milnor, J. W. and Shapley, L. S. [1961], *Values of Large Games II: Oceanic Games*, RM-2649, The Rand Corporation, Santa Monica, CA.

[11] Neyman, A. [1979], *Asymptotic values of Mixed Games* in Game Theory and Related Topics (A. Moeschlin and D. Pallaschke, eds.), North Holland, Berlin and New York, 71-81.

[12] Neyman, A. [1985], *Semi-values of Political Games*, Math. Opns Res., **10**, 390-402.

[13] Neyman, A. [1988], *Weighted Majority Games have Asymptotic Values*, Math. Opns Res., **13**, 556-580.

[14] Owen, G. [1977], *Values of Games with a Priori Unions*, in Essays in Mathematical Economics and Game Theory (R. Hein and O. Moeschlin, eds.), Springer Verlag, Berlin.

[15] Shapiro, N. Z. and Shapley, L. S. [1978], *Values of Large Games I: A Limit Theorem*, Math. Opns Res. **3**, 1-9.

[16] Shapley, L. S. [1953], *A Value for n-Person Games*, in Contributions to the Theory of Games **2**, (H. W. Kuhn and A. W. Tucker, eds.), Princeton University Press, Princeton, NJ, 307-317.

Fields Institute Communications
Volume **23**, 1999

Aumann-Shapley Random Order Values of Non-Atomic Games

Lakshmi K. Raut
Department of Economics
University of Chicago
Chicago, Illinois 60637 USA
raut@spc.uchicago.edu

Abstract. In this paper we report results for a reformulated random order approach to values of non-atomic games. The reformulation is achieved by generating random orders from a fixed subgroup of automorphisms Θ that admits an invariant probability measurable group structure. The resulting Θ-symmetric random order value operator is unique and satisfies all the axioms of a Θ-symmetric axiomatic value operator. For the uncountably large invariant probability measurable group $\left(\breve{\Theta}, \breve{\mathcal{B}}, \breve{\Gamma}\right)$ of Lebesgue measure preserving automorphisms constructed in Raut [1997], $\breve{\Theta}$-symmetric random order value exists for most games in BV and it coincides with the fully symmetric Aumann-Shapley axiomatic value on large subspaces of games in pNA. Thus by restricting the set of admissible orders and the space of games suitably, it is possible to circumvent the Aumann-Shapley Impossibility Principle for the random order approach to values of non-atomic games.

1 Introduction

Given a fixed set of players, called the grand coalition, and an algebra of its subsets as the possible coalitions, a cooperative game is a set function that assigns a real number to a coalition, called the worth of the coalition, to each coalition. The worth of a coalition is the payoff which the players can earn cooperatively. A basic problem in cooperative game theory is to find rules, called solution concepts, for dividing the worth of the grand coalition among the players so that certain desiderata are satisfied. Mathematically, the problem is to find a mapping or an operator satisfying pre-specified conditions from the space of all set functions to the space of additive set functions. The Shapley value is one such solution concept. The axiomatic characterization provided by Shapley is typically viewed as expressing fairness properties; these may account for its wide commendation.

1991 *Mathematics Subject Classification.* Primary 90D12; Secondary 28C10,90A08.

Using the linear vector space structure of the space of games, Shapley [1953] proved the existence and uniqueness of the operator satisfying his axioms and provided a formula to compute the value of individual games. The solution thus obtained is known as *axiomatic value*. Shapley also postulated an alternative set of fairness properties which come to be known as the *random order value*. In this approach, a player is given his expected marginal contribution in a random ordering of players, each ordering being equally likely among all possible orderings of the players. Shapley showed that the formulas for value from both approaches coincide.

Aumann and Shapley [1974] extended the concept of axiomatic value to games with a continuum of players and proved the existence and uniqueness of an axiomatic value operator on the spaces of games such as pNA and bv′ NA (definitions of unknown terms in the introduction can be found in subsequent sections). They also provided a "diagonal formula" for games in pNA. In an attempt to extend the random order value to continuum case, Aumann and Shapley considered the set Ω of all orderings of players satisfying some measurability condition and derived an impossibility principle: *There does not exist a measure structure on Ω with respect to which a random order value could be assigned to games in pNA.*

There have been several developments in the axiomatic value over the past several years. One direction of research has been aimed at finding larger spaces of games on which an axiomatic value, possibly a unique one, exists. The space pNA is economically the most important one since it contains smooth market games and fair cost allocation schemes. However, non-smooth games that arise from markets with strong complementarity do not belong even to the space ASYMP.[1] Mertens [1988] introduced a very large space of games, known as *Mertens space*. This space includes these non-smooth games and also all the above spaces. Using his extension of the diagonal formula Mertens proved the existence of an axiomatic value on Mertens space.

The most important fairness properties of the Shapley value are derived from the symmetry axiom. This axiom was originally specified with respect to the whole group $\mathcal{G}$ of automorphisms of the players set. More generally, we view each axiomatic value with symmetry restricted to a subgroup $\Theta \subset \mathcal{G}$ as representing a particular fairness principle. We shall refer to a value with symmetry restricted to a subgroup Θ as *Θ-symmetric axiomatic value* and the mapping which assigns to each game its Θ-symmetric axiomatic value as a *Θ-symmetric axiomatic value operator*. Some economic applications may dictate the use of particular subgroups Θ. For a given non-atomic measure μ on the unit interval, let us denote by $\mathcal{G}(\mu)$ the group of μ-measure preserving automorphisms. Monderer [1986, 1989] showed that the non-atomic games that arise from smooth market economies have certain characteristics in which symmetry group could be restricted to the group $\mathcal{G}(\mu)$ and the non-atomic measure μ is determined by the data of the economy. For such applications the domain of the value operator can be restricted to the space pNA(μ), the linear subspace of pNA generated by the polynomials in non-atomic probability measures which are absolutely continuous with respect to the above measure μ.

---

[1]The space ASYMP contains the space pNA and it is the largest space on which value was shown to exist in the Aumann and Shapley [1974] monograph. The technique used to prove this result (see Kannai [1966] and also Aumann and Shapley [1974] for details) was to approximate a continuum game with a sequence of finite games and then to use a limiting argument on the sequence of Shapley values of these finite games to arrive at the Shapley value of the continuum game. The space ASYMP consists of games for which this limiting argument was valid.

Monderer showed that any $\mathcal{G}(\mu)$-symmetric axiomatic value operator on pNA(μ) is also symmetric with respect to the full group of automorphisms $\mathcal{G}$.

Aumann and Shapley [1974] proved that there does not exist $\mathcal{G}$-symmetric axiomatic value operator on all of BV. Thus to have a value on all of BV, the symmetry axiom must be restricted to a proper subgroup. Ruckle [1982] has shown that when the symmetry is restricted to any "locally finite" subgroup of automorphisms, there exists a value operator on all of BV. This result is further refined by Monderer and Ruckle [1990].

It is important to note that the main fairness property of the random order value arises from the fact that each player has an equal chance of forming a coalition with a set of players of any size and any names. The random order value assigns to each player the average of his marginal contributions over all coalitions which he may join. In the finite player case, the group of automorphisms of the players set and the set of orderings of players generated by them are identical, and the unweighted averaging of the marginal contributions of a player over all orderings generated by a given group of automorphisms basically symmetrizes the random order value for the group of automorphisms. Raut [1993] shows that for the average marginal contribution to be symmetric for all games and all players it is necessary and sufficient that the randomness of the set of orders that are induced by the symmetry group assigns equal likelihood to each order, i.e., a random order has uniform distribution. I used these insights from finite games to reformulate a Θ-symmetric random order value for continuum case (Raut [1995, 1997]), which can be briefly described as follows.

We begin with a fixed group of automorphisms $\Theta \subset \mathcal{G}$ with an invariant probability measurable group structure in it. We take the set of orders Ω as those which are determined by Θ and induce a uniform distribution on Ω from the action of the invariant probability measurable group Θ. We first show that the expected value of the marginal contribution function on Ω is unique (i.e., independent of any particular invariant measurable group structure on Θ) and it satisfies all the axioms of Θ-symmetric value; we name such a mathematical expectation operator as *Θ-symmetric random order value operator* on any subspace of games on which it is well defined. In particular then the $\mathcal{G}$-symmetric random order value operator is also a $\mathcal{G}$-symmetric axiomatic value operator. The set of orders induced by $\mathcal{G}$ is, however, a proper subset of the set of Aumann-Shapley measurable orders (see Raut [1995, Proposition 1] for details). Could we not then hope to have a possibility result for the Aumann-Shapley Impossibility Principle with respect to the $\mathcal{G}$-symmetric random order value operator? We do not have an answer to this question yet. However, with the help of even more restricted set of orders induced from a suitably constructed uncountably large invariant probability measurable group $\breve{\Theta}$ of Lebesgue measure preserving (l.m.p.) automorphisms, we have proved a partial possibility theorem to the Aumann-Shapley Impossibility Principle. More specifically, we have shown that there exists a unique $\breve{\Theta}$-symmetric random order value operator on a large space of games (we named it as NBV) which contains most of the games in BV and some of the non-smooth games that did not belong even to ASYMP. For any non-atomic probability measure μ, let us define the space lpNA(μ) as the linear space of games generated by the polynomials in μ. Most economic applications consider games in pNA which also belong to lpNA(μ) for some non-atomic probability measure μ. We have also derived a diagonal formula for the $\breve{\Theta}$-symmetric random order value which coincides with the diagonal value formula for the $\mathcal{G}$-symmetric Aumann and Shapley axiomatic value for games in the space lpNA(μ). Thus, when the set of admissible orders Ω are suitably restricted we have a measure structure on Ω with respect to which the random order value exists for large subspaces of games in pNA.

In Section 2 we lay out the basic framework for our random order approach, discuss the relevant issues and state the main results. In Section 3, we discuss the issues concerning choice of a suitable symmetry group, and sketch the construction of invariant probability measurable group $\breve{\Theta}$ which was introduced in Raut [1997]. Finally in Section 4 we put together further remarks concerning the random order approach to values for future research.

2 The basic framework

To denote a Borel σ-algebra of a topological space X (i.e., the σ-algebra generated by the class of open sets of X), we use the subscripted notation $\mathcal{B}_X$ and to denote any general σ-algebra, we do not use a subscript. Let $I = [0,1] \subset \Re$ be the set of players. Let $\mathcal{B}_I$ be the Borel σ-algebra of I. The elements of $\mathcal{B}_I$ are the set of admissible coalitions. A *game* is a set function $V : \mathcal{B}_I \to \Re$ such that $V(\emptyset) = 0$. Let G_I be the set of all games. Let FA be the set of finitely additive set functions on $(I, \mathcal{B}_I)$. A *measure* is a countably additive set function. One can check easily that G_I and FA are linear vector spaces. A game V is *monotonic* if $V(S) \leq V(T)$ for any $S, T \in \mathcal{B}_I$, $S \subset T$. A *Borel automorphism* is a measurable map $\theta : (I, \mathcal{B}_I) \to (I, \mathcal{B}_I)$ such that it is one-one, onto and θ^{-1} is also measurable. Let $\mathcal{G}$ be the set of all Borel automorphisms on $(I, \mathcal{B}_I)$. One can check that $\mathcal{G}$ is a non-commutative (also known as non-abelian) group with composition of functions as group multiplication operation and identity function as the group identity.

For each $\theta \in \Theta$, define the linear operator $\theta^* : G_I \to G_I$ by

$$(\theta^* V)(S) = V(\theta^{-1}(S)), \ \forall S \in \mathcal{B}_I.$$

Given a subgroup of automorphisms, $\Theta \subset \mathcal{G}$, a linear subspace $Q \subset G_I$ is said to be Θ-*symmetric* if $\theta^* Q \subset Q$ for all $\theta \in \Theta$.

Let Q be a linear subspace of G_I. An operator $\Phi : Q \to FA$ is said to be *linear* if $\Phi(\alpha V_1 + V_2) = \alpha\Phi(V_1) + \Phi(V_2) \ \forall \ V_1, V_2 \in Q, \alpha \in \Re$. The operator Φ is said to be *positive* if (ΦV) is monotonic for any monotonic V in the domain of Φ, and *efficient* if $\Phi V(I) = V(I) \ \forall \ V \in Q$. For a Θ-symmetric space Q, the operator $\Phi : Q \to FA$ is said to be Θ-*symmetric* if $\Phi\theta^* V = \theta^*\Phi V, \ \forall \ \theta \in \Theta, V \in Q$.

A Θ-*symmetric axiomatic value operator* on a Θ-symmetric linear space of games Q is a positive, linear, efficient, and Θ-symmetric operator $\Phi : Q \to FA$. When Θ is the full group $\mathcal{G}$, this operator is the *Aumann-Shapley axiomatic value operator*. Aumann and Shapley [1974] proved the existence and uniqueness of this operator axiomatically.

Although in our analysis of random order value we do not use any topological structure on the space of games, to relate our results to the literature, we adopt the following topological concepts from Aumann and Shapley [1974]. A game V is *of bounded variation* if there exist monotonic games U and W such that $V = U - W$. Denote by BV the set of all games of bounded variation. It is known that BV is a linear vector space over $\Re$. Define a map $\| \,.\, \|_{BV}$: BV $\to \Re$ by

$$\| V \|_{BV} = \inf \ \{U(I) + W(I) \mid V = U - W, \ U \text{ and } W \text{ are monotonic games}\}$$

for each $V \in$ BV. It can be shown that $\| \,.\, \|_{BV}$ is a well defined norm, on BV and with this norm, BV is a Banach space (see Aumann and Shapley [1974, Corollary

4.2, and Proposition 4.3]. The following notation is standard in the literature:

$$\begin{aligned}
\mathrm{NA} &= \text{set of non-atomic measures on } (I, \mathcal{B}_I)\\
\mathrm{NA}^1 &= \text{set of probability measures in NA}\\
\mathrm{pNA} &= \|.\|_{BV} - \text{closure of linear space spanned by powers of } \mu \in \mathrm{NA}\\
\mathrm{bv'NA} &= \|.\|_{BV} - \text{closure of linear space spanned by } f \circ \mu, \text{ where } f : I \to \Re\\
&\quad \text{is of bounded variation, continuous at 0 and 1, and } f(0) = 0, \text{and}\\
&\quad \mu \text{ is a non-negative, non-atomic probability measure on } (I, \mathcal{B}_I).
\end{aligned}$$

It is known that FA, NA and pNA are all closed subspaces of BV.

2.1 Generation of random orders. Two main features of the random order approach to values of games with finite sets of players are that (1) each automorphism [2] generates a distinct ordering of players, i.e., the set of orders is the same as the group of automorphisms; (2) for all games, the mathematical expectation of the random marginal contribution set function is symmetric with respect to the group of automorphisms if and only if each random ordering of players is equally likely (see Raut [1993]). In the finite players case, the main reason why the expected marginal contribution set function becomes symmetric for any game and with respect to the full group of permutations is that every player is equally likely to form a coalition with a set of players of any size and names in a random order. We want to adopt these two features to the continuum case.

Each $\theta \in \Theta$ generates a binary relation, $\succ_\theta \subset I \times I$ defined by

$$\text{for any } s, t \in I, \; s \succ_\theta t \;\Leftrightarrow\; \theta(s) > \theta(t).$$

Recall that an order $\succ$ on a set X is a *linear order*, which is also known as *total order* if for any $x, y \in X$, $x \neq y$, either $x \succ y$ or $y \succ x$, for no $x \in I$, $x \succ x$, and for any $x, y, z \in I$, $x \succ y$, $y \succ z \Rightarrow x \succ z$. A total order is a particular type of preference order. We will refer to a total order in this paper simply as an *order*. It is easy to verify that the binary relation $\succ_\theta$ generated by an automorphism θ is an order on I and that each $\theta \in \Theta$ generates a distinct order. Let $\bar{I} = I \cup \{\infty\}$, and for all $\theta \in \mathcal{G}$, we assume that $\theta(\infty) = \infty$. For an order $\succ_\theta$, $\theta \in \mathcal{G}$, and a $s \in \bar{I}$, define *an initial segment* $I(s, \theta)$ by $I(s, \theta) = \{t \in I \mid \theta(s) > \theta(t)\}$. We view $I(s, \theta)$ as the set of players who are before player s in the order $\succ_\theta$.

In the continuum case two Borel automorphisms of I may generate the same ordering of I. For instance, $\theta \in \mathcal{G}$, defined by $\theta(x) = x^2$, $x \in I$ and the identity element $e \in \mathcal{G}$, defined by $e(x) = x, x \in I$, both generate the standard order $\succ_e$. Thus the set of orderings of players and the group of Borel automorphisms of players are not the same set. We derive the set of orders Ω that are generated by a given group of automorphisms Θ as follows:

Define an equivalence relation $\sim$ on $\Theta \times \Theta$ by,

$$\theta_1 \sim \theta_2, \text{ for } \theta_1, \theta_2 \in \Theta \Leftrightarrow \theta_1, \theta_2 \text{ generate the same order on } I.$$

Let $\Theta_e = \{\theta \in \Theta | \theta \sim e\}$. It can be easily shown that Θ_e is a subgroup of Θ and the set of distinct orders, Ω, generated by the automorphisms in Θ is the set of right cossets given by

$$\Omega \equiv \Theta/\Theta_e \equiv \{\Theta_e\theta | \theta \in \Theta\}$$

In the finite player case, the set of automorphisms of players is finite and for finite sets the concept of equal likelihood is intuitive. In the continuum case,

[2] In the finite players case an automorphism is known as permutation.

however, the set of automorphisms of the players is uncountable. An analogue of the equal likelihood in the continuum set is the following concept of an invariant measurable group or invariant measure. This concept requires the underlying space to have a group structure:

Definition 1 A measure space $(\Theta, \mathcal{A}, \Gamma)$ is said to be an *invariant measurable group* if Θ is a group, the map $(\theta_1, \theta_2) \to \theta_1\theta_2^{-1}$ from $(\Theta \times \Theta, \mathcal{A} \times \mathcal{A})$ onto $(\Theta, \mathcal{A})$ is measurable, and Γ is $\sigma-$finite, not identically zero, and right invariant, i.e., $\Gamma(E\theta) = \Gamma(E)$, for all $E \in \mathcal{A}$, and $\theta \in \Theta$, where $E\theta \equiv \{\sigma\theta | \sigma \in E\}$. Γ is known as *right invariant measure.*[3] Furthermore, when Γ is a probability measure, a measurable group $(\Theta, \mathcal{A}, \Gamma)$ is said to be a *right invariant probability measurable group.*

In general Θ_e is not a normal[4] subgroup of Θ and hence Ω is not necessarily a group. To see this, let $\theta \in \mathcal{G}$ and $\theta_e \in \mathcal{G}_e^*$ be defined by

$$\theta(x) = \begin{cases} 1-x & \text{if } 0 \leq x < 1/2 \\ x - 1/2 & \text{if } 1/2 \leq x \leq 1. \end{cases}$$

$$\theta_e(x) = \begin{cases} .01x & if\ 0 \leq x < .8 \\ .008 + 4.96(x - .8) & if\ .8 \leq x \leq 1 \end{cases}$$

Let $t = .4$ and $s = .3$. Thus $\theta_e(s) < \theta_e(t)$, but $(\theta^{-1}\theta_e\theta)(s) = .507 > .506 = (\theta^{-1}\theta_e\theta)(t)$, thus $\theta^{-1}\theta_e\theta \notin \mathcal{G}_e$.

Thus we do not have the group structure on Ω that we needed to extend the concept of equal likelihood of orderings in Ω. But Ω is a homogeneous space acted on by the group Θ, and for homogeneous spaces there is a natural concept of invariant measure (see Parthasarathy [1977, Section 55]; or Segal and Kunz [1978, Section 7.4]). In our set-up, however, we can use the *natural map* $\Pi : \Theta \to \Omega$ defined by $\Pi(\theta) = \Theta_e\theta$ to induce an invariant probability measure structure $(\Omega, \mathcal{B}, \mu)$ on the homogeneous space Ω of induced orders. $(\Omega, \mathcal{B}, \mu)$ will be referred to as a *set of random orders.*

2.2 Θ-symmetric random order value operator. Given a game V, and an order $\succ_\theta$, $\theta \in \mathcal{G}$, a *marginal contribution set function,* $\phi^\theta V$ on $(I, \mathcal{B}_I)$ is a measure on $(I, \mathcal{B}_I)$ such that

$$\left(\phi^\theta V\right)(I(s,\theta)) = V(I(s,\theta)),\ \forall\ s \in \bar{I} \tag{2.1}$$

Notice that for any $\theta, \theta' \in \Theta$, such that $\theta \sim \theta'$, we have $I(s,\theta) = I(s,\theta')$; hence it follows from (2.1) that $\phi^\theta V(S) = \phi^{\theta'} V(S)$ for all $S \in \mathcal{B}_I$. This allows us to unambiguously define $(\phi^\omega V)(S) = (\phi^\theta V)(S)$ where θ is such that $\omega = \Theta_e\theta$.

The *expected marginal contribution set function for a game* V is a set function $\Phi_\Gamma V$ defined by

$$\begin{aligned}(\Phi_\Gamma V)(S) &= \int_\Omega (\phi^\omega V)(S) d\mu(\omega) \\ &= \int_\Theta (\phi^\theta V)(S) d\Gamma(\theta),\ S \in \mathcal{B}_I.\end{aligned} \tag{2.2}$$

[3]When Θ is a locally compact topological group, and $\mathcal{A}$ is the Borel σ-algebra, such that $\Gamma(U) > 0$, for every non-empty open set $U \subset \Theta$, then the Borel measure Γ is known as Haar Measure.

[4]N is a normal subgroup of G if for all $\theta \in$ G, we have $\theta^{-1}\nu\theta \in$ N for all $\nu \in$ N.

The second equality follows from the change of variable formula for Lebesgue integrals and the facts in the previous paragraph. Let us define the space of games:

$$L1(\Theta,\Gamma) = \left\{V \in G_I \mid \phi^\theta V(S) \text{ in (2.2) is integrable for all } S \in \mathcal{B}_I\right\}. \tag{2.3}$$

For the operator Φ_Γ defined in (2.2) to yield a random order value operator in the continuum case, it should have three basic properties (which are shown to hold in Raut [1995]): **First**, for any game V and any order $\succ_\theta$, $\theta \in \Theta$, if there exists a measure $\phi^\theta V$ satisfying (2.1), it should be unique so that for each $S \in \mathcal{B}_I$, $\phi^\theta V(S)$ is a function of θ. **Second**, in order for the operator Φ_Γ to be Θ-symmetric with respect to a given subgroup of automorphisms, $L1(\Theta,\Gamma)$ defined in (2.3) must be a Θ-symmetric linear subspace of G_I. **Third**, the approach is of little use if for a given symmetry group of automorphisms, Θ, two different probability measurable structures assign two different finitely additive set functions to a game. The second part of Raut [1995, Theorem 1] shows that the mathematical expectation in (2.2) depends only on the fixed subgroup of automorphisms, Θ, but not on a specific invariant probability measurable group structure on Θ.

For a given fixed subgroup of automorphisms $\Theta \subset \mathcal{G}$ and a Θ-symmetric linear subspace of games $Q \subset G_I$, the operator $\Phi_\Gamma : Q \to FA$ defined in (2.2) with respect to an invariant probability measurable group structure $(\Theta, \mathcal{A}, \Gamma)$ on Θ such that $Q \subset L1(\Theta,\Gamma)$ is said to be a *Θ-symmetric random order value operator* on Q. Raut [1995, Theorem 1] assures that when such an operator exists, it is independent of a specific invariant probability measurable group structure on Θ and it coincides with the Θ-symmetric axiomatic value operator on Q. In fact, a Θ-symmetric random order value operator is a particular characterization of the Θ-symmetric axiomatic value operator. In particular, therefore, if we take Θ to be the full automorphism group $\mathcal{G}$, then the existence of an Aumann-Shapley axiomatic value operator on Q can be reduced to the question of the existence of an invariant probability measurable group structure, $(\mathcal{G}, \mathcal{B}, \Gamma)$, with the property that $Q \subset L1(\mathcal{G},\Gamma)$.

In our random order approach, the set of orders induced by $\mathcal{G}$ is a proper subset of the set of measurable orders considered by Aumann and Shapley in their random order approach. Therefore we may recast the Aumann-Shapley impossibility issue and expect a possibility result in our framework:

Question 1: *Could there exist an invariant probability measurable group structure on $\mathcal{G}$, such that* pNA $\subset L1(\mathcal{G},\Gamma)$?

Indeed, on any group Θ, there always exists a right invariant probability measurable group structure, for instance, the trivial, coarsest σ-algebra, $\mathcal{B} = \{\emptyset, \Theta\}$ with a trivial probability measure that assigns 0 to empty set and 1 to the whole set. The coarser the σ-algebra, the more meager are the sets of measurable and integrable functions, and hence fewer games belong to $L1(\Theta,\Gamma)$ which may not include games in pNA. If the answer to the above question is still negative, we may find a positive answer when we restrict the question to other possibly smaller classes of economically important games than pNA. I conjecture the proof of Aumann-Shapley impossibility theorem could be adopted to the present framework to produce a negative answer to the above question. Very little is known about the structure of group $\mathcal{G}$ that can shed light on the above issues, and I have not pursued these issues any further in this paper either. Instead, I have carried my analysis along the following line:

Question 2: *If we restrict the symmetry of the value operator to a "reasonable" proper subgroup of automorphisms Θ, could we construct a probability measurable* *group structure*, $(\Theta, \mathcal{A}, \Gamma)$ *on* Θ *such that* $L1(\Theta, \Gamma)$ *contains a large set of games including* pNA *and on a large class of economically important games from* pNA, the Θ-*symmetric random order value operator coincides with the Aumann-Shapley axiomatic value operator?*

We carried out the above type of inquiry for the $\breve{\Theta}$-symmetric random order value operator $\Phi_{\breve{\Gamma}}$ for the uncountably large subgroup $\breve{\Theta}$ of the group of l.m.p. automorphisms with the invariant probability measurable group structure, $\left(\breve{\Theta}, \breve{\mathcal{B}}, \breve{\Gamma}\right)$. This was introduced in Raut [1997], and we explain its structure in Section 3 .

In the context of restricting the symmetry group, two types of issues are addressed in the axiomatic approach to value. **First**, The largest subspace of BV on which Aumann-Shapley established the existence and uniqueness of $\mathcal{G}$-symmetric axiomatic value operator is bv′NA, and they also showed that there cannot exist a $\mathcal{G}$-symmetric axiomatic value operator on all of BV. Ruckle [1982] addressed the question: If the symmetry of the value operator is restricted to a proper subgroup, could there exist an axiomatic value operator on all of BV? Ruckle has shown that if Θ is taken to be any "locally finite" group then there exists a Θ-symmetric axiomatic value operator on all of BV. A group is said to be *locally finite* if for any finite number of elements from the group there is a finite subgroup which contains these elements. Monderer and Ruckle [1990] gave further refinement of this result. We have an analogous result. To describe it we need the following definition:

Definition 2 A set function $V \in G_I$ is said to be *normalized* if (i) $V(A_n) \to 0$ as $n \to \infty$ for any sequence of sets, $A_n \in \mathcal{B}_I$, $A_n \downarrow \emptyset$ as $n \to \infty$, and (ii) $V(A_n) \to V(A)$ as $n \to \infty$ for any sequence of sets, $A_n \in \mathcal{B}_I$, $A_n \uparrow A$ as $n \to \infty$, where $A \in \mathcal{B}_I$.

Let us denote by NBV = the set of normalized set functions from BV. It can be easily seen that NBV is a linear space. The result is as follows:

Theorem 1 *There exists a unique* $\breve{\Theta}$-*symmetric random order value* $\Phi_{\breve{\Gamma}}$ *on* NBV.

Second, Monderer [1986, 1989] addressed the question: If Θ is a proper subgroup of $\mathcal{G}$, then on which linear subspace of pNA, does there exist a unique Θ -symmetric value operator? Or equivalently, on which subspace of games in pNA, Θ-symmetric value operator coincides with the Aumann-Shapley axiomatic value operator? He has shown that when Θ is taken to be the group of μ-measure ($\mu \in NA$) preserving automorphisms and the linear space of games restricted to pNA(μ) (the closed subspace of games in pNA spanned by polynomials of non-atomic probability measures which are absolutely continuous with respect to the given non-atomic probability measure μ).[5] Then there exists a unique Θ-symmetric axiomatic value operator satisfying dummy axiom on pNA(μ). With a little modification in the proof in Raut [1995, Theorem 3]) we have the following analogous result for our random order approach.

Theorem 2 *Let* $f : I \to \Re$ *be absolutely continuous, and* $\lambda \in NA^1$, *then the unique* $\breve{\Theta}$-*symmetric random order value of the scalar measure game* $f \circ \lambda$ *yields*

[5]See his economic justifications for restricting the symmetry group and the class of games in this particular way so that his theory applies to a large class of market games.

the following diagonal formula:

$$\Phi_{\breve{\Gamma}}[f \circ \lambda](S) = \lambda(S) \int_0^1 f'(t) d\lambda.(t) \tag{2.4}$$

Thus, the $\breve{\Theta}$-symmetric random order value operator $\Phi_{\breve{\Gamma}}$ coincides with the Aumann-Shapley axiomatic value operator on all of lpNA(λ).

With symmetry and generation of random orders restricted to a subgroup of μ-measure preserving automorphisms $\breve{\Theta}(\mu)$,[6] it is easy to modify our analysis so that the above result is true for lpNA(μ). I conjecture that the above result holds for pNA(μ). At this point, I have not been able to prove it.

3 On the symmetry subgroup $\breve{\Theta}$

The essence of the success of the random order approach to characterize the Shapley value of games with finite set of players ask the following question: *Does there exist an uncountably large subgroup of automorphisms in $\mathcal{G}$ that can be equipped with an invariant probability measurable group structure, $(\Theta, \mathcal{A}, \Gamma)$, such that in a randomly selected order every player is equally likely to have placed before it a set of players of any size containing players from anywhere in the unit interval?* This notion beckons to the strongly mixing automorphisms. To fix ideas, let us consider mixing with respect to Lebesgue measure λ. A Lebesgue measure preserving automorphism, $\theta \in \mathcal{G}$, is said to be *strongly mixing* if $\lim_{n\to\infty} \lambda\left(\theta^{-n} E \cap F\right) = \lambda\left(E \cap F\right)$, for all $E, F \in \mathcal{B}_I$. In essence a strongly mixing automorphism, θ, allows thorough mixing of any set of players $t \in E$ with any other player in the unit interval I by producing an orbit $\mathcal{Q}(E) = \{\theta^n t | t \in E, n = 0, 1, ..\}$ which is dense and uniformly spread allover I. A l.m.p. automorphism is *weakly mixing* if $\lim_{n\to\infty} \frac{1}{n} \sum_{j=1}^n \lambda\left(\theta^{-j} E \cap F\right) = \lambda\left(E \cap F\right)$. It is known that with respect to the "weak topology", the set of such automorphisms is of the first category and the set of weakly mixing automorphisms is of the second category. This means that generically a measure preserving automorphism is a weakly mixing but not strongly mixing. Aumann [1967] [7] has shown that it is impossible to find an invariant probability measurable group structure on the whole group of l.m.p. automorphisms which satisfy further conditions that the real valued function $f(\theta) \equiv \lambda\left(E \cap \theta F\right)$ is measurable for all $E, F \in \mathcal{B}_I$.

Thus in our approach, we seek to achieve thorough mixing of players with the help of actions of an uncountably large subgroup $\breve{\Theta}$ of l.m.p. automorphisms. This we obtain as a (projective) limit of an increasing sequence of "carefully constructed" increasing finite subgroups, Θ_n, $n \geq 0$ of l.m.p. automorphisms. We describe the construction of $\breve{\Theta}$ briefly in this section, (for details, see Raut [1997]). It is, however, interesting to note that in our construction we get thorough mixing with the help of automorphisms in $\Theta_n' s$ which are all recurrent of period n.

We also want to equip $\breve{\Theta}$ with a fine enough measurable group structure $\left(\breve{\Theta}, \breve{\mathcal{B}}, \breve{\Gamma}\right)$ so that it allows a sufficiently rich set of games in $L1(\breve{\Theta}, \breve{\Gamma})$. We impose the following separability requirement on our measurable group structure:

---

[6]This subgroup is, in fact, a translation of the group $\breve{\Theta}$ with an automorphism $\xi \in \mathcal{G}$, where ξ is such that $\mu = \lambda\xi^{-1}$. That there exists such an automorphism ξ can be found in Parthasarathy [1977].

[7]I am grateful to Professor Robert Aumann for drawing my attention to this result.

Definition 3 A measurable group $(\Theta, \mathcal{A}, \Gamma)$ is *separated* [8] if $\forall \theta \in \Theta,\ \theta \neq e$, there exists $E \in \mathcal{A}$ such that $0 < \Gamma(E) < \infty$ and $\Gamma(E\theta\Delta E) > 0$, where Δ is the symmetric difference operator between two sets.

We define the increasing sequence of finite groups, Θ_n, $n \geq 0$ each containing l.m.p. automorphisms that are discontinuous at only a finite number of points of I. In the n-th automorphism group, Θ_n, the discontinuities of the automorphisms are at the points $\frac{k}{2^n}, k = 1, ..., 2^n - 1$. These $2^n - 1$ points in I determine 2^n dyadic subintervals of I: $I_k = \left[\frac{k}{2^n}, \frac{k+1}{2^n}\right]$, $k = 0, 1, ..., 2^n - 1$. In order for the automorphisms to be Lebesgue measure preserving, we assume that in each subinterval I_k, the automorphisms are linear with slope $\pm$ 1.

Let $N_n = \{0, 1, 2, ..., 2^n - 1\}$. We represent each automorphism in Θ_n by a pair of functions, π_n and $\mathcal{O}_n$ such that $\pi_n : N_n \longrightarrow N_n$ is a permutation of N_n and $\mathcal{O}_n : N_n \rightarrow \{-1, 1\}$ is a map as follows: For each $k \in N_n$, $\pi_n(k)$ specifies which subinterval of the unit interval the image of the kth subinterval be mapped to, and $\mathcal{O}_n \circ \pi_n(k)$ specifies the slope of the automorphism that the image subinterval will take. Sometimes we will refer to $\mathcal{O}_n$ as the *slope map*. We denote such an automorphism as described above by the symbol

$$\theta_n = (\pi_n(k), \mathcal{O}_n \circ \pi_n(k))_{k=0}^{2^n-1}. \tag{3.1}$$

An equivalent description of the above automorphism that we will often use is the following:

$$\theta_n(x) = \begin{cases} \frac{\pi_n(k)}{2^n} - \frac{k}{2^n} + x & \text{if } x \in I_k \text{ and } \mathcal{O}_n(\pi_n(k)) = +1 \\ \frac{\pi_n(k)+1}{2^n} - \frac{k}{2^n} - x & \text{if } x \in I_k \text{ and } \mathcal{O}_n(\pi_n(k)) = -1 \\ & k = 0, 1, ...2^n - 1 \end{cases} \tag{3.2}$$

We use θ_n to mean the representation (3.1) and $\theta_n(.)$ or $\theta_n(x)$ to mean the representation (3.2) of an element in Θ_n.

Let $\Im = \{+1, -1\}$ denote the set of slopes. With the usual multiplication operation of real numbers and with $+1$ as the identity element, it is trivial to show that $\Im$ is a group.

Let

$$\Theta_n = \left\{ \begin{array}{c} \theta_n : I \rightarrow I, \text{ defined by (3.2)} \mid \pi_n \text{ is a permutation of } N_n \\ \text{and } \mathcal{O}_n \text{ is an orientation of the } 2^n \text{sub-intervals.} \end{array} \right\}$$

There are $(2^{2^n}.2^n!)$ total number of elements in Θ_n. It can be easily seen that Θ_n is a subgroup of l.m.p. automorphisms. For instance, when $n = 0$, there is no subdivision of I, and thus

$$\Theta_0 = \{\theta_0 = (\pi_0(0), \mathcal{O}_0(0)) \mid \mathcal{O}_0(0) \in \Im\}.$$

Note that Θ_0 has only two elements.

For further illustration of the above concepts, in panel (a) of Figure 1, we have shown the graph of $\theta_2(x)$ corresponding to the permutation, $\pi_2(1) = 2$, $\pi_2(2) = 4$, $\pi_2(3) = 1$ and $\pi_2(4) = 3$, and the orientation, $\mathcal{O}_2(1) = +1$, $\mathcal{O}_2(2) = -1$, $\mathcal{O}_2(3) = +1$, and $\mathcal{O}_2(4) = -1$.

[8]This separation notion for measurable groups is the analogue of Hausdorff separation axiom for topological spaces, see Halmos [1950, pp. 273].

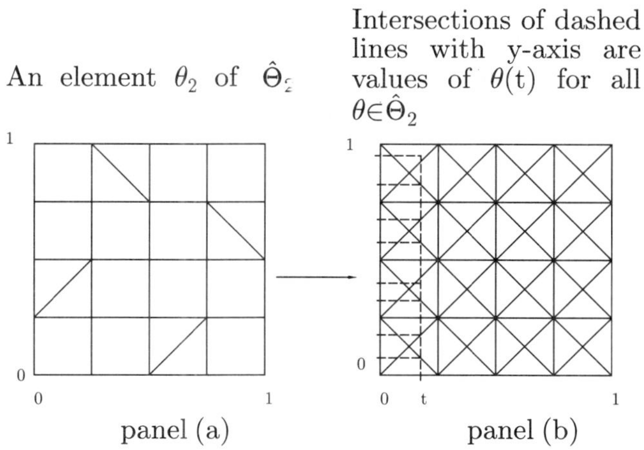

Figure 1

Let us examine the kind of randomization of the players that are performed by the random orders in Θ_n for large n. For illustration purpose, let us consider the automorphism $\theta_2 \in \Theta_2$ that is depicted in panel (a) of Figure 1. Note that the set of players that are placed before player t, $t \in I$, in the random order $\theta_2 \in \Theta_2$ is given by

$$I(\{t\}, \theta_2) = \begin{cases} [0, t) \cup I_3 & \text{if } t \in I_1 \\ I_1 \cup (t, \frac{1}{2}) \cup I_3 \cup I_4 & \text{if } t \in I_2 \\ [\frac{1}{2}, t) & \text{if } t \in I_3 \\ I_1 \cup I_3 \cup (t, 1] & \text{if } t \in I_4 \end{cases}$$

Note that the nature of randomization produced by an element of Θ_n depends on the permutation π and the orientation $\mathcal{O}$. Let us fix a $t \in I$ and suppose $t \in I_1$. Consider the initial segments of player t in each of the random orders $\theta_n \in \Theta_n$, that has the same value for t, say $\theta_n(t) = t_0$. All of these random orders will have either positive orientation or negative orientation. Let us assume that they have positive orientation. Let us denote by [x], $x \in \Re$, as the greatest integer in x. The way the initial segments are randomized by these random orders is that $[\frac{2^n}{t_0}]$ sub-intervals from the set of all sub-intervals except I_1 are randomly selected and then placed before the set of points [0,t) in all possible permutations. For very large n, size of each sub-intervals is very small, and hence for large n, all these $\theta_n' s$ with fixed value of $\theta_n(t) = t_0$ are placing almost any infinitesimally small subintervals of I that can fit in an interval of size $[0, t_0]$. The size of the interval also varies as we vary t_0 in the orbit $\mathcal{Q}_n = \{\theta(t_0) \mid \theta \in \Theta_n\}$.

In panel (b) of Figure (1), we have graphed all the elements of Θ_2. The set $\mathcal{Q}_2$ is shown as the intersection of the dash lines with the y-axis. It is trivial to note that $\Theta_n \subset \Theta_{n+1}$ and as $n \to \infty$, the number of elements in $\mathcal{Q}_n$ becomes large and are spread uniformly over I.

It can be easily shown that the limiting set, $\Theta \equiv \cup_{n=1}^{\infty}\Theta_n = \lim_{n\to\infty}\Theta_n$ is a group of l.m.p. automorphisms, but contains only countable number of elements, and thus cannot admit an invariant probability measurable structure. It is interesting to note, however, that Θ is locally finite. Thus we have an example of a locally finite symmetry group Θ such that there does not exist a Θ-symmetric random order value operator on any space of games at all.

We can, however, construct a large subgroup of l.m.p. automorphisms with a different kind of limit, namely, the projective limit of the above sequence of Θ_n limiting further the elements in each of them as follows: We use $\hat{\Theta}_n$ to denote the modified Θ_n. $\hat{\Theta}_n, n \geq 0$ are recursively constructed as follows:

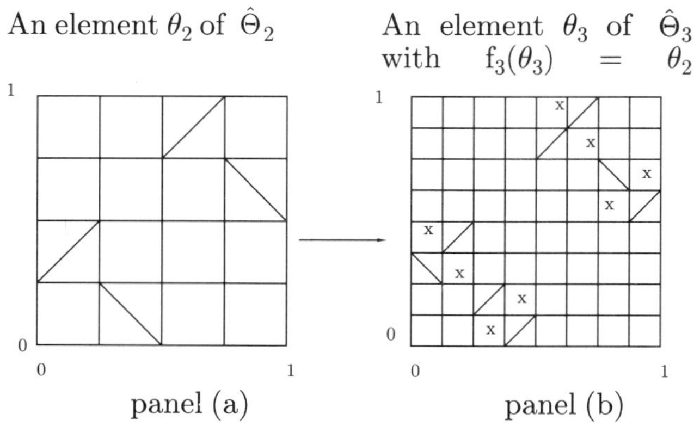

Figure 2

For $n = 0$, take $\hat{\Theta}_0 = \Theta_0$. To define $\hat{\Theta}_1$, note that we have two dyadic sub-intervals of I denoted as I_0 and I_1. Each $\theta_0 \in \hat{\Theta}_0$, induces a unique permutation π_{1,θ_0} of $N_1 = \{0, 1\}$ defined by

$$\pi_{1,\theta_0}(j) = i \;\text{ if for all } x \in I_j,\; \theta_0(x) \in I_i,\; i, j \in N_1.$$

Given $\theta_0 \in \hat{\Theta}_0$, let us denote by

$$A_1(\theta_0) = \left\{\theta_1 = (\pi_{1,\theta_0}(k), \mathcal{O}_1(\pi_{1,\theta_0})(k))_{k=0}^{1} \mid \mathcal{O}_1(\pi_{1,\theta_0})(j) \in \Im, j = 0, 1\right\}.$$

We now define $\hat{\Theta}_1$ by

$$\hat{\Theta}_1 = \bigcup_{\theta_0 \in \hat{\Theta}_0} A_1(\theta_0).$$

Note that each $A_1(\theta_0)$ has $2 \times 2 = 4$ elements and hence $\hat{\Theta}_1$ has $2 \times 4 = 8$ elements.

Let us suppose that we have already defined $\hat{\Theta}_{n-1}$. We now define $\hat{\Theta}_n$ from $\hat{\Theta}_{n-1}$.

Let the 2^n dyadic sub-intervals at stage n be denoted as $I_0,\dots\ I_{2^{n-1}}$. Each $\theta_{n-1} \in \hat{\Theta}_{n-1}$ induces a unique permutation $\pi_{n,\theta_{n-1}}$ of the set N_n defined by

$$\pi_{n,\theta_{n-1}}(j) = i \text{ if for all } x \in I_j,\; \theta_{n-1}(x) \in I_i,\; i, j \in N_n. \tag{3.3}$$

For each $\theta_{n-1} \in \hat{\Theta}_{n-1}$, we define

$$A_n(\theta_{n-1}) = \left\{ \begin{array}{c} \theta_n = \left(\pi_{n,\theta_{n-1}}(k), \mathcal{O}_n(\pi_{n,\theta_{n-1}})(k)\right)_{k=0}^{2^n-1} \mid \\ \mathcal{O}_n(\pi_{n,\theta_{n-1}})(i) \in \Im \; \forall \; i \in N_n \end{array} \right\}$$

and

$$\hat{\Theta}_n = \bigcup_{\theta_{n-1} \in \hat{\Theta}_{n-1}} A_n(\theta_{n-1}).$$

For each $n \geq 1$, we define the multiplication operation in $\hat{\Theta}_n$ as the composition of functions, namely, for $\hat{\theta}_n$, $\hat{\theta}'_n \in \hat{\Theta}_n$, we define $\hat{\theta}_n \hat{\theta}'_n = \hat{\theta}_n\left(\hat{\theta}'_n(x)\right)$, $x \in I$, and for each $n \geq 1$, we define the projection maps $f_n : \hat{\Theta}_n \to \hat{\Theta}_{n-1}$, $f_n(\theta_n) = \theta_{n-1}$, where θ_{n-1} is such that $\theta_n \in A_n(\theta_{n-1})$.

To get an idea about these projection maps, in panel (b) of Figure 2, we have drawn a $\theta_3 \in \hat{\Theta}_3$ whose projection using the map f_3 is $\theta_2 \in \hat{\Theta}_2$.

Denote by

$$\breve{\Theta} = \left\{\breve{\theta} = (\theta_0, \theta_1, \theta_2, ...) \mid \theta_n \in \hat{\Theta}_n, \forall n \geq 0 \text{ and } f_n(\theta_n) = \theta_{n-1}, \forall n \geq 1\right\}.$$

For any two elements $\breve{\theta} = (\theta_0, \theta_1, \theta_2, ...)$ and $\breve{\theta}' = (\theta'_0, \theta'_1, \theta'_2, ...)$ from $\breve{\Theta}$, define the multiplication operation $\breve{\theta} \circ \breve{\theta}'$ by

$$\breve{\theta} \circ \breve{\theta}' = (\theta_0\theta'_0, \theta_1\theta'_1, \theta_2\theta'_2, ...)$$

By Raut [1997, Proposition 3], $f_n(\theta_n\theta'_n) = f_n(\theta_n)f_n(\theta'_n) \in \hat{\Theta}_{n-1}$. Hence, $\breve{\theta} \circ \breve{\theta}' \in \breve{\Theta}$. With $\breve{\theta}^{-1} = (\theta_0^{-1}, \theta_1^{-1}, \theta_2^{-1}, ...)$ as the inverse of $\breve{\theta} = (\theta_0, \theta_1, \theta_2, ...)$, and with $\breve{e} = (e_0, e_1, ...)$, where e_n is the identity element of $\hat{\Theta}_n$ as the unit element, we note that $\breve{\Theta}$ is a group.

Define for $n \geq 0$ the projection maps $\pi_n : \breve{\Theta} \to \hat{\Theta}_n$ by

$$\pi_n(\breve{\theta}) = \theta_n, \quad \text{where} \quad \breve{\theta} = (\theta_0, \theta_1, \theta_2, ...).$$

Let $\mathcal{F} = \cup_{n=1}^{\infty} \pi_n^{-1}(\mathcal{B}_n)$. It can be easily shown that $\mathcal{F}$ is a Boolean algebra. Let $\breve{\mathcal{B}}$ be the $\sigma-$algebra generated by $\mathcal{F}$. $\left(\breve{\Theta}, \breve{\mathcal{B}}\right)$ is called the *projective limit* of the sequence of measure spaces, $(\hat{\Theta}_n \mathcal{B}_n)$, $n \geq 0$ through the maps $f_n, n \geq 1$.

The following theorem summarizes the main results from Raut [1995, 1997] for the projective limit group $\left(\breve{\Theta}, \breve{\mathcal{B}}, \breve{\Gamma}\right)$:

Theorem 3 *There exists a unique right invariant probability measure, $\breve{\Gamma}$ on the projective limit, $(\breve{\Theta}, \breve{\mathcal{B}})$ of the sequence of measurable groups, $\left(\hat{\Theta}_n, \mathcal{B}_n, \Gamma_n\right)_0^{\infty}$, through the sequence of homomorphisms, $\{f_n\}_0^{\infty}$, such that*

(i) $\breve{\Gamma}\pi_n^{-1} = \Gamma_n$

(ii) $\left(\breve{\Theta}, \breve{\mathcal{B}}, \breve{\Gamma}\right)$ *is an uncountably large separated probability measurable group.*

(iii) *For each $\breve{\theta} = (\theta_0, \theta_1, ..., \theta_n, ...) \in \breve{\Theta}$, the limit $\breve{\theta}(t) = \lim_{n\to\infty} \theta_n(t)$ exists for all $t \in I$ and the limit function $\theta : I \to I$ is a Lebesgue measure preserving automorphism.*

(iv) *The projective limit group $\left(\breve{\Theta}, \breve{\mathcal{B}}, \breve{\Gamma}\right)$ is isomorphic to the unit interval with Lebesgue measure, $(I, \mathcal{B}_I, \lambda)$.*

Let us examine the limitations that are imposed on the pattern of randomization by the projective limit group as compared to the limit group Θ that we discussed earlier. Let us fix a $t \in I$ and suppose $t \in I_1$. Consider the initial segments of player t in each of the random orders $\theta_n \in \breve{\Theta}_n$, that has the same value for t, say $\theta_n(t) = t_0$. From Figure 2 it is clear that all these random orders place a particular type of sets of Lebesgue measure t_0 before player t; and the type of the sets depends on t_0, and the order of the dyadic subdivision, n. For example, suppose $t_0 < 1/2$, then all these random orders do not place any players from the interval [1/2,1]. Thus for a given size t_0, $0 \le t_0 \le 1$, the random orders in $\breve{\Theta}$ allow t to form coalition only with certain sets of players of size t_0 but not every (Borel) set of players whose size is t_0. On the other hand, t gets to have any given player placed before him in a suitable random order and a suitable size t_0.

For games in which the worth of a coalition depends only through its size but not through any of their other identities, i.e., for anonymous games for instance such as the games in lpNA(λ), we expect and we have shown formally in Raut [1995] that the expected marginal contribution of a player with respect to the group of random orders, $\left(\breve{\Theta}, \breve{\mathcal{B}}, \breve{\Gamma}\right)$, coincides with the axiomatic value for these games as characterized by Aumann and Shapley [1974]. See Theorem 2 for details.

4 Final remarks

Remark 1 There are economically important non-smooth games which neither belong to bv′NA, MIX, nor even to ASYMP. Mertens [1988] extended the diagonal formula for value to a very powerful closed subspace of games in BV, known as *Mertens space*, on which the extended diagonal formula provides a value operator of norm 1 and the Mertens space was shown to include all well known spaces such as bv′NA, ASYMP, DIFF and DIAG. Both J. F. Mertens and Abraham Neyman suggested to me to examine if Mertens space belongs to $L1(\breve{\Theta}, \breve{\Gamma})$. I have not tried to get a general answer to this question, instead I show that the $\breve{\Theta}$-symmetric random order value exists for the non-smooth game of "n-handed gloves markets" considered in example 19.2 of Aumann and Shapley [1974, p.136]: $V(S) = \min\{\mu_1(S), \mu_2(S), ..., \mu_n(S)\}$, $\mu_i \in NA^1$, $i = 1, 2, ...n$, and $S \in \mathcal{B}_I$. This kind of non-smooth games arise in economies with strong complementarities. Aumann and Shapley showed that this game did not belong even to ASYMP when $n > 2$. One of the motivations for Mertens [1988] to extend the diagonal formula to the Mertens space was to include such games in the space. Notice that V is of bounded variation. Since each μ_i is a non-atomic probability measure, the game $V(S)$ is normalized and hence belongs to NBV. Thus there exists a unique $\breve{\Theta}$-symmetric random order value for V.

Remark 2 An important issue regarding our reformulated random order approach is: What characteristics of the group $\breve{\Theta}$ that makes the random order value coincides with the axiomatic value on lpNA(λ)? In Section 3, I argued that a random order generated according to the probability model $\left(\breve{\Theta}, \breve{\mathcal{B}}, \breve{\Gamma}\right)$ has the characteristics that the random set of players that is placed before any given player is equally likely to be of any size $s \in [0, 1]$; the anonymous games in which worth of a coalition depends only through its size not names such as games in lpNA(λ), each player gets the average of the set of all possible marginal contributions and thus average is fully symmetrized in the sense that the value thus obtained is symmetric with respect to the full group of automorphisms. Locally finite groups of automorphisms may not do the job, as we have illustrated in Section 3. The games that arise in most economic applications are anonymous. However, for a wider applicability of the present approach, we must construct a larger invariant probability measurable group structure $(\Theta, \mathcal{B}, \Gamma)$ than $\left(\breve{\Theta}, \breve{\mathcal{B}}, \breve{\Gamma}\right)$, so that the random order value $\Phi_\Gamma(V)$ with respect to it also fully symmetrizes non-anonymous games.

Remark 3 Robert Aumann pointed out to me that for an alternative reformulation of random order approach to value, one might give up the measure theoretic model of the player set, i.e., $(I, \mathcal{B}_I)$, and instead consider the player space to be torus or other topological spaces with more well-behaved automorphism groups. It should be noted that there can exist only two orders on any topological space that is connected. This, for instance, will greatly simplify our analysis of random order value. We do not know, however, what kind of fairness such a symmetry group entails and what kind of economic situations are appropriate for such models; most of the economic models with a continuum of agents, however, have employed measure theoretic structures for the space of agents, and thus we must begin to imagine the nature and study the implications of economic models with a topological space of agents.

Acknowledgements

An earlier draft of this paper was presented at the conference on "New Directions in the Theory of Markets and Games in honour of Robert J. Aumann", Toronto, Canada, October 19-23, 1995. The main work of this paper was done during my stay at the University of California-San Diego. I got the idea of using invariant measure to characterize random order values of games with a continuum of players (Raut [1981]) when I was a junior research fellow at the Indian Statistical Institute. I have benefitted from delightful introduction to game theory by Lloyd Shapley and to measure theory by K. R. Parthasarathy, and from research advising by T. Parthasarathy during my stay there. Comments from Robert Aumann, Ori Haimanek, J. F. Mertens, Abraham Neyman, Joel Sobel and Myrna Wooders were very useful in writing this revised draft. My gratitude goes to all.

References

Aumann, Robert J. [1967], *Random Measure preserving Transformations*, Proceedings of the Fifth Berkeley Symposium on Mathematical Statistics and Probability, vol. II, Part II, 321-326, University of California Press, Berkeley.

Aumann R. J. and Shapley L. S. [1974], *Values of non-atomic games*, Princeton University Press, Princeton, New Jersey.

Halmos, P. R. [1950], *Measure Theory*, Affiliated East West Pvt. Ltd., New Delhi.

Kannai, Y. [1966], *Values of games with a continuum of players*, Israel J. Math., vol. 4, 54-58.

Mackey, G. W. [1957], *Borel structures in groups and their duals*, Trans. Amer. Math. Soc., vol. 85, 134-165.

Mertens, J. F. [1988], *The Shapley value in the non-differential case*, Int. J. of Game Theory, vol.17(1), 1-65.

Monderer, D. [1986], *Measure based values of non-atomic games*, Mathematics of Operations research, vol.11(2), 321-335.

Monderer, D. [1989], *Weighted majority games have many μ-values*, Int. J. of Game Theory, vol.18, 321-325.

Monderer, D. and Ruckle, W. H. [1990], *On the symmetry axiom for values of non-atomic games*, Int. J. of Math. and Math. Sci., vol.13(1), 165-170.

Parthasarathy, K. R. [1977], *Introduction to Probability and Measure*, Springer-Verlag, New York.

Raut, L. K. [1981], *Random Order Approach to Semi-values of Non-atomic games*, Indian Statistical Institute Discussion Paper no. 8108, New Delhi, India.

Raut, L. K. [1993], *Random Order Approach to Shapley-value of Games and Haar measure*, University of California-San Diego, Discussion Paper.

Raut, L. K. [1995], *A Reformulation of Aumann-Shapley random order values of non-atomic games using invariant measures*, University of California-San Diego, Discussion Paper.

Raut, L. K. [1997], *Construction of Haar measure on projective limit group and random order values of non-atomic games*, Journal of Mathematical Economics, Vol. 27, Issue 2, pp. 229-250.

Ruckle, W. [1982], *Projection in certain spaces of set functions*, Math. Operations Res. vol. 7, 314-318.

Rudin, W. [1966], *Real and Complex Analysis*, McGraw-Hill, New York.

Segal, I. E. and Kunze, R. A. [1978], *Integrals and Operators*, Springer-Verlag, New York.

Shapley, L. S. [1953], *A value for n-person games in Contributions to the Theory of Games, Part II*, Princeton University Press, Princeton, New Jersey.

Economics and Social Choice

Fields Institute Communications
Volume **23**, 1999

On the Existence of Core Allocations in a Large Economy with Incentive-Compatibility Constraints

Beth Allen
Department of Economics
University of Minnesota
Minneapolis, MN 55455 USA
and
Federal Reserve Bank of Minneapolis
sallyoss@atlas.socsci.umn.edu

Abstract. For an atomless exchange economy with finitely many states of the world, possibly asymmetric initial information, and suitably dispersed initial endowment vectors for (almost) every state-dependent utility function, the *NTU* incentive-compatible core is nonempty, where (ex ante) blocking allocations must satisfy incentive-compatibility conditions. Such incentive-compatible core allocations are obtained by demonstrating the existence of "competitive equilibria" despite the non-convexities in agents' consumption sets induced by the presence of incentive-compatibility considerations and then observing that the "equilibrium" allocations belong to the core.

1 Introduction

This paper considers the possibilities for the existence of core allocations in a large exchange economy in which all allocations, including those employed in (ex ante) blocking, must satisfy incentive-compatibility constraints.[1,2] Previous work has demonstrated that the incentive-compatible core may be empty (for an explicit

1991 *Mathematics Subject Classification.* Primary 90D12; Secondary 90A36 .

This research was supported by NSF Grant SBR-9309854, the Curtis Carlson Chair in Economics at the University of Minnesota, and the Federal Reserve Bank of Minneapolis. The views expressed herein are those of the author and not necessarily those of the Federal Reserve Bank of Minneapolis or of the Federal Reserve System.

It's a pleasure to thank Leo Hurwicz, Akira Yamazaki, and Bill Zame for helpful discussions on this topic.

[1]Note that, for the Wilson [**21**] fine core, blocking occurs ex post so that his core concept with incentives is quite different from mine.

[2]Koutsougeras and Yannelis [**17**] ask whether their core concepts with asymmetric information satisfy incentive compatibility, but they do not require that blocking be achieved using only incentive-compatible allocations.

example, see Allen [**3**]),[3] although replication guarantees the nonemptiness of approximate cores (see Allen [**5**]). A remaining issue is whether cores are nonempty in large economies with incentives.

Traditionally, two distinct methods have been used in economic theory to prove nonemptiness of the core of an economy. The game-theoretic approach considers the games (with or without transferable utility) generated by economies. Balancedness of these games guarantees that the core is nonempty.[4,5] Alternatively, one can work directly with the economies and exploit the fact that, under very mild assumptions, competitive equilibria belong to the core. For this strategy, the nontrivial step is to demonstrate existence of competitive equilibrium. The inclusion of core allocations in the set of competitive equilibrium allocations, like the first welfare theorem, requires few hypotheses; the relation remains valid even with nonconvexities in preferences or consumption sets.

Here the existence of competitive equilibrium is difficult, even in large economies.[6] While the presence of a nonatomic continuum of agents can eliminate the effects of nonconvex preferences, the incentives problem does not reduce to this case. However, there are results about existence with nonconvex consumption sets that can be utilized, as the technicalities are similar. Note that one need not be concerned with the precise interpretation of competitive equilibrium under incentive-compatibility constraints; it suffices that the equilibrium concept be such that it is contained in the core.

The solution proposed in this paper to the existence problem for competitive equilibrium with consumption set nonconvexities caused by incentive-compatibility constraints is to hypothesize a mild dispersion condition on traders' initial endowments. Yamazaki [**22**] shows that, in the presence of nonconvexities in consumption sets, such suitably distributed endowments guarantee that pure exchange economies (without asymmetric information or incentive constraints) have competitive equilibria. However, my paper appears to constitute the first time that dispersion has been used to obtain nonemptiness of the core as well as the first use of dispersion in the context of incentives problems.

The remainder of this paper is organized as follows: Section 2 presents the model and states the dispersion condition. Incentive-compatibility is formalized in Section 3, while incentive-compatible competitive equilibria and the incentive-compatible core are defined in Section 4. Main results are stated and proved in Section 5. Extensions appear in Section 6, while some properties of the incentive-compatible core are examined in Section 7. Finally, Section 8 concludes with some remarks.

---

[3]This contrasts with some previous literature on incentive-compatible cores with (exclusively) idiosyncratic risk. See Prescott and Townsend ([**19**], [**20**]), Boyd and Prescott [**11**], Boyd, Prescott, and Smith [**12**], Marimon [**18**], and Berliant [**9**], who tend to focus on situations of adverse selection, moral hazard, signaling, financial intermediation, and income taxation.

[4]A good reference for this material is the textbook by Hildenbrand and Kirman ([**14**], [**15**]).

[5]The concept of balancedness is used to demonstrate nonemptiness of the core with asymmetric information in the papers by Wilson ([**21**]; he considers balancedness in a related game in which the players consist of traders in various date-event pairs to show that his coarse core is nonempty) and Allen ([**1**], [**2**], [**3**], [**4**]; see also [**5**]). Yannelis [**23**] examines the private, strong and coarse cores; his nonemptiness proof deals directly with core allocations in exchange economics rather than using the device of deriving the associated games and checking for balancedness.

[6]Contrast this with the situation of nonconvexities in preferences (only) as examined in Aumann's [**7**] paper on existence of competitive equilibria in economies with an atomless continuum of agents. See also the monograph by Hildenbrand [**13**].

2 The Model

This paper analyzes a pure exchange economy with a finite number of goods in each state of the world, a finite number of states or payoff-relevant events, and a continuum of traders, albeit they are restricted to finitely many utility-types. I use the term "utility-type" to avoid possible confusion with the concept of "type" in games with incomplete information. To fix notation, let the finite set $I = \{1, \dots, n\}$ with typical element $i \in I$ denote the set of utility-types present in the economy, and suppose that there are ℓ commodities in the economy (in each state of the world).

The uncertainty in the model is represented by an abstract set Ω of states of the world and a σ-field $\mathcal{F}$of measurable subsets of Ω. Elements of $\mathcal{F}$ are termed events. Write ω for a typical element of Ω. Then $(\Omega, \mathcal{F})$ is a measurable space; endow it with a probability measure μ so that $(\Omega, \mathcal{F}, \mu)$ is a probability space. The probability μ indicates the likelihoods of various events. Let the sub-σ-field $\mathcal{G}_i$ of $\mathcal{F}$, for $i \in I$, be the information of each agent of utility-type i. Then $\mathcal{G} = \sigma(\bigcup_{i \in I} \mathcal{G}_i)$ includes all of the relevant information in the economy in the sense that, if two states of the world cannot be distinguished based on the pooled information of all agents, then trades or allocations that are different in the two states can never arise. For technical reasons, I must require that $\mathcal{G}$ is finite. This is automatic if Ω is a finite set; the condition is obviously equivalent to finiteness of $\mathcal{G}_i$ for every $i \in I$.

2.1. Assumption. The sub-σ-field $\mathcal{G} = \sigma(\bigcup_{i \in I} \mathcal{G}_i)$ is finite.

2.2. Simplification. The set Ω of states of the world is finite. Moreover, each state occurs with strictly positive probability $\big(\mu(\omega) > 0$ for all $\omega \in \Omega\big)$ and the pooled information of all agents suffices to distinguish any two distinct states $\big(\sigma(\bigcup_{i \in I} \mathcal{G}_i) = 2^{\Omega}\big)$.

Having specified the information of each agent, I must now specify their preferences and endowments to complete the description of the economy. Because uncertainty is present, it is appropriate to specify preferences in terms of cardinal utility functions. Risk aversion is meaningful and, because I assume that traders maximize their state-dependent expected utilities conditional on the information that the trader may use, is equivalent to concavity of the (state-dependent) utility functions.

More formally, for each $i \in I$, let the utility function $u_i \colon \mathbb{R}^{\ell}_{++} \times \Omega \to \mathbb{R}$ for utility-type i be such that u_i is continuous on $\mathbb{R}^{\ell}_{++}$ and $\mathcal{F}$-measurable on Ω, so that it is jointly measurable. Assume also that for μ-almost all $\omega \in \Omega$, $u_i(\cdot\,;\omega)$ is strictly concave and strictly monotone. (Note that $u_i(\cdot\,;\cdot) = u_j(\cdot\,;\cdot)$ for some $i, j \in I$ with $i \neq j$ is permitted.)

To incorporate dispersion of initial endowment vectors for each utility-type, let the economy contain a nonatomic continuum of traders of each utility-type. Let λ_i denote the distribution on $\mathbb{R}^{\ell}_{++}$ summarizing the initial endowments of agents of utility-type $i \in I$. Do not require that the λ_i be probabilities but only that they be nonnegative Borel (σ-additive) measures on $\mathbb{R}^{\ell}_{++}$ such that the measure λ on $\mathbb{R}^{\ell}_{++}$, defined by $\lambda(S) = \sum_{i \in I} \lambda_i(S)$ for all Borel-measurable subsets S of $\mathbb{R}^{\ell}_{++}$, is a probability. Note that $\lambda_i(\mathbb{R}^{\ell}_{++}) \leq 1$ gives the proportion of agents that are of utility-type $i \in I$. In addition, $\operatorname{supp}(\lambda) = \bigcup_{i \in I} \operatorname{supp}(\lambda_i)$, which is a compact subset of $\mathbb{R}^{\ell}_{++}$ by the following assumption. Dispersion can now be formalized.

2.3. Assumption. For each $i \in I$, the distribution λ_i on $I\!R^\ell_{++}$ is absolutely continuous with respect to ℓ-dimensional Lebesgue measure and has compact support.

2.3′. Assumption. For each $i \in I$, the distribution λ_i on $I\!R^\ell_{++}$ has compact support and the associated wealth distribution $\lambda_{i,p}$ on $I\!R$, defined for Borel subsets $B \subseteq I\!R$ and for all $p \in \bar{\Delta} = \{p \in I\!R^\ell_+ \mid \sum_{j=1}^{\ell} p_j = 1\}$, by $\lambda_{i,p}(B) = \lambda_i(\{e_i \in I\!R^\ell_{++} \mid p \cdot e_i \in B\})$ is absolutely continuous with respect to 1-dimensional Lebesgue measure on $I\!R$ for each $p \in \bar{\Delta}$.

2.4. Observation. Assumption 2.3′ implies Assumption 2.3.

Remark. Alternatively, I could have defined marginal distributions on $C(I\!R^\ell_{++}, I\!R)$ and $\mathcal{F}^*$, where $\mathcal{F}^*$ denotes the space of all sub-σ-fields of $\mathcal{F}$. With appropriate measurability restrictions—which involve endowing $C(I\!R^\ell_{++}, I\!R)$ and $\mathcal{F}^*$ with topologies and using the Borel subsets—one could specify a Borel probability measure (the distribution of agents' characteristics) on the product space $\mathcal{F}^* \times C(I\!R^\ell_{++}, I\!R) \times I\!R^\ell_{++}$. Then dispersion could be defined, as in Yamazaki [**22**], in terms of the marginal distributions on $I\!R^\ell_{++}$. However, this approach seems more awkward and complicated than necessary, in view of the fact that finiteness (Assumption 2.1) of $\sigma(\bigcup_{i\in I} \mathcal{G}_i)$ would still be needed, so that the assumption of finitely many information-types would be needed in addition to a dispersion assumption in terms of the marginal wealth distribution. More importantly, Yamazaki [**22**] postulates the same nonconvex consumption set for all agents, whereas the consumption sets here are defined by incentive-compatibility constraints, so that each utility-type has a potentially distinct consumption set. Because I apply the Yamazaki [**22**] aggregation technology to each utility-type separately, I must have dispersion of endowments or wealths for each particular utility-type. However, I could weaken the dispersion in Assumption 2.3 to the form of the assumption that he uses.

The n-tuple $(\lambda_1, \dots, \lambda_n)$ of measures on $I\!R^\ell_{++}$ can be converted to a probability measure $\tilde{\lambda}$ on $\widetilde{\mathcal{F}}^* \times \widetilde{\mathcal{U}} \times \operatorname{supp}(\lambda)$, where

$$\begin{aligned}\widetilde{\mathcal{F}}^* &= \{\mathcal{G} \subseteq \mathcal{F} \mid \mathcal{G} \text{ is a sub-}\sigma\text{-field and } \mathcal{G} = \mathcal{G}_i \text{ for some } i \in I\}\\ &= \bigcup_{i\in I}\{\mathcal{G}_i\}\end{aligned}$$

and

$$\widetilde{\mathcal{U}} = \{u \colon I\!R^\ell_{++} \times \Omega \to I\!R \mid u = u_i \text{ for some } i \in I\} = \bigcup_{i\in I}\{u_i\}$$

[note that $\#\widetilde{\mathcal{F}}^* \leq \#I$ and $\#\widetilde{\mathcal{U}} \leq \#I$] by defining $\tilde{\lambda}(S) = \sum_{i\in I} \tilde{\lambda}_i(\operatorname{slice}_i(S))$, where

$$\operatorname{slice}_i(S) = \{\mathcal{G} \times u \times e \in \widetilde{\mathcal{F}}^* \times \widetilde{\mathcal{U}} \times \operatorname{supp}(\lambda) \mid \mathcal{G} = \mathcal{G}_i \text{ and } u = u_i\}$$

and $\tilde{\lambda}_i$ is the positive measure on (Borel subsets of) $\{\mathcal{G}_i\} \times \{u_i\} \times \operatorname{supp}(\lambda_i)$ defined by $\tilde{\lambda}_i(S) = \lambda_i(\operatorname{proj}_{I\!R^\ell} S)$. Observe that $\tilde{\lambda}(\widetilde{\mathcal{F}}^* \times \widetilde{\mathcal{U}} \times S) = \lambda(S)$ for all Borel measurable subsets S of $I\!R^\ell_{++}$. I identify the extension of measures on $\operatorname{supp}(\lambda_i)$ to $\operatorname{supp}(\lambda)$ and $I\!R^\ell_{++}$ in the obvious way with no change of notation.

Because $\tilde{\lambda}$ is a Borel probability measure on an (uncountable) complete separable metric space $(\widetilde{\mathcal{F}}^* \times \widetilde{\mathcal{U}} \times \operatorname{supp}(\lambda))$, it has a standard representation; see Aumann and Shapley [**8**] or Hildenbrand [**13**]. Therefore, if T denotes the closed unit interval $[0, 1]$ equipped with the Borel σ-field $\mathcal{T}$ and 1-dimensional Lebesgue measure τ, there is a measurable mapping $\mathcal{E} \colon T \to \widetilde{\mathcal{F}}^* \times \widetilde{\mathcal{U}} \times \operatorname{supp}(\lambda)$ such that the image measure $\tau \circ \mathcal{E}^{-1}$ is the same as the distribution of $\tilde{\lambda}$; i.e., $\tau \circ \mathcal{E}^{-1} = \tilde{\lambda}$ as probability measures on the Borel subsets of $\widetilde{\mathcal{F}}^* \times \widetilde{\mathcal{U}} \times \operatorname{supp}(\lambda)$. Hence I can equally well consider $\mathcal{E}\colon T \to \widetilde{\mathcal{F}}^* \times \widetilde{\mathcal{U}} \times \operatorname{supp}(\lambda)$ as the definition of the economy, where an "individual agent" with "name" $t \in T$ is described by the characteristics $\mathcal{E}(t) = (\mathcal{G}_t, u_t, e_t) \in \widetilde{\mathcal{F}}^* \times \widetilde{\mathcal{U}} \times \operatorname{supp}(\lambda)$.

With this new notation, Assumption 2.3 and Assumption 2.3′ can be reformulated. Assumption 2.1 and Simplification 2.2 remain in force.

2.3S. Assumption. The measurable mapping $\mathcal{E}_3\colon T \to \operatorname{supp}(\lambda)$ defined by $\mathcal{E}_3 = \operatorname{proj}_{I\!R^\ell} \circ \mathcal{E}$ induces a distribution $\tau \circ \mathcal{E}_3^{-1}$ on $I\!R^\ell_{++}$ which is absolutely continuous with respect to ℓ-dimensional Lebesgue measure and has compact support.

2.3S′. Assumption. In the notation of Assumption 2.3S, the distribution $\tau \circ \mathcal{E}_3^{-1}$ has compact support and, for each $p \in \bar{\Delta} = \{p \in I\!R^\ell_+ \mid \sum_{j=1}^{\ell} p_j = 1\}$, the distribution $(\tau \circ \mathcal{E}_3^{-1})_p$, defined for Borel subsets B of $I\!R$ by

$$(\tau \circ \mathcal{E}_3^{-1})_p(B) = \tau \circ \mathcal{E}_3^{-1}(\{e \in I\!R^\ell_{++} \mid p \cdot e \in B\}),$$

is absolutely continuous with respect to 1-dimensional Lebesgue measure on $I\!R$.

3 Incentive Compatibility

The strategy in this paper is first to define a family of incentive-compatible core concepts, for any information structures available to agents in coalitions (other than the grand coalition), and then to examine a "auxiliary" core concept. The auxiliary core is defined to have the property that its nonemptiness implies nonemptiness for the core concept of interest. Moreover, the dispersion assumption guarantees that a competitive equilibrium exists. Since it belongs to the auxiliary core, the incentive-compatible core is indeed nonempty.

Specify the information that may be used by a trader (not a utility-type) in each coalition to which he or she belongs by fixing sub-σ-fields $\mathcal{G}_t^S$ if $t \in S$ and S is a measurable subset (or coalition) of T by a measurable mapping $\mathcal{E}_1^S\colon T \to \widetilde{\mathcal{F}}^*$, where $\mathcal{E}_1^S(t) = \mathcal{G}_t^S$. It is not necessary that the same utility-types receive the same information in every coalition to which they belong—i.e., $\mathcal{G}_t^S \neq \mathcal{G}_{t'}^S$ is permitted even when $\mathcal{G}_t = \mathcal{G}_{t'}$ and $u_t = u_{t'}$. If $S = T$ (so that S is the grand coalition), set $\mathcal{G}_t^S = 2^\Omega$ for all $t \in T$. Note that I invoke the simplifying assumption (Assumption 2.2) of the previous section. It is natural—but not necessary—to assume that $\mathcal{G}_t \subseteq \mathcal{G}_t^S$ whenever $t \in S$. Each collection $\{\mathcal{G}_t^S\}$ of these information sub-σ-fields or each collection $\{\mathcal{E}_t^S\}$ of measurable functions defines an associated core. This is formalized in the next section after incentive compatibility is defined. Note that $\mathcal{E}_1^S$ is defined for all $t \in T$ (or, equivalently, that $\mathcal{G}_t^S$ is defined even if $t \notin S$), provided that S is a measurable subset of T, although altering $\mathcal{E}_1^S$ or the $\mathcal{G}_t^S$ off of S (in a measurable way) does not alter the concepts in this paper.

A state-dependent allocation (or a state-dependent net trade) for agent $t \in T$ is a function $x_t\colon \Omega \to I\!R^\ell_{++}$ (or $z_t\colon \Omega \to I\!R^\ell$). A state-dependent allocation function (or state-dependent net trade function) is a measurable function $X\colon T \to (I\!R^\ell_{++})^\Omega$ (or $Z\colon T \to (I\!R^\ell)^\Omega$). Write $x_t = X(t)$ and $z_t = Z(t)$.

3.1. Definition. A state-dependent allocation $x_t\colon \Omega \to I\!R^\ell_{++}$ (or a state-dependent net trade $z_t\colon \Omega \to I\!R^\ell$) for trader $t \in T$ satisfies *incentive compatibility* if for almost all $\omega' \in \Omega$, $u_t(x_t(\omega'); \omega') \geq u_t(x_t(\omega''); \omega')$ for all $\omega'' \in \Omega$ or $u_t(e_t + z_t(\omega'); \omega') \geq u_t(z_t(\omega''); \omega')$ for all $\omega'' \in \Omega$ respectively. A state-dependent allocation function X or a state-dependent net trade function Z satisfies *incentive compatibility* if for almost all $t \in T$, x_t or z_t satisfies incentive compatibility.

Incentive compatibility requires a trader to prefer (weakly) his allocation or net trade in state $\omega' \in \Omega$ (for almost all $\omega' \in \Omega$) to the allocation or net trade for state $\omega' \neq \omega''$ when the state of the world is $\omega' \in \Omega$, except possibly for a null subset of traders. It means that the agent has no reason to lie and say that state ω'' has occurred when the true state is ω'. Note that the definition fails to take account of the information available to the agent. The concept does not restrict allocations or net trades to functions depending only on information available to the agent. Moreover, it requires that the self-selection constraints be satisfied even if the trader cannot distinguish state ω' from state ω''. Note that the existence of feasible incentive-compatible allocations for every agent in every coalition is guaranteed, as $e_t \in I\!R^\ell_{++}$ does not depend on $\omega \in \Omega$.

The proof that competitive equilibria belong to the core ordinarily requires that preferences satisfy a monotonicity or local nonsatiation property when consumption sets are convex. With incentive-compatibility constraints, the substitute for this property requires the existence of some nearby incentive-compatible allocation which is strictly better than a given potential competitive equilibrium allocation. Further assumptions are needed to demonstrate this condition; they are imposed at this time to emphasize their role in obtaining a version of local nonsatiation. These additional assumptions can be viewed as a very weak form of some standard smoothness and full rank conditions. Note that they can be weakened from global requirements to the statement that the first two parts of the assumption hold at one competitive equilibrium allocation which lies in the interior of $I\!R^\ell_+$, where $\ell \geq \#\Omega$.

3.2. Assumption. The economy satisfies the following three conditions:

(a) *Differentiable strict monotonicity.* For almost all $t \in T$, u_t is C^1 on $I\!R^\ell_{++}$ and $Du_t(x;\omega) \gg 0$ for all $x \in I\!R^\ell_{++}$ and almost every $\omega \in \Omega$.

(b) *Boundary condition—indifference curves through individually rational allocations don't cut the boundary of the positive orthant.* For all $i \in I$ and almost all $e_j \in \operatorname{supp} \lambda_i$ (which is a compact subset of $I\!R^\ell_{++}$),

$$\operatorname{cl}\big(\{y \in I\!R^{\#\Omega\ell}_{++} \mid \sum_{\omega\in\Omega} u_i(e_i;\omega) \leq \sum_{\omega\in\Omega} u_i\big(y_i(\omega);\omega\big)\}\big) \cap \partial I\!R^\ell_+ = \emptyset.$$

(c) *Full rank.* The number of commodities is not less than the number of nonnull states minus one ($\ell \geq \#\Omega$) and the matrix $(Du_t(x(\omega_1);\omega_1), \dots, Du_t(x(\omega_{\#\Omega});\omega_{\#\Omega}))$ is of full rank almost everywhere on $I\!R^{\#\Omega\ell}_{++}$.

3.3. Theorem For almost all $t \in T$, almost every incentive-compatible individually rational (and hence almost every possible incentive-compatible competitive equilibrium) state-dependent allocation function $\hat{x}_t \colon \Omega \to I\!R^\ell_{++}$ satisfies the following version of local nonsatiation: Given $\hat{x}_t \in I\!R^{\#\Omega\ell}_{++}$ as above and given $\varepsilon > 0$, there exists $\bar{x}_t \in I\!R^{\#\Omega\ell}_{++}$ with $\|\hat{x}_t - \bar{x}_t\|_{I\!R^{\#\Omega\ell}} < \varepsilon$ such that $\bar{x}_t \colon \Omega \to I\!R^\ell_{++}$ is incentive compatible and, for (almost) all $\omega \in \Omega$, $u_t\big(\hat{x}_t(\omega);\omega\big) < u_t\big(\bar{x}_t(\omega);\omega\big)$.

Remark. This local nonsatiation condition is due to Koopmans ([**16**], pp. 47–49).

Proof Let $\varepsilon > 0$ and $\hat{x}_t$ be as in the statement of the theorem. Incentive compatibility implies that for (almost) all $\omega', \omega'' \in \Omega$, $u_t\big(\hat{x}_t(\omega');\omega'\big) \geq u_t\big(\hat{x}_t(\omega'');\omega'\big)$. If some of these $\#\Omega\times(\#\Omega - 1)$ inequalities are satisfied with strict inequality, make

$\varepsilon > 0$ smaller if necessary so that if $\tilde{\omega}, \tilde{\tilde{\omega}} \in \Omega$ are such that $u_t(\hat{x}_t(\tilde{\omega}); \tilde{\omega}) > u_t(\hat{x}_t(\tilde{\tilde{\omega}}); \omega)$, then $u_t(\bar{x}_t(\tilde{\omega}); \tilde{\omega}) > u_t(\bar{x}_t(\tilde{\tilde{\omega}}); \omega)$ for all $\bar{x}_t \in I\!R_{++}^{\#\Omega\ell}$ with $\|\hat{x}_t - \bar{x}_t\|_{I\!R^{\#\Omega\ell}} < \varepsilon$; such an $\varepsilon > 0$ exists, for any given $\hat{x}_t$, by the continuity of each $u_t(\cdot; \omega)$ on $I\!R_{++}^{\ell}$. Renormalize the utility functions $u_t(\cdot; \omega_1)\colon I\!R_{++}^{\ell} \to I\!R, \ldots, u_t(\cdot; \omega_{\#\Omega})\colon I\!R_{++}^{\ell} \to I\!R$ by adding a constant to each so that $u_t(\hat{x}_t(\omega_1); \omega_1) = 0, \ldots, u_t(\hat{x}_t(\omega_{\#\Omega}); \omega_{\#\Omega}) = 0$. This implies that $u_t(\hat{x}_t(\omega'); \omega') = 0$, $u_t(\hat{x}_t(\omega''); \omega') = 0$, etc., for all $\omega', \omega'' \in \Omega$ as before. There remain at most $(\#\Omega)^2 \le \#\Omega\ell$ equalities of the form $u_t(\hat{x}_t(\omega'); \omega') = 0$ or $u_t(\hat{x}_t(\omega''); \omega') = 0$ for $\omega' \neq \omega''$ in the variables $(\hat{x}_t(\omega_1), \ldots, \hat{x}_t(\omega_{\#\Omega})) \in I\!R_{++}^{\#\Omega\ell}$. Replace each zero on the right hand side of these equations by some sufficiently small $\delta > 0$ and solve, by the inverse function theorem, for $\bar{x}_t \in I\!R_{++}^{\#\Omega\ell}$ with $\|\hat{x}_t - \bar{x}_t\|_{I\!R^{\#\Omega\ell}} < \varepsilon$. Then $\bar{x}_t\colon \Omega \to I\!R_{++}^{\ell}$ satisfies incentive compatibility and, for (almost) all $\omega \in \Omega$, $u_t(\bar{x}_t(\omega); \omega) = \delta > u_t(\hat{x}_t(\omega); \omega) = 0$ (after the renormalization of utility functions), as desired. □

4 Cores and Competitive Equilibria

This section defines two core concepts and an auxiliary competitive equilibrium concept which appear in the proof of the main theorem. The auxiliary core also serves as an auxiliary concept in the proof strategy.

4.1. Definition. A state-dependent allocation function X belongs to the *auxiliary core* if the state-dependent allocations are feasible $\left(\int_T x_t(\omega)\, d\tau(t) = \int_T e_t\, d\tau(t)\right.$ for (almost) all $\left.\omega \in \Omega\right)$ and incentive compatible $\left(u_t(x_t(\omega'); \omega') \ge u_t(x_t(\omega''); \omega')\right.$ for almost all $\omega', \omega'' \in \Omega$ and almost all $\left.t \in T\right)$ and if there is no coalition S and alternative state-dependent allocation function X' (defining $x'_t\colon \Omega \to I\!R_{++}^{\ell}$ for $t \in S$) such that $\tau(S) > 0$, $\int_S x'_t(\omega)\, d\tau(t) = \int_S e_i\, d\tau(t)$ for almost all $\omega \in \Omega$,

$$\tau\Bigl(\Bigl\{t \in S \mid \int_\Omega u_t(x'_t(\omega); \omega)\, d\mu(\omega) \ge \int_\Omega u_t(x_t(\omega); \omega)\, d\mu(\omega)\Bigr\}\Bigr) = \tau(S)$$
$$= \tau\bigl(\{t \in S \mid x'_t \text{ is incentive compatible for } t\}\bigr),$$

and

$$\tau\Bigl(\Bigl\{t \in S \mid \int_\Omega u_t(x'_t(\omega); \omega)\, d\mu(\omega) > \int_\Omega u_t(x_t(\omega); \omega)\, d\mu(\omega)\Bigr\}\Bigr) > 0.$$

4.2. Definition. Given $\{\mathcal{E}_1^S\}$ for all measurable subsets S of T, a state-dependent allocation function X (where the $x_t\colon \Omega \to I\!R_{++}^{\ell}$ are necessarily $\mathcal{G}$-measurable state-dependent allocations for $t \in T$) belongs to the *incentive-compatible core* if

$$\tau\bigl(\{t \in T \mid u_t(x_t(\omega'); \omega') \ge u_t(x_t(\omega''); \omega') \text{ for almost all } \omega' \in \Omega \text{ and all } \omega'' \in \Omega\}\bigr) = 1$$

and

$$\int_T x_t(\omega)\, d\tau(t) = \int_T e_t\, d\tau(t)$$

for almost all $\omega \in \Omega$, and if there is no (measurable) coalition $S \subseteq T$ and allocations $x'_t\colon \Omega \to I\!R_{++}^{\ell}$ for $t \in S$ such that $\tau(S) > 0$, $\int_S x'_t(\omega)\, d\tau(t) = \int_S e_t\, d\tau(t)$ for almost all $\omega \in \Omega$,

$$\tau\bigl(\{t \in S \mid x'_t \text{ is incentive compatible and } \mathcal{G}_t^S\text{-measurable}\}\bigr)$$
$$= \tau(S) = \tau\Bigl(\Bigl\{t \in S \mid \int_\Omega u_t(x'_t(\omega); \omega)\, d\mu(\omega) \ge \int_\Omega u_t(x_t(\omega); \omega)\, d\mu(\omega)\Bigr\}\Bigr)$$

and

$$\tau\left(\{t\in S \mid \int_{\Omega} u_t\big(x_t'(\omega);\omega\big)d\mu(\omega) > \int_{\Omega} u_t\big(x_t(\omega);\omega\big)d\mu(\omega)\}\right) > 0.$$

4.3. Definition. A state-dependent allocation function X^* is an *incentive-compatible competitive equilibrium allocation function* if there is a price vector $p^* \in \triangle = \{p \in I\!R^{\#\Omega\ell}_{++} \mid \sum_{\omega}\sum_{j=1}^{\ell} p_{j,\omega} = 1\}$ such that $\int_T x_t^*(\omega)\,d\tau(t) = \int_T e_t\,d\tau(t)$ for almost all $\omega \in \Omega$, $\lambda\big(\{t \in T \mid p^* \cdot (e_t, \dots, e_t) = p^* \cdot \big(x_t^*(\omega_1), \dots, x_t^*(\omega_{\#\Omega})\big)$, $u_t\big(x_t^*(\omega');\omega'\big) \geq u_t\big(x_t^*(\omega'');\omega'\big)$ for almost all $\omega' \in \Omega$ and all $\omega'' \in \Omega$, and if $\bar{x}_t \colon \Omega \to I\!R^{\ell}_{++}$ satisfies

$$\int_{\Omega} u_t\big(\bar{x}_t(\omega);\omega\big)\,d\mu(\omega) \geq \int_{\Omega} u_t\big(x_t^*(\omega);\omega\big)d\mu(\omega)$$

and

$$u_t\big(\bar{x}_t(\omega');\omega'\big) \geq u_t\big(\bar{x}_t(\omega'');\omega'\big)$$

for almost all $\omega' \in \Omega$ and all $\omega'' \in \Omega$, then

$$p^* \cdot \big(\bar{x}_t(\omega_1), \dots, \bar{x}_t(\omega_{\#\Omega})\big) > p^* \cdot (e_t, \dots, e_t).$$

An *incentive-compatible competitive equilibrium* is an incentive-compatible competitive equilibrium allocation function (or a collection of associated state-dependent allocations generated by an incentive-compatible competitive equilibrium allocation function) combined with the above price vector p^*.

An incentive-compatible competitive equilibrium is simply a competitive equilibrium in which traders' consumption sets are constrained to satisfy incentive compatibility, which causes these consumption sets to be nonconvex and potentially different for each utility-type. An auxiliary core allocation is a feasible and incentive-compatible allocation that cannot be blocked ex ante by any nonnull coalition using feasible and incentive-compatible allocations. Finally, given a collection $\{\mathcal{G}_t^S\}$, or more precisely a collection of measurable maps $\mathcal{E}_1^S$ for all measurable subsets S of T, the incentive-compatible core consists of feasible and incentive-compatible allocations for the grand coalition which cannot be blocked ex ante by any nonnull coalition S using allocations that are feasible in each state of the world and which are $\mathcal{G}_t^S$-measurable and incentive compatible for (almost) all $t \in S$.

Remark. Notice that I have defined blocking in terms of incentive-compatible state-dependent allocation functions for the grand coalition in place of the usual definition that specifies blocking allocations only for members of the blocking coalition. This device was chosen to avoid the introduction of additional notation and definitions. The choice does not affect the underlying idea, as blocking allocations are resource constrained (or required to be feasible) for only the blocking coalition. Indeed, any blocking allocation according to the standard usage can always be extended to a blocking (incentive-compatible state-dependent) allocation (function) by specifying that all members of the complementary coalition keep their initial endowments, which are necessarily incentive compatible and measurable with respect to the desired information. Conversely, the restriction to the blocking coalition of a blocking allocation according to my definition yields a blocking allocation in the standard sense.

5 Results

This section presents statements and proofs of the main results of this paper. More precisely, I verify that there is an incentive-compatible competitive equilibrium for an economy—as described by my model—satisfying Assumption 2.1, Simplification 2.2, and Assumption 2.3S or Assumption 2.3S′. Then I show that any incentive-compatible competitive equilibrium necessarily belongs to the auxiliary core. Finally, since the incentive-compatible core contains the auxiliary core, this proves that the incentive-compatible core is nonempty.

5.1. Proposition. There exists an incentive-compatible competitive equilibrium.

Proof This follows readily from Yamazaki's [**22**] theorem. Aggregation over dispersed endowments for each utility-type yields an aggregate excess demand satisfying Yamazaki's [**22**] properties, which are preserved by addition over finitely many utility-types, each having the same nonconvex consumption set defined by the incentive-compatibility self-selection constraints given the utility-type. Therefore, Yamazaki's [**22**] fixed point argument applies, and there is an incentive-compatible competitive equilibrium. □

5.2. Proposition. Any incentive-compatible competitive equilibrium allocation function belongs to the auxiliary core.

Remark. The standard proof, as in Aumann [**6**] or Hildenbrand [**13**], applies to my model, even with nonconvex consumption sets caused by incentive compatibility. As the argument is short, I include it here for completeness.

Proof Suppose that $X^*\colon T \to (I\!R^\ell_{++})^\Omega$ is an incentive-compatible competitive equilibrium allocation function that does not belong to the auxiliary core. Then it can be blocked; there is a measurable subset S of T with $\tau(S) > 0$ and there is an incentive-compatible state-dependent allocation function $\bar{X}\colon T \to (I\!R^\ell_{++})^\Omega$ such that, for almost all $\omega \in \Omega$,

$$\int_S \bar{x}_t(\omega)\, d\tau(t) = \int_S e_t\, d\tau(t),$$

$$\tau\left(\{t \in S \mid \int_\Omega u_t\big(\bar{x}_t(\omega);\omega\big)\, d\mu(\omega) \geq \int_\Omega u_t\big(x^*_t(\omega);\omega\big)\, d\mu(\omega)\}\right) = \tau(S),$$

and if $\bar{S}$ is defined by

$$\bar{S} = \{t \in S \mid \int_\Omega u_t\big(\bar{x}_t(\omega);\omega\big)\, d\mu(\omega) > \int_\Omega u_t\big(x^*_t(\omega);\omega\big) d\mu(\omega)\},$$

then $\tau(\bar{S}) > 0$. Let p^* be a (equilibrium) price vector that makes X^* an incentive-compatible competitive equilibrium allocation function. Because all agents in S weakly prefer $\bar{X}$ to X^* and all of those in $\bar{S}$ strictly prefer $\bar{X}$ to X^*; $p^* \cdot \bar{X}(t) > p^* \cdot (e_t, \ldots, e_t)$ for $t \in \bar{S}$ and $p^* \cdot \bar{X}(t) \geq p^* \cdot (e_t, \ldots, e_t)$ for all $t \in S$. This contradicts the requirement that $\int_S \bar{x}_t(\omega)\, d\tau(t) = \int_S e_t\, d\tau(t)$ for almost all $\omega \in \Omega$, which says that S cannot block X^*. Hence X^* belongs to the auxiliary core. □

5.3. Corollary. The auxiliary core is nonempty.

Proof Combine Proposition 5.1 with Proposition 5.2. □

5.4. Proposition. The incentive-compatible core is nonempty.

Proof The difference between the auxiliary core and the incentive-compatible core is that potential state-dependent blocking allocations for subcoalitions S in the incentive-compatible core must be $\mathcal{G}_t^S$-measurable. (By Simplification 2.2, $\mathcal{G}_t^T = 2^\Omega = \mathcal{G}$ for all $t \in T$, so that this restriction reduces to the same requirement for the grand coalition according to both core concepts.) Hence blocking is harder for the incentive-compatible core while the same "commodity space" of incentive-compatible state-dependent allocation functions—the feasible allocations for the grand coalition—is used in both cores. This implies that the incentive-compatible core contains the auxiliary core, so that nonemptiness of the auxiliary core implies the same property for the incentive-compatible core. □

Remark. One cannot directly show that a corresponding competitive equilibrium allocation concept is contained in the incentive-compatible core because the information measurability conditions implicitly say that an agent's commodity space or consumption set depends on the coalition to which he or she belongs. This destroys the implication that preferred allocations for (sub)coalitions cannot be afforded at competitive equilibrium prices, since they may have failed to be chosen as part of a competitive equilibrium because the measurability constraint, rather than the budget constraint, is not satisfied.

6 Extensions

The same proof strategy can be employed to examine some alternative incentive-compatible core concepts. For this technique to be valid, one needs a notion of incentive compatibility that is independent of the particular coalition to which an agent belongs, so that there is an auxiliary core containing the appropriate incentive-compatible competitive equilibrium allocations. Moreover, the assumptions that $\mathcal{G}_t^T = \mathcal{F}$ for all $t \in T$ [which guarantees balancedness of the associated cooperative game with nontransferable utility for the case of a finite economy without incentive-compatibility restrictions] and that $\mathcal{G}_t^S \supseteq \mathcal{G}_t$ for all $t \in S$ and all $S \subseteq T$ imply that the modified incentive-compatible core in which one is interested indeed contains the auxiliary core, since the imposition of additional information measurability conditions renders blocking more difficult.

The preceding section considered incentive-compatibility constraints stating that no trader would wish to lie in any state of the world (or in any event in $\mathcal{G}_t^T = \mathcal{F}$ for all $t \in T$). A weaker but extremely plausible version of incentive compatibility would instead impose self-selection constraints that are expressed in terms of conditional expected utilities given each agent's initial private information $\mathcal{G}_t$ only. The interpretation involves the willingness to reveal initial private information rather than an unwillingness to upset a proposed realization of a (state-dependent) random allocation. A disadvantage of this modification is that it leaves open the possibility that traders may wish to lie by saying that they possess more information than they actually do—i.e., they could strictly prefer the outcome of one state over another in some event which they know has occurred. In this case, they would exaggerate their private information by claiming that they know the first state rather than just the event containing both states. The assumptions of my model imply that other agents would then be able to distinguish between the two states, although such information is helpful only if it can be verified by some referee or planner.

To be specific, replace Definitions 4.1, 5.1, and 5.2 for the remainder of this section by concepts that have been modified as described above.

6.1. Definition. A state-dependent allocation function $X\colon T \to (\mathbb{R}^\ell_{++})^\Omega$ or a state-dependent net trade function $Z\colon T \to (\mathbb{R}^\ell)^\Omega$ is *privately incentive compatible* if for almost all $t \in T$ and for all $\omega', \omega'' \in \Omega$ (where ω' is the true state, t knows $\mathcal{G}_t(\omega')$ but he attempts to lie by stating that ω'' is the state),

$$\int_\Omega u_t\big(x_t(\omega);\omega\big)\,d\mu\big(\omega|\mathcal{G}_t(\omega')\big) \geq \int_\Omega u_t\big(x_t(\omega'');\omega\big)\,d\mu\big(\omega|\mathcal{G}_t(\omega')\big)$$

or if

$$\int_\Omega u_t\big(e_t + z_t(\omega);\omega\big)\,d\mu\big(\omega|\mathcal{G}_t(\omega')\big) \geq \int_\Omega u_t\big(e_t + z_t(\omega'');\omega\big)\,d\mu\big(\omega|\mathcal{G}_t(\omega')\big)$$

respectively.

6.2. Definition. A privately incentive-compatible state-dependent allocation function $X^*\colon T \to (\mathbb{R}^\ell_{++})^\Omega$ is a *privately incentive-compatible competitive equilibrium allocation* if $\int_T x_t(\omega)\,d\tau(t) = \int_T e_t\,d\tau(t)$ for almost all $\omega \in \Omega$ and if there is $p^* \in \{p \in \mathbb{R}^{\#\Omega\ell}_{++} \mid \sum_{\omega\in\Omega}\sum_{j=1}^{\ell} p_j(\omega) = 1\}$ for which $\sum_\omega p^*(\omega)\cdot x_t(\omega) = \sum_\omega p^*(\omega)\cdot e_t$ for almost all $t \in T$ such that, for almost all $t \in T$, if there is a privately incentive-compatible state-dependent allocation $\hat{x}_t\colon \Omega \to \mathbb{R}^\ell_{++}$ with $\int_\Omega u_t\big(\hat{x}_t(\omega);\omega\big)\,d\mu(\omega) > \int_\Omega u_t\big(x^*_t(\omega);\omega\big)\,d\mu(\omega)$, then $\sum_\omega p^*(\omega)\cdot\hat{x}_t(\omega) > \sum_\omega p^*(\omega)\cdot e_t$.

6.3. Definition. A privately incentive-compatible state-dependent allocation function $X\colon T \to (\mathbb{R}^\ell_{++})^\Omega$ belongs to the *modified auxiliary core* if $\int_T x_t(\omega)\,d\tau(t) = \int_T e_t\,d\tau(t)$ for almost all $\omega \in \Omega$ and if there is no privately incentive-compatible state-dependent allocation function $\hat{X}\colon T \to (\mathbb{R}^\ell_{++})^\Omega$ for which there is a measurable subset S of T with $\tau(S) > 0$ such that

$$\tau(S) = \tau\Big(\Big\{t \in S \mid \int_\Omega u_t\big(\hat{x}_t(\omega);\omega\big)\,d\mu(\omega) \geq \int_\Omega u_t\big(x_t(\omega);\omega\big)\,d\mu(\omega)\Big\}\Big)$$

and

$$\tau\Big(\Big\{t \in S \mid \int_\Omega u_t\big(\hat{x}_t(\omega);\omega\big)\,d\mu(\omega) > \int_\Omega u_t\big(x_t(\omega);\omega\big)d\mu(\omega)\Big\}\Big) > 0,$$

and

$$\int_S \hat{x}_t(\omega)\,d\tau(t) = \int_S e_t\,d\tau(t)$$

for almost all $\omega \in \Omega$.

6.4. Definition. A privately incentive-compatible state-dependent allocation function $X\colon T \to (\mathbb{R}^\ell_{++})^\Omega$ belongs to the *privately incentive-compatible core* if for all $\omega \in \Omega$, $\int_T x_t(\omega)\,d\tau(t) = \int_T e_t\,d\tau(t)$ and if there is no alternative privately incentive-compatible state-dependent allocation function $\bar{X}\colon T \to (\mathbb{R}^\ell_{++})^\Omega$ for which there is a measurable subset S of T with $\tau(S) > 0$ such that

$$\tau(S) = \tau\Big(\Big\{t \in S \mid \int_\Omega u_t\big(\bar{x}_t(\omega);\omega\big)\,d\mu(\omega) \geq \int_\Omega u_t\big(x_t(\omega);\omega\big)\,d\mu(\omega)\Big\}\Big),$$

$$\tau\Big(\Big\{t \in S \mid \int_\Omega u_t\big(\bar{x}_t(\omega);\omega\big)\,d\mu(\omega) > \int_\Omega u_t\big(x_t(\omega);\omega\big)\,d\mu(\omega)\Big\}\Big) > 0,$$

$$\int_S \bar{x}_t(\omega)\,d\tau(t) = \int_S e_t\,d\tau(t)$$

for almost all $\omega \in \Omega$, and for almost all $t \in S$, $x_t(\cdot)$ is $\mathcal{G}^S_t$-measurable.

Exactly as in Section 5, one can prove the following results.

6.5. Proposition. There exists a privately incentive-compatible competitive equilibrium.

6.6. Proposition. Privately incentive-compatible competitive equilibrium allocations belong to the modified auxiliary core.

6.7. Proposition. The privately incentive-compatible core contains the auxiliary core.

6.8. Corollary. The privately incentive-compatible core is nonempty.

However, this proof strategy does not apply to yet another potential incentive-compatible core concept with an "average" incentive-compatibility restriction conditioned on each trader's conditional expected utility in each coalition. The reason is that, in this case, incentive compatibility would be more restrictive for coalitions in which an agent has more information, so that the containment of various cores based on the relative difficulty of blocking would be the reverse of the desired inclusions. Note that such incentive compatibility could not be incorporated into a corresponding auxiliary core, as its nonemptiness could not be derived from an associated competitive equilibrium because equilibrium notions do not permit an agent to have different consumption sets in different coalitions.

7 Properties

This section considers the possibilities for the incentive-compatible core (or the privately incentive-compatible core) in my model to possess certain desirable features commonly exhibited by cores of economies.

7.1. Equal Treatment.

Even though my model has finitely many utility-types, it does not make sense to ask whether the equal treatment property holds. No collection of identical agents can have positive measure due to the dispersion assumption, which implicitly requires a continuum of endowment types. Note, however, that in a related model (see Allen [**5**]), the incentive-compatible ε-cores of finite replica economies necessarily contain allocations satisfying the equal treatment property.

7.2. Uniqueness.

One cannot expect the incentive-compatible core or the privately incentive-compatible core to be unique—or even to lead to unique outcomes in expected utility terms—in this model. Since these cores contain the corresponding competitive equilibrium allocations, uniqueness fails because competitive equilibria need not be unique.

7.3. Core Equivalence?

One may wonder about the prospects for obtaining a core equivalence result, given that the model features a nonatomic continuum of agents. Here two major obstacles arise.

The first problem is directly associated with the incentive-compatibility constraints and may, in fact, prevent equality of the set of auxiliary core allocations and the set of incentive-compatible competitive equilibrium allocations. The standard proof (i.e., see Hildenbrand [**13**] or Aumann [**6**]) uses monotonicity of preference relations to argue, after the separating hyperplane (through the origin) step, that dividing any strictly positive commodity bundle among the agents would lead to an allocation that could be blocked, so that any feasible allocation that fails to be a competitive equilibrium cannot belong to the core. However, for the model in this paper, incentive compatibility renders it easily possible for the addition of a strictly positive vector to destroy incentive compatibility. While this argument destroys the monotonicity property, a weaker local nonsatiation condition could well be satisfied that would guarantee auxiliary core equivalence.

However, auxiliary core equivalence is not an economically interesting property as the auxiliary core was proposed only for its role as an intermediate concept to guarantee nonemptiness of the appropriate incentive-compatible core. Since the incentive-compatible core, as defined in Section 4, contains the auxiliary core but is expected to be generally strictly larger than the auxiliary core, auxiliary core equivalence would be necessary but certainly not sufficient for equivalence of the incentive-compatible core allocations and the incentive-compatible competitive equilibrium allocations because information measurability constraints render blocking more difficult.

Similar remarks apply to the potential for other asymmetric information core concepts to possess valid equivalence results. In general, the presence of asymmetric information moves the model away from the case of cores that are directly associated with competitive equilibria. A possible exception concerns the private information competitive equilibria, either with or without incentive compatibility. The major difference is derived from the fact that private information measurability essentially constitutes a restriction of each agent's consumption set to a subset of a linear subspace. In the absence of incentive compatibility, core equivalence follows from a slight variation on the standard argument. With incentive compatibility, the prospect for a positive result depends on the demonstration of a suitable form of local nonsatiation, which remains an open question.

8 Concluding Remarks

Finiteness of the set of states of the world or finiteness of the information partition for each utility-type of trader was needed for technicalities associated with the nonconvexities arising from incentive compatibility. Extension to a genuinely infinite state space is nontrivial and, I conjecture, probably impossible. While one could use ℓ^1 or $\mathcal{L}^1$ to model the commodity space as states vary in a countable or uncountable set (respectively), the analogue of Yamazaki's [**22**] theorem is likely to fail because of the nonconvexity problem, although compactness is unlikely to raise difficulties in such a framework. In fact, Yamazaki's [**22**] lemma on the countability of the set of problematic tangent planes (i.e., those tangent planes that are tangent to the consumption set at a nonconvex set of points) to any nonconvex consumption set appears to be a property of Euclidean space, although one could instead attempt to demonstrate that the set of such problematic price vectors is of measure zero in some appropriate simplex of normalized prices. Note also that containment of competitive equilibrium allocations in the core of an economy with an infinite dimensional commodity space generally holds and, in fact, the standard proof mimicking the proof of the first welfare theorem extends automatically to this setting; for example, check Aumann's [**6**] proof or see Bewley [**10**]. However, existence of competitive equilibria can fail in this framework with infinitely many states and incentive compatibility.

References

[1] Allen, B. [1991a], *Market games with asymmetric information and nontransferable utility: Representation results and the core*, CARESS Working Paper #91-09, Department of Economics, University of Pennsylvania.

[2] Allen, B. [1991b], *Transferable utility market games with asymmetric information: Representation results and the core*, CARESS Working Paper #91-16, Department of Economics, University of Pennsylvania.

[3] Allen, B. [1991c], *Incentives in market games with asymmetric information: The core*, CARESS Working Paper #91-38, Department of Economics, University of Pennsylvania.

[4] Allen, B. [1992], *Market games with asymmetric information: The private information core*, Econom. Theory, forthcoming.

[5] Allen, B. [1994], *Incentives in market games with asymmetric information: Approximate NTU cores in large economies*, Chapter 11 in Social Choice, Welfare, and Ethics: Proceedings of the Eighth International Symposium in Economic Theory and Econometrics (William A. Barnett, Hervé Moulin, Maurice Salles, and Norman J. Schofield, eds.), Cambridge University Press.

[6] Aumann, R. J. [1964], *Markets with a continuum of traders*, Econometrica **32**, 39-50.

[7] Aumann, R. J. [1966], *Existence of competitive equilibrium in markets with a continuum of traders*, Econometrica **34**, 1-17.

[8] Aumann, R. J. and Shapley, L. S. [1974], *Values of Non-Atomic Games.* Princeton University Press.

[9] Berliant, M. [1992], *On income taxation and the core*, J. Econom. Theory **56**, 121-141.

[10] Bewley, T. F. [1973], *The equality of the core and the set of equilibria in economies with infinitely many commodities and a continuum of agents*, Internat. Econom. Rev. **14**, 383-394.

[11] Boyd, J. H. and Prescott, E. C. [1986], *Financial intermediary-coalitions*, J. Econom. Theory **38**, 211-232.

[12] Boyd, J. H., Prescott, E. C. and Smith, B. D. [1988], *Organizations in economic analysis*, Canad. J. Econom. **21**, 477-491.

[13] Hildenbrand, W. [1974], *Core and Equilibria of a Large Economy.* Princeton University Press.

[14] Hildenbrand, W. and Kirman, A. P. [1976], *Introduction to Equilibrium Analysis.* North-Holland, Amsterdam.

[15] Hildenbrand, W. and Kirman, A. P. [1988], *Equilibrium Analysis.* North-Holland, Amsterdam.

[16] Koopmans, T. C. [1957], *Three Essays on the State of Economic Science.* McGraw-Hill, New York.

[17] Koutsougeras, L. and Yannelis, N. C. [1992], *Incentive compatibility and information superiority of the core of an economy with differential information*, mimeo, Department of Economics, University of Illinois at Urbana-Champaign.

[18] Marimon, R. [1989], *The core of private information economies*, UAB/IAE Discussion Papers W.P.131.90, Universitat Autónoma de Barcelona.

[19] Prescott, E. C. and Townsend, R. M. [1984a], *Pareto optima and competitive equilibria with adverse selection and moral hazard*, Econometrica **52**, 21-45.

[20] Prescott, E. C. and Townsend, R. M. [1984b], *General competitive analysis in an economy with private information*, Internat. Econom. Rev. **25**, 1-20.

[21] Wilson, R. [1978], *Information, efficiency and the core of an economy*, Econometrica **46**, 807-816.

[22] Yamazaki, A. [1978], *An equilibrium existence theorem without convexity assumptions*, Econometrica **46**, 541-555.

[23] Yannelis, N. C. [1991], *The core of an economy with differential information*, Econom. Theory **1**, 183-198.

Fields Institute Communications
Volume **23**, 1999

Economic Applications of Probabilistic Cheap Talk

Bhaskar Chakravorti
The Monitor Corporation
25 First Street
Cambridge, MA 02141 USA
bhaskar-chakravorti@notes.monitor.com

John P. Conley
Department of Economics
University of Illinois
Champaign, IL 61820 USA
jpconley@uiuc.edu

Bart Taub
Department of Economics
University of Illinois
Champaign, IL 61820 USA
b-taub@uiuc.edu

Abstract. In this paper, we apply the concept of "Probabilistic Cheap Talk" (PCT), introduced in [**3**] to a variety of simple economic problems in public economics and industrial organization. We focus on problems which have a single Nash (often dominant strategy) equilibrium which is Pareto dominated by other outcomes. We show that adding a PCT structure to the pre-existing game transforms these problems into multi-stage games with a single, Pareto dominant, efficient, subgame perfect equilibrium. Nevertheless, the game remains essentially one-shot. Payoffs are distributed once and only once, and in equilibrium, strategies are chosen once and for all. Our focus is on resolving inefficient outcomes when agents' strategy spaces and outcome rules are given by pre-existing institutions, and the choices of agents are unverifiable by a third-party. By way of example, we show how PCT can be successfully applied problems involving voluntary provision of public goods, market failure due to externalities, and failures of collusion among oligopolists.

1 Introduction

In this paper, we apply the concept of "probabilistic cheap talk" (PCT) to a variety of simple economic problems in public economics and industrial organization. This concept was introduced in [**3**] , and explored in the context of abstract prisoners' dilemma games in [**4**].

1991 *Mathematics Subject Classification.* Primary 90D80; Secondary 90D10, 90D20.

By way of illustration, consider the case of joint ventures among regulated firms. Managers of such firms engage in collaborative ventures in the form of investments in mutually beneficial projects. For example, computer or tele-telecommunication firms might jointly develop new technological standards and banks might jointly petition to be allowed to offer new types of financial services. Due to the long-lasting nature of these projects, their horizons are often limited to the venture in question. Firms receive payoffs only when the regulator grants permission to bring the output to market. Managers know that approval will be given eventually, but each time they go before the regulatory body, there is a chance that it will be refused. The actions of each firm may be revised prior to the granting of regulatory permission. Once approval is granted, the product is put on the market and the game ends. Managers often have an incentive to free-ride on the investments of others, or follow strategies which would give them an inequitable advantage were the approval to be given. Nevertheless, we still see successful joint ventures despite the one-shot nature of the payoffs. We argue below that it is precisely the uncertain timing of the arrival of payoffs which makes punishment for non-cooperative behavior possible, and provides an explanation for the observed cooperative behavior.

More abstractly, imagine any one shot game. Now suppose that we transform the game as follows. Agents simultaneously and secretly commit to strategies. With some probability, the game ends and payoffs are distributed according to the strategies agents choose. If the game does not end, then the round of play is ex post cheap talk and each agent's (now payoff irrelevant) strategy choices are publicly revealed. We then move to a new round of play and the process is repeated until eventually the game ends. The probability with which the game ends in any given round might be determined by some random element in nature (messengers being comprised or communications networks failing) or chosen by a mediator. Note that the game ends and payoffs arrive with certainty if we aggregate these probabilities over the entire time horizon; however, payoffs are received once and only once. This implies the important fact that the standpoint of economic resources used, therefore, the game is one-shot. This is in sharp contrast to other types of repeated games.

We consider several economic problems which have some fundamental features in common. In each application, there is a single Nash equilibrium (which in some cases is also a dominant strategy equilibrium) which is Pareto dominated by other feasible allocations. We show that through the use of a PCT mechanism, the equilibrium set can be expanded to include Pareto efficient outcomes. Better still, we show that the probability of cheap talk is the variable which can be manipulated to generate an equilibrium set with a single Pareto dominant, efficient, individually rational, subgame perfect allocation. In most cases, the equilibrium set can be reduced to consist of only the intuitively appealing Pareto efficient allocation, and the one-shot Nash equilibrium allocation. Moreover, this is accomplished without the necessity of a third-party who can observe the strategy choices of the agents. Thus, the PCT mechanisms allow efficient outcomes to be achieved in environments in which binding contracts cannot be written.

It is instructive to compare our approach to the one usually taken in the related literature on the theory of implementation. Implementing efficient allocations in both public and private goods has a long history.[1]

[1]For a recent example in the public goods context, see [**6**].

The central premise of implementation theory is that agents' preferences (or other crucial payoff-relevant parameters) are unobservable to the mechanism designer. The designer is a social planner who invites the agents to play a game whose equilibrium outcomes are contained in the set of "desirable" outcomes. The notion of what constitutes desirability and, of course, the equilibrium outcomes would depend on the actual preferences of the agents. The appealing aspect of the designer's game is that this containment relationship holds regardless of what the agents' true preferences really are. The game generally involves the agents sending messages to the designer who, in turn, allocates resources as a function of the messages.

This function and the message space are chosen by the designer.

Our motivation is quite different from that of the implementation literature. We recognize that certain institutions, and therefore the underlying game as well (including strategies and outcome functions), are already in place. The problem is that the existing structure makes it impossible for the agents to achieve an efficient outcome. The PCT mechanism takes the underlying game as historically given and provides the agents a means of communicating with each other prior to playing the game. The game induced by this procedure can be operated by a designer who cannot observe the strategies played by the agents.

Our mechanism has three important features that are lacking in a standard implementation mechanism:

(i) The PCT mechanism preserves the pre-existing institution or game and uses it as a building block to induce a new game. In contrast, implementation mechanisms are typically complex and do not generally correspond to any existing economic institution. Also, it is clearly more difficult to find a designer with the power to determine not only the strategy space from which agents must make choices, but also the outcome function. It is much more plausible to imagine that the designer merely modifies certain rules of play in a pre-existing game.

(ii) Many actions taken by economic agents are unverifiable to third-parties. The implementation literature presumes that a "message space" can be invented by the designer who can also observe the actual messages chosen. Our method simply requires the agents to choose from the set of actions available in the pre-existing institution; and these actions need not necessarily be observable to an outsider. Thus, the mediator, in our context, can simply be agreed upon by the agents as some form of "noise". For example, the agents could enter their strategies into computer terminals hooked up to a network that goes down with a given frequency, thereby halting the transmission of the entries to the payoffs with some probability. The downtime frequency could be agreed upon by the agents.

(iii) Finally, many implementation schemes have non-unique equilibria without a clear indication of how the agents would resolve the coordination problem over the Pareto unranked outcomes that would result. Thus, we are left with a generalized Battle of the Sexes problem. Our mechanism also has multiple equilibria, but there is a very simple resolution to the coordination problem.

On the other hand, our mechanisms have certain weaknesses vis-a-vis the implementation mechanisms: (i) In general, our focus is on problems where strategy choice, rather than private information on preferences, is unobservable to third-parties. (ii) In the case of implementation mechanisms, *every* equilibrium is efficient. Our mechanisms do not have such a strong property. We show that, beginning with a game in which there is only one inefficient equilibrium, the game can be modified to expand the equilibrium set in an extremely attractive manner:

the expanded set now includes (in addition to the status quo outcome) a unique Pareto dominant, efficient outcome. There are no problems relating to coordination over Pareto unranked outcomes, and it is generally expected that the agents will find this dominant efficient outcome focal. Of course, strict Pareto dominance is not a guarantee that agents will choose such an outcome (for example, if they use coalition-proofness a la Bernheim, Peleg and Whinston [**2**] as a solution concept). However, most economic agents will agree that any mechanism that transforms a problem with a single inefficient outcome to one that also has a unique Pareto dominant, efficient outcome transforms the opportunity set in a most dramatic and highly desirable manner.

To summarize, our approach yields results that are weaker in some aspects and stronger in others as compared to those in standard implementation theory. Also, the class of problems that the two approaches address are quite different.

Before proceeding, we note that we do not explicitly introduce a discount factor. This may be interpreted in different ways depending on the application. Often, it is appropriate to think of the delay in the realization of payoffs as occurring either in non-real time or within a period of time that is so small that there is no discounting. If, on the other hand, we were to apply our models to environments with strong discounting of the future, recall that while the payoffs are not realized, the underlying economic resources may be assumed to grow (for example, through the accumulation of interest earnings in an escrow account). Thus, it may be assumed that the rate of growth of the resources exactly neutralizes the discounting.

The remainder of the paper is organized as follows. In Section 2, we describe a general PCT mechanism. In Section 3, we describe a simple public goods problem. We show how the PCT mechanism can be used to generate equal sharing of cost by all agents as the unique Pareto dominant, efficient, subgame perfect equilibrium. In Section 4, we look at an externality problem. We show how efficient provision of a positive externality, or abatement of a negative externality, can be generated through the use of a PCT mechanism. In Section 5, we look at Bertrand oligopoly. We show that all firms setting price at the monopoly level is the unique Pareto dominant, efficient, subgame perfect equilibrium of a PCT extension of a one-shot game. Section 6 contains concluding remarks.

2 Probabilistic cheap talk mechanisms

In this section we define an abstract PCT mechanism. First, consider a one-shot game $G \equiv < N, M, v >$. Let N be the set of agents. Let M^i be the set of *moves* available to agent $i \in N$. Let $v^i : M^1 \times \cdots \times M^n \equiv M \to \Re^n$, be the payoff function for $i \in N$. We use m to denote $(m^i)_{i \in N}$ and m^{-i} to denote $(m^j)_{j \in N \setminus \{i\}}$.

Agents simultaneously choose moves. The designer is a third party (a "mediator", a computer with random down-time, or just plain noise) with the ability to delay the payoffs based on the outcomes of a randomization device with two realizations: "Cheap Talk"(CT) and "DeadLine" (DL). If DL is realized, the payoffs are generated for the moves the agents chose. If CT is realized, then the payoff distribution is put on hold, the chosen moves are treated as cheap talk, and the process is repeated. In all such repetitions, each agent has two options: i) An agent may opt out of the communications process by not taking any further actions after the first choice. This is interpreted as a willingness to replay the same move chosen in the previous round of (now payoff irrelevant) play. ii) An agent may choose a new move to replace the previous one. If all agents "opt out" in the first round of play, we refer to this as *event* (α), otherwise, we have *event* (β). Note that payoffs are received only once. If event (α) occurs, actions are taken only once and, de facto, there is no communication. Agents "observe" all ex post cheap play but not the current moves of the other agents. The designer can only distinguish between event (α) and event (β). The ex post cheap talk is unobservable to the designer.

Denote the number of rounds of ex post cheap talk by $t \in \{1, 2 \dots\}$. The history of talk at t is denoted by h_t. Let $\mathcal{H}$ be the set of all possible histories over all $t \in \{1, 2, \dots\}$. Let $\mathcal{H}_t$ be the space of all possible histories at time t. We shall set $h_1 = \emptyset$. A *strategy profile* for $i \in N$ is a mapping $s^i = \{s^i_t : \mathcal{H}_t \to M^i\}_{t=1}^{\infty}$. Let S^i be the class of all possible strategy profiles.

If for some $t' < t$ the agents have chosen $m \in M$ in round t', we shall say that the resulting history h_t *contains* m at t'. We shall write this as $m \in_{t'} h_t$. A history $h \in \mathcal{H}$ is said to be *stationary* if

$$\exists\, m \in M \text{ s.t. } \forall t \in \{1, 2, \dots\}, \text{ and } \forall t' \leq t, m \in_{t'} h_t.$$

If $h \in \mathcal{H}$ is stationary, then the restriction of h to the first t rounds, h_t, is also said to be stationary. Note that event α generates a stationary history. Let $\mathcal{H}^\alpha$ denote the sub-class of stationary histories generated by event α. Let $\mathcal{H} \setminus \mathcal{H}^\alpha = \mathcal{H}^\beta$.

We now define two subclasses of strategies. Let $S_M \in S$ be the class of *stationary trigger strategy profiles* in which agents play a stationary strategy, and respond to any deviation from stationary by going to a "punishment" move the next round:

$$S_M \equiv \{s_{\tilde{m}} \in S\}$$

where for all $h \in \mathcal{H}$, and all $t \in \{1, 2, \dots\}$

$$s_{\tilde{m},t}(h_t) \equiv \begin{cases} s_{\tilde{m},t-1}(h_{t-1}), & \text{if } i)\ \exists m \in M \text{ s. t. } m \in_{t-2} h^t, \text{ and } m \in_{t-1} h^t \\ & \text{or } ii)\ t = 2 \\ \tilde{m}, & \text{otherwise.} \end{cases}$$

Let $S_M^M \subset S_M$ be the class of *enforcing trigger strategy profiles* in which agents play a *particular* stationary strategy and respond to any deviation from this strategy by going to a punishment move next round. The difference between these two is that in stationary trigger strategies, an agent may or may not require that other agents make a particular move in the first round. However, he always goes to the punishment move if any agent makes a different move in any subsequent round. In an enforcing trigger strategy, punishment is induced not only by nonstationary play, but also by deviation from a specific move by the other agents, even in the first round. Formally:

$$S_M^M \equiv \{s_{\tilde{m}}^{\bar{m}} \in S_M\}$$

where for all $h \in \mathcal{H}$, and all $t \in \{1, 2, \dots\}$,

$$s_{\tilde{m},t}^{\bar{m}}(h_t) = \begin{cases} \bar{m}, & \text{if } i)\ \bar{m} \in_{t-1} h_t; \\ & \text{or } ii)\ t = 1 \\ \tilde{m}, & \text{otherwise.} \end{cases}$$

Our convention is to have the superscripted move (if any) to be the proposed stationary one, while the subscripted move (if any) is the punishment move.

Although our attention will focus on enforcing stationary trigger strategies, the former class is needed for technical reasons in the proofs of the lemmata that follow.

For any $i \in N$, and any $m \in M$ let $m^{i,opt}$ denote the set of *optimal one-shot defections from m for agent i*

$$m^{i,opt} \equiv \{\hat{m}^i \in M^i \mid \forall \bar{m}^i \in M^i, v^i(m^1,\ldots,\hat{m}^i,\ldots,m^n) \geq v^i(m^1,\ldots,\bar{m}^i,\ldots,m^n)\}.$$

In the applications given below, this will always be single valued. In the general case, it is sufficient to take any element of this set to prove the lemmata below. Note that if $\tilde{m}$ is a Nash equilibrium, then $\tilde{m}^{i,opt} = \tilde{m}^i$.

We are now able to give a formal definition the mechanism.[2] A PCT mechanism is a profile $\delta = \{\delta_t : h_t \to [0,1]\}_{t=1}^{\infty}$ such that for all $t \in \{1,2,\ldots\}$,

$$\delta_t(h_t) = \begin{cases} \delta & \text{if } h_{t-1} \in \mathcal{H}^\alpha \\ 1 & \text{if } h_{t-1} \notin \mathcal{H}^\alpha. \end{cases}$$

Given a history h_t, $\delta_t(h_t)$ is the probability that DL is realized at round t. Thus, if the game continues to round t, and the history involves agents opting out up until then (that is, up to $t-1$,) then the game ends with probability δ. However, the game ends with certainty, and the payoffs are distributed, on the first round after any agent is seen revising his strategy. Note that at each t, δ_t is measurable with respect to the information partition $\{\mathcal{H}^\alpha, \mathcal{H}^\beta\}$. Let $\Gamma(G,\delta)$ denote the multi-stage game induced by the PCT mechanism, δ, given the underlying one-shot game, G.

For any agent $i \in N$, participating in a strategy profile $s \in S$ yields the following expected payoff at time t, given history h:

$$\delta_t(h_t)v^i(s_t(h_t)) + (1-\delta_t(h_t))\delta_{t+1}(h_{t+1})v^i(s_{t+1}(h_{t+1}))$$
$$+(1-\delta_t(h_t))(1-\delta_{t+1}(h_{t+1}))\delta_{t+2}(h_{t+2})v^i(s_{t+2}(h_{t+2})) + \ldots .$$

Therefore, define the *expected value of strategy* s *to agent* i *at time* t, $E_t^i : S \times \mathcal{H}_t \mapsto \mathfrak{R}_+$ as

$$E_t^i(s,h_t) \equiv \left\{ \delta_t(h_t)v^i(s_t(h_t)) + \sum_{k=t+1}^{\infty} \left[\delta_k(h_k)v^i(s_k(h_k)) \prod_{r=t}^{k-1}(1-\delta_r(h_r)) \right] \right\}$$

where the histories after t are generated by equilibrium play of s given h_t. A strategy profile $s \in S$ is *subgame perfect equilibrium (SPE) of the game* $\Gamma(G,\delta)$ if

$$\forall t \in \{1,2,\ldots\}, \forall i \in N, \forall \bar{s}^i \in S^i, \text{ and } \forall h \in \mathcal{H},$$
$$E_t^i(s,h_t) \geq E_t^i(s^1,\ldots,\bar{s}^i\ldots,s^n,h_t).$$

The following lemmata will be useful in proving the results in subsequent sections. The first of these simply says that if a strategy profile is an SPE of a game, then it must be a stationary trigger strategy. This means that the only equilibrium histories are stationary histories.

Lemma 2.1 *Let* $\tilde{m} \in M$ *be the unique Nash equilibrium of a one-shot game* G. *If a strategy,* s, *is an SPE of* $\Gamma(G,\delta)$, *then there is* $s_{\tilde{m}} \in S_M$ *such that* $s \equiv s_{\tilde{m}}$.

Proof First, in any SPE strategy, it must be the case that $\tilde{m}$ is played on the round after any nonstationary move occurs. This is because by construction of the PCT mechanism, the game ends with certainty in the next round, and by hypothesis, $\tilde{m}$ is the only Nash equilibrium in such a subgame.

[2]We use δ to mean both the profile of maps from histories into probabilities, and the actual probability with which the game ends if the history is stationary. This is a slight abuse of notation, but it should not cause any confusion in context. We immediately restrict attention to the class of profiles that enforce stationary play.

Second, in any SPE, all agents must make the same move each round if the history has been stationary. Suppose instead that it was not optimal to play a stationary strategy. Then suppose that the history h_{t+2} is stationary up to t, with agents playing $\bar{m} \in M$ each round. However at round $t+1$ the equilibrium move is $\hat{m} \neq \bar{m}$. The expected payoff to agent j from abiding by the "equilibrium" strategy in this subgame is:

$$\delta v^j(\hat{m}) + (1-\delta)v^j(\tilde{m}).$$

This is because if the PCT game continues until round t, there is a probability of δ that the game ends in round $t+1$. If the game does not end in round $t+1$, the mediator ends the game with certainty in $t+2$ due to the nonstationary play. Thus the probability that the game ends at $t+2$ is $1-\delta$. Assume that $\hat{m} \neq \tilde{m}$.

Then the expected payoff to agent j from deviating optimally in round $t+1$ is

$$\delta v^j(\hat{m}^1, \ldots \hat{m}^{j,opt}, \ldots, \hat{m}^n) + (1-\delta)v^j(\tilde{m}).$$

But since $\tilde{m}$ is the only one-shot Nash equilibrium, for at least one $j \in N$

$$\delta v^j(\hat{m}^1, \ldots \hat{m}^{j,opt}, \ldots, \hat{m}^n) + (1-\delta)v^j(\tilde{m}) > \delta v^j(\hat{m}) + (1-\delta)v^j(\tilde{m}).$$

and so the strategy could not have been an SPE.

The argument is similar if it happens that $\hat{m} = \tilde{m}$. In this case, agent j can improve his expected payoff by deviating in round t. Since he then receives the Nash equilibrium payoffs in $t+1$ anyway, there is no incentive not to defect in the previous round t, and so s could not have been an SPE. □

The next lemma shows that for any enforcing trigger strategy, $s^{\bar{m}}_{\tilde{m}}$, it is a best response for agents to invoke the punishment move, $\tilde{m}$, if there is any deviation from the move $\bar{m}$. Note that this lemma does not say that $\bar{m}^i$ is a best response to $\bar{m}^{-i}$.

Lemma 2.2 *Let $\tilde{m} \in M$ be the unique Nash equilibrium of a one-shot game G, and consider any enforcing trigger strategy $s^{\bar{m}}_{\tilde{m}} \in S\bar{m}_{\tilde{m}}$. For any $t \in \{1, 2, \ldots\}$, suppose also that $h_{t+1} \in \mathcal{H}_{t-1}$ is such that for all $t' < t$ $\bar{m} \in_{t'} h_{t+1}$, but $\bar{m} \neq \hat{m} \in_t h_{t+1}$.*

Then for all $i \in N$, it is a best response in this subgame to abide by the trigger strategy and play $\tilde{m}^i$ in round $t+1$ and all future rounds.

Proof For $t = 1$, under the hypothesis, $h_2 = \{\hat{m}\}$ where for at least one agent j, $\hat{m}^j \neq \bar{m}^j$. This history is not consistent with the equilibrium play of the trigger strategy, but is trivially stationary. However, under the trigger strategy, all agents other than j respond to this history by playing $\tilde{m}^{-j}$ in the second round. If the deadline happens not to hit, and the game does not end in the second round, then it certainly ends in the third due to the non-stationary play. Then clearly it is a best response for all agents $i \in N$ to abide by this trigger strategy and play $\tilde{m}^i$ in the second and third rounds.

Finally for any $t > 2$, suppose h_t satisfies the hypothesis. Then the history is stationary at $\bar{m}$ up until $t-2$, but at $t-1$ at least one agent j, makes the move $\hat{m}^j \neq \bar{m}^j$. Again, if the deadline happens not to hit in round $t-1$, it certainly ends in round t due to the non-stationary play. Then clearly it is a best response all agents $i \in N$ to abide by this trigger strategy and play $\tilde{m}^j$ in this last round. □

The next lemma shows that all equilibrium histories can be generated by enforcing trigger strategies, and so attention can be restricted to this class.

Lemma 2.3 *Let $\tilde{m} \in M$ be the unique Nash equilibrium of a one-shot game, G. Let $s_{\tilde{m}} \in S_M$ be an SPE of $\Gamma(G, \delta)$. Then $h \in \mathcal{H}^\alpha$, with $\hat{m} \in_1 h$, is a possible equilibrium history generated by this strategy if and only if $s_{\tilde{m}}^{\hat{m}} \in S_M^M$ is also an SPE of $\Gamma(G, \delta)$.*

Proof It is immediate that $h \in \mathcal{H}^\alpha$ with $\hat{m} \in_1 h$ is a possible equilibrium history generated by some strategy $s_{\tilde{m}} \in S_M$ if $s_{\tilde{m}}^{\hat{m}} \in S_M^M$ is also an SPE since $s_{\tilde{m}}^{\hat{m}} \in S_M$.

To see the reverse implication, suppose that $h \in \mathcal{H}^\alpha$, with $\hat{m} \in_1 h$, is a possible equilibrium history generated by some strategy $s_{\tilde{m}} \in S_M$ but that $s_{\tilde{m}}^{\hat{m}} \in S_M^M$ was not an SPE. Then for some $t \in \{1, 2, \dots\}$, and some $\bar{h}^t \in \mathcal{H}$, there is an agent $i \in N$ and a strategy s_t^i such that

$$E_t^i(s_{\tilde{m},t}^{\hat{m},1}, \dots, s_t^i, \dots, s_{\tilde{m},t}^{\hat{m},n}, \bar{h}_t) > E_t^i(s_{\tilde{m}}^{\hat{m}}, \bar{h}_t).$$

Suppose first that for all $t' < t$, $\hat{m} \in_{t'} \bar{h}^t$. But since the future play of both $s_{\tilde{m}}^{\hat{m}}$ and $s_{\tilde{m}}$ are the same given this history, it follows that:

$$\begin{aligned} E_t^i(s_{\tilde{m},t}^{\hat{m},1}, \dots, s_t^i, \dots, s_{\tilde{m},t}^{\hat{m},n}, \bar{h}_t) &= E_t^i(s_{\tilde{m},t}^1, \dots, s_t^i, \dots, s_{\tilde{m},t}^n, h_t) \\ > E_t^i(s^{\tilde{m}}, h_t) &= E_t^i(s_{\tilde{m}}^{\hat{m}}, \bar{h}_t), \end{aligned}$$

which contradicts the hypothesis that $s_{\tilde{m}}$ is an SPE.

Suppose instead that for some $t' < t, \bar{m} \in_{t'} h^t$, and $\bar{m} \neq \hat{m}$. Without loss of generality, assume that round t' is the first time any move other than $\hat{m}$ is seen. Then by Lemma 2.2, it is a best response for agent i to abide by the trigger strategy and invoke the punishment move in all rounds after t'.

Thus, $s_{\tilde{m}}$ is an SPE, then $s_{\tilde{m}}^{\hat{m}}$ also has all the agents playing a best response in every subgame, and is therefore an SPE as well. □

We close this section with the observation that the construction of δ is such that in equilibrium there is not cheap talk. The agents choose strategies in the first round and then opt out of communication.

3 A hand raising mechanism for the provision of discrete public goods

Consider the following one-shot game $G^{hr} \equiv \langle N, M, v \rangle$. Each agent $i \in N$, chooses between two moves: $M^i = \{HO, FR\}$ (Help Out, or Free Ride). Let $\#HO : M \mapsto \{0, \dots, n\}$ be a function that gives the number of agents that agree to help out in any given profile of moves. Let B^i be the benefit that agent i stands to receive if the public project is undertaken, and C be the cost of the project. Then for all $i \in N$, and all $m \in M$,

$$v^i(m) = \begin{cases} B^i - \frac{C}{\#HO(m)} & \text{if } m^i = HO \\ B^i & \text{if } m^i = FR \text{ and } \#HO(m) \neq 0 \\ 0 & \text{if } \#HO(m) = 0. \end{cases}$$

We assume that for all $i \in N, C > B^i > 0$, and $C < \sum_{i=1}^n B^i$. This means that no single agent would be willing to build the project on his own, but the sum of the benefits to all agents exceeds the cost.

This is a very simple model of a mechanism to produce and pay for a discrete level of a public good. To fix the idea, suppose that a group of neighbors is considering the construction of a community playground for the local children. By raising his hand and agreeing to help out with the playground, an agent promises to show up at the proposed site and share in the effort needed to complete the task.

The difficulty, of course, is in overcoming the free rider problem. In the one-shot game, in which the moves are simultaneous, it is a dominant strategy to free ride. Also notice that building a playground is fundamentally a one-shot problem, or at best, a finitely repeated one. It does not make sense to think about building an infinity of playgrounds. We therefore cannot expect agents to arrive at an efficient outcome by using a repeated game argument.

Consider the incentive problem facing any particular agent i when all the other agents play their part of an arbitrary enforcing trigger strategy $s_{\tilde{m}}^{\bar{m}} \in S_M^M$. Since $\{FR, \dots FR\}$ is the only Nash equilibrium of G^{hr}, by Lemma 2.1, we can assume that the punishment move is $\tilde{m} \equiv (FR, \dots, FR)$. If $\bar{m}^i = FR$, then i is being asked to play his one-shot dominant strategy. Then it is trivially a best response in every subgame for him to play his part in the trigger strategy equilibrium. On the other hand, if $\bar{m}^i = HO$, he finds that playing his part of this strategy is a best response if and only if the following condition is met:

$$B^i - \frac{C}{\#HO(\bar{m})} \geq \delta B^i.$$

The left-hand side of this expression is i's payoff from playing along with the trigger strategy. By so doing the agent i guarantees himself this payoff whenever the deadline happens to fall. The right-hand side is his payoff from the optimal deviation, free riding from the first round of the subgame. If the deadline hits in round one (which is a probability δ event), then he gets the full benefit of the public good, B^i, without paying any of the cost of the project. If the game does not end, then all other agents free ride in the following round. This nonstationary play induces the mediator to end the game with certainty in the third round. Thus, the only possible benefit from defecting from the trigger strategy is free riding in the first round in the event that the deadline happens to hit. Then clearly the agent will abide by the trigger strategy if and only if:

$$\delta \leq 1 - \frac{C}{B^i \#HO(\bar{m})}.$$

First, suppose all agents are identical and so $\forall i \in N, B^i = B$. Let

$$\delta^* = 1 - \frac{C}{Bn}$$

and $\bar{m} \equiv (HO, \dots, HO)$.

Lemma 3.1 *Suppose all agents in the game G^{hr} are identical. Then $s_{\tilde{m}}^{\bar{m}}$ and $s_{\tilde{m}}^{\tilde{m}}$ are the only SPE of $\Gamma(G^{hr}, \delta^*)$.*

Proof Clearly, $s_{\tilde{m}}^{\tilde{m}}$ is an SPE for any δ since $\tilde{m}$ is the only Nash equilibrium of the one-shot game. To see that $s_{\tilde{m}}^{\bar{m}}$ is also an SPE, consider any $i \in N$ and any $h \in \mathcal{H}$. This history could have evolved in one of two ways.

Suppose first that the game has not ended at any given t, and suppose that all agents have been playing HO each round. Then by construction of δ^* it is a best response for i to play HO in round t.

Suppose on the other hand that at some $t \in \{1, 2, \dots\}$, some agent j plays FR. Then by Lemma 2.2, it is a best response for all agents to abide by the trigger strategy and play FR in all remaining rounds.

Finally, suppose there was another trigger strategy $s_{\tilde{m}}^{\hat{m}}$ that is an SPE. But then $\#HO(\bar{m}) > \#HO(\hat{m})$ and so

$$\delta^* B = B - \frac{C}{\#HO(\bar{m})} > B - \frac{C}{\#HO(\hat{m})}.$$

Thus it would be optimal for every agent to defect from any other trigger strategy. □

Theorem 3.2 *Suppose all agents in the game G^{hr} are identical. Then the only SPE payoffs of $\Gamma(G^{hr}, \delta^*)$ are: $\{(B - \frac{C}{n}, \ldots, B - \frac{C}{n}), (0, \ldots 0)\}$.*

Proof By Lemma 3.1, $s^{\bar{m}}_{\bar{m}}$ and $s^{\tilde{m}}_{\tilde{m}}$ are the only SPE. Then by Lemma 2.3,

$$\bar{h} \equiv \{h \in \mathcal{H} \mid \forall t \in \{1, 2, \ldots\}, \bar{m} \in_t h\},$$

and

$$\tilde{h} \equiv \{h \in \mathcal{H} \mid \forall t \in \{1, 2, \ldots\}, \tilde{m} \in_t h\},$$

are the only SPE histories. Thus

$$\{(B - \frac{C}{n}, \ldots, B - \frac{C}{n}), (0, \ldots 0)\}$$

are the only SPE payoffs.

Thus, if all agents derive the same benefit from the project, we can choose δ high enough so that everyone helping out, and everyone trying to free ride (and the project not being undertaken) are the only an SPE outcomes. Then since everybody helping out strongly Pareto dominates everybody free riding, it is a focal equilibrium which on which all agents share an interest in coordinating.

Now consider the case of nonidentical agents. We know that an agent is better off helping out than free riding in a trigger strategy enforcing $\bar{m}$ if:

$$B^i - \frac{C}{\#HO(\bar{m})} \geq \delta B^i.$$

Assume that the agents are ordered so that agents with a lower index place more value on the project. For simplicity, assume that all agents receive different benefits. Thus $B^1 > B^2 > \cdots > B^n$.

For any $j \in \{1, \ldots, n\}$, let $m^{(j)}$ denote the move in which the first j agents help out and the rest free ride:

$$m^{(j)} \equiv (m^{(j),1}, \ldots, m^{(j),j}, m^{(j),j+1}, \ldots, m^{(j),n},) = (HO, \ldots HO, FR, \ldots FR).$$

For any $j \in \{1, \ldots, n\}$, define:

$$\delta^{(j)} = 1 - \frac{C}{B^{(j)} \#HO(m^{(j)})}.$$

Finally, define j^* to be:

$$j^* \equiv \{j \in \{1, \ldots, n\} \mid \forall i \in \{1, \ldots, n\}, \delta^{(j)} \geq \delta^{(i)}\}.$$

Note that j^* may not always be a singleton, but is generically unique. We will assume in the following that j^* is unique.

Lemma 3.3 *Suppose that j^* is unique. Then the strategies $s^{m^{(j^*)}}_{\tilde{m}}$ and $s^{\tilde{m}}_{\tilde{m}}$ are the only SPE of $\Gamma(G^{hr}, \delta^{(j^*)})$.*

Proof Clearly, $s_{\tilde{m}}^{\tilde{m}}$ is an SPE for any δ since $\tilde{m}$ is the only Nash equilibrium of the one-shot game.

For all agents $i > j^*$, playing strategy $s_{\tilde{m}}^{m^{(j^*)},i}$ is obviously a best response in any subgame since they play FR in each round. So consider any agent $j \leq j^*$ and any $h \in \mathcal{H}$. This history could have evolved in one of two ways.

Suppose the game has not ended at any given t, and suppose all other agents have been playing the move given by $m^{(j^*)}$ each round. Then by construction of $\delta^{(j^*)}$ it is a best response for j to play HO.

Suppose on the other hand that at some $t \in \{1, 2, \dots\}$, some agent $j \leq j^*$ plays FR. Then by Lemma 2.2 it is a best response for all agents to abide by the trigger strategy and play FR in all remaining rounds.

Next suppose there is another trigger strategy $s_{\tilde{m}}^{m^{(i)}}$ that was an SPE. To see that this could not be observe that for any $i \neq j^*$

$$\delta^{(j^*)}B^i > \delta^{(i)}B^i = B^i - \frac{C}{\#HO(m^{(i)})}.$$

Thus it would be optimal for agent i to defect from this strategy.

It only remains to consider trigger strategies in which a coalition of agents that is not contiguous in the benefit ordering agrees to help out. That is, we must show that a trigger strategy in which some of the agents who free ride value the pubic good more than other agents who help out is not an SPE. Let $s_{\tilde{m}}^{\hat{m}}$ be any such noncontiguous trigger strategy. Suppose that $\#HO(s_{\tilde{m}}^{\hat{m}}) = i$. But we know from the above that $s_{\tilde{m}}^{m^{(i)}}$ is not an SPE because it is optimal for i to defect. Let $i' \in N$ be the agent who values the public good the least, but is supposed to help out in the trigger strategy $s_{\tilde{m}}^{\hat{m}}$. Then by construction $B^i > B^{i'}$. Therefore,

$$\delta^{(j^*)}B^i > \delta^{(i)}B^i > \delta^{(i)}B^{i'} = B^{i'} - \frac{C}{\#HO(\hat{m})}.$$

Thus it would be optimal for agent i' to defect from this strategy. □

Intuitively, what this is saying is that if the benefits are unequal, and drop off rapidly, then for the critical δ (which induces the shortest game in expectation), only the highest benefitting agents will raise their hands in an SPE. This description of the efficient SPE makes it possible to look at the real world for verification of the model. For example, if parents receive relatively high benefits from the playground, and non-parents receive only a little, it is likely that only the parents will share in the cost in equilibrium.

From a social efficiency, or egalitarian standpoint, it may be important to know how close the benefits of the agents must be to make unanimous hand raising the only Pareto efficient SPE. Below we give a sufficient bound on the rate of decline of the benefit profile for this to be so.

Lemma 3.4 *Suppose that for all $i \in N$, $B^{\frac{i+1}{B^i}} > \frac{i}{i+1}$, then $s_{\tilde{m}}^{m^{(n)}}$ and $s_{\tilde{m}}^{\tilde{m}}$ are the only SPE of $\Gamma(G^{hr}, \delta^{(n)})$.*

Proof In this case $j^* = n$. To see this note that for all $i \in N$:

$$\delta^{(i)} = 1 - \frac{C}{B^i \#HO(m^{(i)})} < 1 - \frac{C}{B^{i+1}(\#HO(m^{(i)}) + 1)} = \delta^{(i+1)}.$$

Then apply Lemma 3.3. □

4 Self-enforcing optimal urban planning

Consider the following one-shot game: $G^{up} \equiv< N, M, v >$. Here each agent $i \in N$, chooses between two moves: $M^i = \{CC, MB\}$ (Comply with the Code, or Maximize private Benefit). Let $\#CC : M \mapsto \{0, \dots, n\}$ be a function that gives the number of agents who agree to comply with the code in any given set of moves. Let B^i be the external benefit that agent i receives when any agent undertakes the socially beneficial action (complying with the code), and C^i be the private cost of undertaking this action himself. Then for all $i \in N$, and all $m \in M$,

$$v^i(m) = \begin{cases} B^i\#CC(m) - C^i & \text{if } m^i = CC \\ B^i\#CC(m) & \text{if } m^i = MB. \end{cases}$$

We assume that for all $i \in N, C^i > B^i$. This means that the only Nash equilibrium of the one-shot game is for each agent to maximize the private benefit of his property. In fact, this is a dominant strategy equilibrium.

This is a model of an urban planning game. If an agent complies with the code, he generates benefits for himself and his neighbors. However, the private cost of compliance is higher than the private benefit. To fix the idea, we can imagine a group of developers filing plans with the building commission. The random deadline aspect might be generated by uncertainty over when the building commission meets. The game described above is substantially different from the first two. The problem before was to divide the cost of a discrete level of public good. This game is directed toward assuring the provision of an efficient level of a positive externality, or the abatement of a negative externality. The attraction of thinking about this specifically as an urban planning problem is that putting up a development is fundamentally a one-shot proposition characterized by a high probability of delay in finalizing decisions due to bureaucratic procedures.

Let us first consider the case of identical agents. Let

$$\delta^* = \frac{Bn - C}{B(n-1)}$$

$\tilde{m} \equiv (\tilde{m}^1, \dots, \tilde{m}^n) = (MB, \dots, MB)$, and $\bar{m} \equiv (\bar{m}^1, \dots, \bar{m}^n) = (CC, \dots, CC)$. Lemma 4.1 says that if all agents are identical, then at the critical δ, everybody undertaking the beneficial action is an efficient SPE which dominates the only other SPE in which no one undertakes the action.

Lemma 4.1 *Suppose all agents in the game G^{up} are identical. Then $s^{\bar{m}}_{\tilde{m}}$ and $s^{\tilde{m}}_{\tilde{m}}$ are the only SPE of $\Gamma(G^{up}, \delta^*)$.*

Proof Clearly, $s^{\tilde{m}}_{\tilde{m}}$ is an SPE for any δ since $\tilde{m}$ is the only Nash equilibrium of the one-shot game.

To see that $s^{\bar{m}}_{\tilde{m}}$ is also an SPE consider any $i \in N$ and any $h \in \mathcal{H}$. This history could have evolved in one of two ways.

Suppose first that the game has not ended at any given t, that the other agents have been playing $\bar{m}^{-i}$ each round. If i makes the optimal defection from $\bar{m}$ and tries to free ride, then his expected payoff is $\delta^*(n-1)B$. On the other hand, not defecting yields a payoff of $Bn - C$ with certainty. But by construction:

$$\delta^* B(n-1) = \frac{Bn - C}{B(n-1)} B(n-1) = Bn - C.$$

Thus, it is a best response for i to play CC since no additional benefit is gained by defecting.

Suppose on the other hand that at some $t \in \{1, 2, \dots\}$, some agent j plays $\hat{m}^j \neq \bar{m}^j$. Then by Lemma 2.2 it is a best response for all agents to abide by the trigger strategy and play FR in all remaining rounds.

Finally, suppose there was another trigger strategy $s^{\hat{m}}_{\tilde{m}}$ that was an SPE. But then $\#CC(\bar{m}) > \#CC(\hat{m})$ and $\delta^* < 1$ so

$$\begin{aligned}\delta^* B(\#CC(\hat{m}) - 1) &= \delta^* B(n-1) - \delta^* B(n - \#CC(\hat{m})) \\ &= Bn - C - \delta^* B(n - \#CC(\hat{m})) > \#CC(\hat{m})B - C.\end{aligned}$$

Thus it would be optimal for every agent to defect from any other trigger strategy. □

Thus we have:

Theorem 4.2 *Suppose that all the agents are identical. Then the only SPE payoffs of* $\Gamma(G^{up}, \delta^*)$*, are:* $\{(Bn - C, \dots, Bn - C), (0, \dots 0)\}$.

Proof By Lemma 4.1, $s^{\bar{m}}_{\tilde{m}}$ and $s^{\tilde{m}}_{\tilde{m}}$ are the only SPE trigger strategies. Then by Lemma 2.3,

$$\bar{h} \equiv \{h \in \mathcal{H} \mid \forall t \in \{1, 2, \dots\}, \bar{m} \in_t \bar{h}\},$$

and

$$\tilde{h} \equiv \{h \in \mathcal{H} \mid \forall t \in \{1, 2, \dots\}, \tilde{m} \in_t \bar{h}\},$$

are the only SPE histories. Thus,

$$\{(Bn - C, \dots, Bn - C), (0, \dots 0)\}$$

are the only SPE payoffs. □

This result also holds if agents are almost identical, or if at least the ratios of costs to benefits increases sufficiently slowly. It is possible to prove a result similar to Lemma 3.3 for this game. We will not do so here because the sufficiency condition is less easy to interpret. But it turns out that if we order the agents so that those with a lower index have a lower ratio of costs to benefits then if for all $i \in \{1, \dots, n-1\}$,

$$1 \leq \left(\frac{C^i}{B^i} - \frac{C^{i+1}}{B^{i+1}}\right) + \frac{C^{i+1}}{B^{i+1}}$$

then for

$$\delta = \frac{B^n n - C^n}{B^n(n-1)}$$

the only SPE trigger strategies have all agents abiding by the code or all agents maximizing private benefits. If the cost-benefit ratio increases faster than this, there may also be other SPE trigger strategies.

5 Bertrand oligopoly

Next we consider the case of constant marginal cost Bertrand oligopolists facing a known demand curve. For simplicity, we will assume that demand is linear, but this is easily generalized. Let the demand be given by

$$Q = \alpha - \beta p,$$

where $\alpha \geq 0$, and $\beta > 0$. Let C^i be the per unit cost to firm i of making good.

Now, consider the one-shot game $G^{bo} \equiv < N, M, v >$ where for all $i \in N$, $M^i \equiv \mathfrak{R}_+$. Define $LPF : \mathfrak{R}^n_+ \mapsto \{I \mid I \subseteq N\}$:

$$LPF(m) \equiv \{I \subseteq N \mid i \in I \text{ iff } \forall j \in N, m^i \leq m^j\}.$$

This correspondence gives the subset of Lowest Priced Firms. The for all $i \in N$ and all $m \in M$ the payoff function is:

$$v^i(m) = \begin{cases} \frac{(\alpha-\beta m^i)(m^i-C^i)}{|LPF(m)|/} & \text{if } i \in LPF(m) \\ 0 & \text{otherwise.} \end{cases}$$

Thus, all of the lowest priced firms share the demand equally. Other firms sell nothing. It is well known that the only Nash equilibrium in the case of identical firms is for each firm to price at cost. We restrict attention to this case. Lemma 5.1 establishes that in equilibrium, all firms will name the same price and this price will be somewhere between cost, and the monopoly price. Thus all firms naming the monopoly price is an efficient SPE that Pareto dominates all other SPE's. Let $\tilde{m} = (C, \ldots, C)$.

Lemma 5.1 *Suppose all firms are identical. Then for* $\delta =\frac{1}{n}$ $s^{\bar{m}}_{\tilde{m}}$ *is an SPE of* $\Gamma(G^{bo}, \delta)$, *if and only if for all* $i, j \in N$ $\bar{m}^i = \bar{m}^j$, *and* $\bar{m} \in [C, \frac{\alpha+\beta C}{2\beta}]$.

Proof First suppose that $\bar{m} = \tilde{m} = (C, \ldots, C)$. Then clearly, $s^{\bar{m}}_{\tilde{m}}$ is an SPE for any δ since it $\tilde{m}$ is the only Nash equilibrium of the one-shot game.

Now suppose that $\bar{m} \in (C, \frac{\alpha+\beta C}{2\beta}]$. To see that $s^{\bar{m}}_{\tilde{m}}$ is also an SPE consider any $i \in N$ and any $h \in \mathcal{H}$. This history could have evolved in one of two ways.

Suppose first that the game has not ended at any given t, and that the other agents have been playing $\bar{m}^{-i}$ each round. By hypothesis, the equilibrium price is below the monopoly price, $\alpha + \beta \mathcal{C} 2\beta$. Thus, optimal defection for any $i \in N$ is to lower the price by ϵ and capture all the demand. His expected payoff in this case is

$$\delta(\alpha - \beta(\bar{m} - \epsilon))(\bar{m} - \epsilon - C).$$

This is because all agents revert to playing C if the game happens not to end in the round that i defects. On the other hand, by not defecting agent i gets a payoff of

$$\frac{(\alpha - \beta\bar{m})(\bar{m} - C)}{n}$$

with certainty. But by construction:

$$\delta(\alpha - \beta(\bar{m} - \epsilon))(\bar{m} - \epsilon - C) < \delta(\alpha - \beta\bar{m})(\bar{m} - C) = \frac{(\alpha - \beta\bar{m})(\bar{m} - C)}{n}.$$

Thus, it is a best response for i to play $\bar{m}$ since no additional benefit is gained by defecting.

Suppose on the other hand that at some $t \in \{1, 2, \ldots\}$, some agent j plays $\hat{m}^j \neq \bar{m}^j$. Then by Lemma 2.2 it is a best response for all agents to abide by the trigger strategy and play $m^i = C$ in all remaining rounds.

Finally, suppose there is another trigger strategy, $s^{\hat{m}}_{\tilde{m}}$, that is an SPE but does not satisfy the hypothesis of the Lemma. Note first that if $s^{\hat{m}}_{\tilde{m}}$ is an SPE, then for all $i, j \in N$, $\hat{m}^i = \hat{m}^j$. This is because any agent who names a price which is not the lowest receives no profit. Thus for all $\delta > 0$, the optimal defection would necessarily have a positive expected value. Therefore, no agent could name a price above the lowest price in equilibrium. Second, clearly $\hat{m} \geq C$. Otherwise defection would give agents zero profits instead of the negative profits they get from offering to sell below cost. Finally, suppose that $\hat{m} > \frac{\alpha+\beta C}{2\beta}$. In this case, it is optimal to defect to the monopoly price. This gives an expected profit of

$$\delta(\alpha - \frac{\alpha + \beta C}{2})(\frac{\alpha + \beta C}{2\beta} - C).$$

But for all $\hat{m}^i > \frac{\alpha+\beta C}{2\beta}$,

$$\delta(\alpha - \frac{\alpha+\beta C}{2})(\frac{\alpha+\beta C}{2\beta} - C) > \frac{(\alpha-\beta\bar{m}^i)(\bar{m}^i - C)}{n},$$

since it is easy to check that $m = \frac{\alpha+\beta C}{2\beta}$ is profit maximizing. □

Thus we have:

Theorem 5.2 *Suppose that all the agents are identical. Then the only SPE payoffs of* $\Gamma(G^{bo}, \frac{1}{n})$ *are* $\left\{\frac{(\alpha-\beta m)(m-C)}{n}, \ldots, \frac{(\alpha-\beta m)(m-C)}{n} \mid m \in [C, \frac{\alpha+\beta C}{2\beta}]\right\}$.

Proof By Lemma 5.1, $s_{\bar{m}}^{\bar{m}}$ for $\bar{m} \in [C, \frac{\alpha+\beta C}{2\beta}]$ are the only SPE trigger strategies. Then by Lemma 2.3,

$$\bar{h} \equiv \left\{h \in \mathcal{H} \mid \forall t \in \{1, 2, \ldots\}, \bar{m} \in_t \bar{h}\right\},$$

for $\bar{m} \in [C, \frac{\alpha+\beta C}{2\beta}]$ are the only SPE histories. Thus

$$\left\{\frac{(\alpha-\beta m)(m-C)}{n}, \ldots, \frac{(\alpha-\beta m)(m-C)}{n} \mid m \in [C, \frac{\alpha+\beta C}{2\beta}]\right\}$$

are the only SPE payoffs. □

Thus, all the mediator needs to know to choose the appropriate δ is the number of firms in the market. Equal sharing of the monopoly profit is the unique Pareto dominant SPE equilibrium payoff.

It may be possible to sharpen this result by requiring that the defecting firm lower his price by at least a fixed ϵ (instead of the arbitrarily small ϵ here) in order to capture the market. This seems to reduce the equilibrium set to three elements: the monopoly price, the competitive price, and ϵ above the competitive price. The problem is that these last two are also equilibria of the one-shot game. We therefore could not use the lemmata proved in the early sections since the hypothesis that there be only one equilibrium in the one-shot game is not met. Consequently, we do not pursue this further in the current paper.

6 Conclusions

In this paper, we show the usefulness of probabilistic cheap talk as a simple mechanism for resolving a variety of market and coordination failures. A possible criticism of our approach is its requirement that there be an infinite horizon with no bound on the stopping time (we emphasize that although the horizon is infinite, the stopping time is finite with certainty). The same criticism would apply to infinite horizon repeated games. In this regard, we are influenced by [**1**], and, more recently, [**7**]. We offer the following passage of Aumann, quoted in Rubinstein as commentary:

> in the notion of a supergame that will be used in this paper, each superplay consists of an infinite number of plays of the original game G. On the face of it, this would seem to be unrealistic, but actually it is more realistic than the notion in which each superplay consists of a fixed finite (large) number of plays of G. ... Of course when looked at in the large, nobody really expects an infinite number of plays to take place; on the other hand, after each play we do expect that there will be more. A. Tucker has

pointed out that this condition is mathematically equivalent to an infinite sequence of plays, so that is what our notion of supergames will consist of.

Rubinstein continues:

> "By using infinite horizon games we do not assume that the real world is infinite. Taking the view that models are not supposed to be isomorphic with reality, I see the infinitely repeated game model as a tool for analyzing situations where players examine a long term situation without assigning a specific status to the end of the world."

Though the previous observations are made in the context of *repeated* games, Aumann's reference to the uncertainty about the possibility of future play translates here into uncertainty about the binding nature of current strategic choices. The "end of the world" in Rubinstein's comment might easily refer to the actual determination of payoffs. The key element in our model which justifies the infinite horizon is the fact that at any given round of strategy choice agents are aware that they operate in an uncertain world, and due to the exogenous circumstances that determine the arrival of payoffs, they may still have the opportunity to revise their actions.

A second issue involves the partial stationarity restriction on the set of equilibrium strategies. Despite our assertion in the introduction that stationarity is compelling in games with PCT, one may wonder what the implications of nonstationary equilibria would be. We conjecture that the unique implementation results achieved herein would no longer hold. This is based on analysis of the behavior of the set of equilibrium payoffs as the discount factor varies in [**8**], and [**9**] for the infinitely repeated prisoners' dilemma. In general, there is no value of the discount factor for which the set of Pareto efficient equilibria is a singleton. Hence, we would expect a similar non-uniqueness to extend to PCT games as well. See [**5**] for more discussion.

Acknowledgements

We are grateful to the participants of the "Incentives, Implementation, and Applications to Industrial Organization" workshop at SUNY, Stony Brook.

References

[1] Aumann, R. [1959], *Acceptable Points in General Cooperative N-Person Games*, in Contributions to the Theory of Games IV (R. D. Luce and A. W. Tucher, eds.), Princeton University Press, Princeton NJ.

[2] Bernheim, D., Peleg, B. and Whinston, M. [1987], *Coalition-Proof Equilibrium*, Journal of Economic Theory, pp. 1-29.

[3] Chakravorti, B., Conley, J. and Taub, B. [1993], *On Resolving the One-shot Prisoners' Dilemma through Probabilistic Cheap Talk*, Mimeo, University of Illinois and Bellcore.

[4] Chakravorti, B., Conley, J. and Taub, B. [1996], *On Uniquely Implementing Cooperation in the One-Shot Prisoners' Dilemma*, Economic Theory **8**, pp. 347-366.

[5] Chakravorti, B., Conley, J. and Taub, B. [1997], *On Uniquely Implementing Cooperation in the Prisoner's Dilemma: Corrigendum*, Economic Theory **9**, pp. 377-378.

[6] Jackson, M, and Moulin, H. [1992], *Implementing a Public Project and Distributing Its Cost*, Journal of Economic Theory, pp. 125-140.

[7] Rubinstein, A. [1992], *Comments on the Interpretation of Repeated Games*, Advances in Economic Theory (J. J. Laffont, ed.), Vol. 1, Cambridge University Press, Cambridge.

[8] Stahl, D. [1991], *The Graph of the Prisoner's Dilemma Supergame Payoffs as a Function of the Discount Factor*, Games and Economic Behavior **3**, pp. 368-384.
[9] van Damme, E. [1991], *Stability and Perfection of Nash Equilibria*, Springer-Verlag, Berlin, New York.

Fields Institute Communications
Volume **23**, 1999

Multilaterally Strategy-Proof Mechanisms in Random Aumann–Hildenbrand Macroeconomies

Peter J. Hammond
Department of Economics
Stanford University
Stanford, CA 94305-6072 USA
hammond@leland.stanford.edu

Abstract. By definition, multilaterally strategy-proof mechanisms are immune to manipulation not only by individuals misrepresenting their preferences, but also by finite coalitions exchanging tradeable goods on the side. Continuum economies are defined in which both agents' identifiers and their privately known characteristics are joint i.i.d. random variables. For such economies, conditions are given for multilateral strategy-proofness to imply decentralization by a budget constraint with linear prices for tradeable goods and lump-sum transfers independent of individual characteristics. Also, adapting Aumann's key proof [**7**] avoids using Lyapunov's theorem or its corollary, Richter's theorem on integrating a correspondence w.r.t. a non-atomic measure.

1 Introduction and outline

A macroeconomy is one that is adequately described by statistical aggregates. Usually macroeconomists look at the means of important variables like income, hours worked, etc. But other moments of the relevant distribution are also important. For example, when different income groups have different marginal propensities to spend on various goods, aggregate demand is influenced by the distribution of income between those groups. Accordingly, the obvious step for an economic theorist to take is to consider the whole distribution. This suggests examining economies with a continuum of agents, i.e., one in which the usual finite set of agents is replaced by a non-atomic measure space.

The earliest paper I recall seeing that uses a continuum of economic agents explicitly is by Vickrey [**58**], in his discussion of the limits to redistribution. Much later, Mirrlees [**51**] used a model like Vickrey's with a one-dimensional continuum of skill levels in order to formulate and solve a problem involving optimal non-linear income taxation. There have been many successors, of course. But the first general equilibrium model of a continuum economy is due to Aumann [**7, 10**]. Such models can be motivated using the words of Aumann himself ([**7**], p. 41):

1991 *Mathematics Subject Classification.* 90A36, 90A14.

> The idea of a continuum of traders ... is no stranger than ... a continuum of "particles" in fluid mechanics. ... [T]he continuum can be considered as an approximation to the "true" situation in which there is a large but finite number of particles .... The purpose of adopting the continuous approximation is to make available the powerful and elegant methods of the branch of mathematics called "analysis," in a situation where treatment by finite methods would be much more difficult or even hopeless (think of trying to do fluid mechanics by solving n-body problems for large n).
>
> There is perhaps a certain psychological difference between a fluid with a continuum of particles and a market with a continuum of traders. Though we are intellectually convinced that a fluid contains only finitely many particles, to the naked eye it still looks quite continuous. The economic structure of a shopping center, on the other hand, does not look continuous at all. But, *for the economic policy maker in Washington, or for any professional macroeconomist*, there is no such difference. He works with figures that are summarized for geographic regions, different industries, and so on; the individual consumer (or merchant) is as anonymous to him as the individual molecule is to the physicist.
>
> (emphasis added)

Indeed, continuum economies have come to play a central role in much recent work in macroeconomics, as well as in general equilibrium theory. Continuum economies capture very well the intuitive idea that individuals have negligible power to affect prices. They also represent how individuals acting alone cannot control the levels of "widespread externalities" like carbon dioxide in the atmosphere, or those associated with learning by doing in models of "endogenous" growth.[1]

Aumann used his general equilibrium model in order to prove first core equivalence and then later the existence of Walrasian equilibrium. These results were soon extended in several articles by Hildenbrand and others (see especially [**40**]). In addition, following an idea he ascribes to Mertens [**50**], Hildenbrand also introduced a metric on the space of preferences. However, Hildenbrand's earlier work [**39**] on finite random economies suggests what may be the most satisfactory interpretation of a continuum economy — as the statistical limit of a sequence of large finite economies in which individuals' identifiers, characteristics, and also their corresponding allocations are all sampled independently from an appropriate common joint probability distribution. By an application of the law of large numbers — or, more precisely, of Glivenko–Cantelli's theorem in the form set out by Parthasarathy ([**52**], Theorem 7.1, p. 53) — the sample distributions converge almost surely to their values in the continuum economy.[2]

In general equilibrium theory, two well known and several less well known results are true under much weaker convexity assumptions about individuals in

[1] For work on general widespread externalities, see for example [**44, 45, 46**], [**34**], and [**31, 32**]. For some of the many recently published surveys and general discussions of endogenous growth theory, see for example [**54**], [**57**], [**38**], [**47**], [**1**], [**12**], and [**24**], as well as in [**35**].

[2] Actually, as Walter Trockel has kindly reminded me, the continuum economy model is just one particular way of representing an economy with many participants. One prominent alternative is the non-standard model with a hyperfinite set of agents. This was pioneered by Brown and Robinson in [**15, 16**] (see also [**3**]). A less well known alternative with finitely additive measures is due to [**60**].

economies with many agents. These include results on the existence of Walrasian equilibrium and the characterization of Pareto efficient allocations as Walrasian equilibria with lump-sum redistribution of wealth. Neither of these results requires preferences to be convex. Typically, both use proofs relying on Richter's theorem [**53**] which states that integrating w.r.t. a non-atomic measure any measurable correspondence with values in $\Re^n$ produces a convex set.[3] Convexity of individual feasible sets, however, still helps avoid difficulties as in Arrow's [**5**] "exceptional case" (see [**30**]).

Aumann's core equivalence theorem, however, falls into a different class of results that, except for economies with many agents, are simply not true in general. Moreover, unlike the existence and second efficiency theorems mentioned above, Aumann's proof [**7**] of core equivalence might seem at first to avoid Lyapunov's theorem. However, it does rely on the range of a non-atomic measure being an interval of the real line; at this step of essentially the same proof, Hildenbrand ([**40**], p. 144; [**41**], p. 844) explicitly invokes Lyapunov's theorem "in one dimension, the proof of which is much easier".[4] In [**40**], on the other hand, Aumann's proof is left as an exercise; in the main text (p. 133), Hildenbrand uses Richter's theorem. Other results similar to core equivalence include [**49**] discussion of the bargaining set for a continuum economy, and Aumann and Shapley's [**11**] treatment of values in non-atomic cooperative games without transferable utility.

For Aumann, as well as in later work on the core of a continuum economy that immediately succeeded his pioneering contribution, coalitions were always sets of positive measure, even if this measure might be arbitrarily small as in [**21**], [**56**] and [**59**]. In the 1980s, however, Mamoru Kaneko, Myrna Wooders and I all became interested in what finite coalitions could achieve in continuum economies. Of course, because a continuum economy involves a non-atomic measure, any finite coalition has zero measure — i.e., it is a null set.

In fact, [**44**] is the first paper on what came to be called the "f-core." Using a related idea, Hammond in [**27**] improved some very preliminary ideas found in [**26**], which had been taken further by Gale in [**19, 20**] and Guesnerie in [**22, 23**]. These concerned strategy-proof allocation mechanisms in a continuum economy which are also immune to manipulation by finite coalitions exchanging on an unofficial or hidden market the goods allocated to them by the official mechanism. Similar results were applied in different contexts by [**42**], [**14**], [**36**], and [**37**].

In the work on multilateral strategy-proofness reported in [**27**] and also in our joint paper [**34**] on an equivalence theorem for the f-core with widespread externalities, the main results were proved using Richter's theorem. It has since become clear to me that both proofs can be made considerably simpler by following Aumann's original approach [**7**]. Moreover, because only finite coalitions are being considered,

---

[3]In symbols, Richter's theorem says that $\int_T F(t)\, d\tau$ is a convex set whenever $F : T \twoheadrightarrow \Re^n$ is a correspondence with a measurable graph and τ is a non-atomic measure. As Hildenbrand's elegant and concise proof ([**40**], Theorem 3, p. 62) makes clear, Richter's theorem is really a corollary of the well known theorem of Lyapunov [**48**] stating that the range of a finite-dimensional vector-valued non-atomic measure is a convex set Note too that Richter's theorem is different from what Clarke in [**17**] calls "Aumann's theorem" stating that the integral of the convex hull of an integrably bounded, measurable and non-empty valued correspondence $F : T \twoheadrightarrow \Re^n$ is identical to the integral of the correspondence — or in symbols, that $\int_T F(t)\, d\tau = \int_T \operatorname{co} F(t)\, d\tau$. Aumann in [**9**] proved this using Lyapunov's theorem. See also [**6**]. Clarke in [**17**] provides a proof of Aumann's result without using Lyapunov's theorem.

[4]Actually, sometimes a measure on a space of one dimension is defined to be non-atomic iff Lyapunov's theorem is true.

one can avoid using Richter's theorem or even Lyapunov's theorem in one dimension. There is a small price to pay, however: the conventional assumption of local non-satiation must be strengthened so that it becomes local non-satiation among suitable vectors with rational coordinates — an assumption that is automatically satisfied if preferences are monotone, for instance.

For the case of the f-core equivalence theorem in an economy with rather general widespread externalities, such a proof is included in [**33**]. The present paper concentrates instead on the multilateral incentive compatibility or strategy-proofness result of [**27**], and extends the key lemma in that paper. In particular, mechanisms that are not anonymous or symmetric can now be considered. Also, a subtle difficulty raised Guesnerie in [**23**] is overcome by considering a random continuum economy. Moreover, some of the earlier assumptions are replaced by a dispersion assumption along the lines suggested by Yamazaki in [**61, 62**]. Finally, exceptional null sets of agents are dealt with more satisfactorily than in [**26, 27**].

In what follows, Section 2 sets out the model of a random economy with a continuum of agents whose identifiers and privately known characteristics are pairs of independently and identically distributed random variables with an unknown joint distribution. For such random continuum economies, Section 3 defines a multilaterally strategy-proof mechanism which is immune to manipulation even when finite coalitions can exchange tradeable goods unofficially on the side. Then Sections 4 and 5 discuss the conditions under which such mechanisms are decentralizable by a budget constraint with linear prices for tradeable goods, but non-linear pricing of non-tradeable goods.

2 Random continuum economies

Let T be a finite set of tradeable goods, and $\Re^T$ the corresponding Euclidean space of net trade vectors with a typical member denoted by x or t. In addition, suppose that allocations of goods which individuals cannot freely exchange are described by the points z of an abstract locally compact and separable metric space Z. The relevant commodity or goods space is therefore the locally compact and separable product metric space $G = \Re^T \times Z$, of which the typical member is (x, z) where $x \in \Re^T$ and $z \in Z$.

The set of agents is assumed to be the measure space $(I, \mathcal{B}, \mu)$. Here I is the interval $[0, 1]$ of the real line, whose members $i \in I$ are individual *identifiers*. Then $\mathcal{B}$ is the real Borel σ-field, and μ is Lebesgue measure.

Let Θ be the domain of possible individual *characteristics*. Each typical member $\theta \in \Theta$ determines a non-empty closed *feasible set* $F(\theta) \subset G$ of net trade vectors, together with a (reflexive, complete and transitive binary weak) preference ordering $\succsim_\theta$ on $F(\theta)$ which is *continuous* in the sense of having a closed graph in $G \times G = G^2$ given by

$$\text{Graph}\,(\succsim_\theta) := \{\, \big((x, z), (x', z')\big) \in F(\theta) \times F(\theta) \mid (x, z) \succsim_\theta (x', z') \,\}.$$

Let $\succ_\theta$ be the strict preference relation corresponding to $\succsim_\theta$. Note that, because $\succsim_\theta$ is reflexive, once the graph of $\succsim_\theta$ is known, so is the feasible set, which must be

$$F(\theta) = \{\, (x, z) \in G \mid \big((x, z), (x, z)\big) \in \text{Graph}\,(\succsim_\theta) \,\}.$$

Following Hildenbrand in [**40**], who credits Mertens [**50**] with the idea, the space Θ of (non-empty) closed subsets of G^2 representing possible consumer characteristics will be given the *closed convergence* topology. This corresponds to the metric ρ on the space Θ defined by $\rho(E,F) = d_H(E \cup \{\infty\}, F \cup \{\infty\})$ for all non-empty closed subsets $E, F \subset G^2$, where d_H denotes the Hausdorff metric, and ∞ denotes an added point at infinity which makes G^2 compact.[5] Because the space G is locally compact and separable, it follows from ([**40**], Theorem 2, p. 19) that Θ is compact. Also, give the product space $I \times \Theta$ any reasonable metric such as $d\big((h,\theta),(i,\eta)\big) := |h - i| + d_H(\theta,\eta)$. Because the spaces I and G are respectively compact and locally compact $I \times \Theta$ is a compact and so separable metric space whose Borel σ-field is equal to the product of the Borel σ-fields for the two component metric spaces.

Each individual $i \in I$ is assumed to have a range $\Theta_i \subset \Theta$ of possible characteristics. Furthermore, the correspondence $i \mapsto\mapsto \Theta_i$ is assumed to have a graph

$$A := \{\, (i,\theta) \in I \times \Theta \mid \theta \in \Theta_i \,\}$$

which is Borel measurable in the product space $I \times \Theta$. Let $\mathcal{A}$ denote the restriction to the measurable set A of the Borel σ-field on $I \times \Theta$. Let $\mathcal{M}(A)$ denote the set of probability measures on the measurable space $(A, \mathcal{A})$. Give $\mathcal{M}(A)$ the topology of weak convergence.

The above formulation has been chosen because it allows a number of interesting different cases of private information, such as when agent i's feasible set F_i is known but i's preference ordering $\succsim_i$ is not. It even allows for the complete absence of any relevant private information when both F_i and $\succsim_i$ are completely known and so Θ_i is a singleton.

Following [**7, 8**] or more especially [**40**], it has become standard practice to define a continuum economy as a measurable mapping $\mathcal{E} : I \to \Theta$ or $i \mapsto \theta_i$ from agents' identifiers to their characteristics. The measurability property appears to be unnecessary, however, as well as rather artificial. Worse, Lusin's theorem states that, except on a set of arbitrarily small measure, any measurable function on a compact domain is equal to some continuous function (see, for instance, [**25**], pp. 242-3, or [**55**], p. 72, Problem 31). More precisely, if $i \mapsto \theta_i$ is measurable, then for every $\epsilon > 0$ there exists a measurable set $I_\epsilon \subset I$ with $\mu(I_\epsilon) < \epsilon$ and a continuous function $i \mapsto \theta_i^\epsilon$ such that $\theta_i^\epsilon = \theta_i$ for all $i \in I \setminus I_\epsilon$. Thus, any measurable $i \mapsto \theta_i$ can be approximated rather well by a sequence of continuous mappings. Moreover, if $i \mapsto \theta_i$ is continuous, then any particular individual h's characteristic θ_h can be inferred from knowledge of θ_i for all $i \in N_h \setminus \{h\}$, where N_h is any neighbourhood of h. As Guesnerie in ([**23**], Section 1.6) points out, this implies that some otherwise unintuitive or rather unappealing first-best mechanisms may be incentive compatible. For example, suppose that the announcements of θ_i for $i \in N_h \setminus \{h\}$ are such that $\theta_h^* := \lim_{h \to i} \theta_i$ exists. Then the mechanism might ignore entirely whatever θ_h is announced by h and simply generate whatever allocation would be appropriate if h's characteristic really were θ_h^*. After all, if the mapping $i \mapsto \theta_i$ really is continuous at $i = h$, and if all individuals $i \in N_h \setminus \{h\}$ are announcing their true characteristics, then θ_h^* must be h's true characteristic. Also, if necessary a severe punishment can be used to deter individuals from announcing a characteristic $\theta_h \neq \theta_h^*$. For such a mechanism, then, truthful revelation by almost all agents is at least a Nash equilibrium whenever $i \mapsto \theta_i$ is continuous. To avoid such awkward possibilities, I shall instead consider random economies in which all individuals' characteristics are i.i.d. random variables. Then, of course, knowing θ_i for all $i \in N_h \setminus \{h\}$ gives no information about θ_h.

---

[5] More formally, $G^2 \cup \{\infty\}$ is the Alexandroff one-point compactification of G^2. For interesting criticism of the closed convergence topology, see ([**4**], p. 448) and Manelli's work cited there.

Accordingly, a *random continuum economy* E is defined as a probability measure $\alpha \in \mathcal{M}(A)$ on the measurable set $A \subset I \times \Theta$ whose marginal distribution $\mathrm{marg}_I \alpha$ on $I = [0, 1]$ is assumed to be uniform — i.e., equal to the Lebesgue measure μ. Let $\mathcal{E}$ be the domain of all such random continuum economies. The intended interpretation is that any $\alpha \in \mathcal{E}$ is the limit as $n \to \infty$ of a finite sample of n individuals whose identifiers and characteristics are drawn independently from the set A according to the joint probability measure α. For $n = 1, 2, \ldots$, let α_n denote the random empirical distribution from a sample of n agents. Because $I \times \Theta$ is a separable metric space, one can apply the Glivenko–Cantelli theorem in the form due to ([**52**], Theorem 7.1, p. 53), as cited also by ([**40**], pp. 52–3). This result is a suitable law of large numbers ensuring that, almost surely as $n \to \infty$, the sequence of measures α_n converges weakly to α.

Note that this formulation of a random economy differs from that due to [**39**] and [**13**] because they assume a fixed set of agents, independent of the state of the world. Here, by contrast, in any approximation to the continuum economy, the set of agent identifiers appearing in the finite random sample is itself random.

As discussed in [**8**], [**18**], [**43**] and [**2**], a continuum of independent random variables θ_i $(i \in I)$ almost never produces a measurable mapping from I to Θ. As an illustration, suppose that each $i \in I$ consists of a (nine-digit) U.S. Social Security number multiplied by 10^{-9}, and that $\theta_i \in \Re_+$ indicates the height of the person with number i. Then the mapping $i \mapsto \theta_i$ surely has a very messy graph, thus suggesting that it would not be measurable in the limit as $n \to \infty$.

For this reason, any realization of a random continuum economy almost surely does not meet the measurability requirements imposed by Aumann or Hildenbrand. This creates no difficulties, however, because there will never be a need to integrate any mapping whose domain is I instead of A; for the latter domain, all the functions that need to be considered will be measurable. In particular, for the example considered above, the mean height of people with U.S. Social Security numbers can be calculated by integrating w.r.t. the joint distribution α the (measurable) projection mapping from $A \subset I \times \Theta$ onto Θ. Equivalently, one could integrate w.r.t. the marginal measure $\mathrm{marg}_\Theta \alpha$ on Θ. Of course, this marginal measure describes the distribution of heights precisely.

3 Multilaterally strategy-proof allocation mechanisms

An *allocation mechanism* $(\mathbf{x}, \mathbf{z})$ is a pair of mappings $\mathbf{x} : \mathcal{E} \times A \to \Re^T$ and $\mathbf{z} : \mathcal{E} \times A \to Z$ with values denoted by $x_i(\alpha, \theta)$ and $z_i(\alpha, \theta)$ respectively which, for all fixed $\alpha \in \mathcal{E}$, are measurable w.r.t. (i, θ). Note that the distribution α is treated as a variable of the allocation mechanism. This reflects the obvious requirement that an economic system should function and produce a feasible allocation even when α can only be discovered by inducing individuals to reveal their privately known true characteristics either directly or indirectly.

Overall feasibility typically requires that the mechanism should satisfy an additional *resource balance* constraint. In the case when z represents a vector of private goods that individuals cannot exchange on the side, Z should be a normed linear space. In this case, the mean net trade vector per head is given by $y(\alpha) := \int_A (x_i(\alpha, \theta), z_i(\alpha, \theta)) d\alpha$. Then the resource balance constraint would be $y(\alpha) = 0$, or alternatively $y(\alpha) \leqq 0$ if free disposal of every good were possible. On the other hand, if each point z represents a collection of public goods, then overall feasibility would require that $z_i(\alpha, \theta) = z(\alpha)$ for α-almost all $(i, \theta) \in A$, and that

$\int_A x_i(\alpha, \theta)\, d\alpha = x(\alpha)$ where $(x(\alpha), z(\alpha)) \in Y$ for some known production possibility set $Y \subset G$. In this case, the space Z does not need a linear structure. No such resource balance constraint plays any role in the subsequent analysis of this paper, however, because only individual feasibility and incentive constraints will be considered. For this reason, say that the mechanism $(\mathbf{x}, \mathbf{z})$ is *feasible* provided that, for all $\alpha \in \mathcal{E}$, one has $(x_i(\alpha,\theta), z_i(\alpha,\theta)) \in F(\theta)$ for α-almost all $(i, \theta) \in A$.

Individual incentive constraints arise because individuals are privately informed about their characteristics. The constraints require that no individual i with true characteristic θ acting alone can benefit from manipulating the mechanism by simulating how i would be expected to behave with characteristic η instead of θ. Hence, for each distribution $\alpha \in \mathcal{E}$ and combination (i, θ, η) with $i \in I$ and $\theta, \eta \in \Theta_i$, the *individual incentive constraint*

$$(x_i(\alpha, \theta), z_i(\alpha, \theta)) \succsim_\theta (x_i(\alpha, \eta), z_i(\alpha, \eta)) \tag{3.1}$$

must be satisfied. Note how (3.1) reflects the fact that one individual acting alone in a continuum economy is powerless to change the apparent distribution α. In fact, these incentive constraints are natural adaptations of those that apply to the anonymous or symmetric mechanisms considered in [**26, 27**], [**19, 20**], and [**22, 23**].

For manipulation to be feasible, it should be true that $(x_i(\alpha, \eta), z_i(\alpha, \eta)) \in F(\theta)$. Otherwise individual i would be unable to carry out the expected net trade, in which case the presumption is that i will be caught in time and required to behave in some other way that does allow the mechanism to allocate a feasible net trade vector. If i does not get caught in time, however, there may still be scope for manipulation unless additional incentive constraints are satisfied, such as those discussed in [**29**]. Of course, the difficulties surrounding this extra restriction disappear if $F(\theta) = F_i$ for all $\theta \in \Theta_i$ because i's feasible set of net trades is known to be F_i.

The ensuing analysis will make use of the notation

$$A(\theta) := \{ (i, \eta) \in A \mid (x_i(\alpha, \eta), z_i(\alpha, \eta)) \in F(\theta) \}$$

for the set of pairs (i, η) which any agent i with true characteristic θ can credibly claim without violating individual feasibility.

After these preliminaries, the allocation mechanism $(\mathbf{x}, \mathbf{z})$ is said to be *almost strategy-proof* provided that, for all fixed $\alpha \in \mathcal{E}$, for α-almost all pairs $(i, \theta) \in A$, and for α-almost all $(i, \eta) \in A(\theta)$, the incentive constraint (3.1) is satisfied. Hence, there must exist a *full* measurable set $A^1_\alpha \subset A$ with $\alpha(A^1_\alpha) = 1$ such that (3.1) is true for all (i, θ, η) satisfying $(i, \theta) \in A^1_\alpha$ and $(i, \eta) \in A(\theta)$.

The above definition is really too weak, however, because a non-null set of potential agents $(i, \theta) \in A^1_\alpha$ might still be able to manipulate by behaving as though they have a characteristic η satisfying $(i, \eta) \in A \setminus A^1_\alpha$. For this reason, say that $(\mathbf{x}, \mathbf{z})$ is *strategy-proof* provided that it is almost strategy-proof, but in addition (3.1) is true for *all* $(i, \eta) \in A(\theta)$ without exception, not just for α-almost all such (i, η). More precisely, (3.1) must be true for all $(\alpha, i, \theta, \eta) \in \mathcal{E} \times I \times \Theta \times \Theta$ which, for some full set $A^1_\alpha \subset A$, satisfy $(i, \theta) \in A^1_\alpha$ and $(i, \eta) \in A(\theta)$. Note that a mechanism is almost strategy-proof if it is equal to a strategy-proof mechanism for α-almost all $(i, \theta) \in A$.

Also, for the game of direct revelation corresponding to the mechanism in which each individual $i \in I$ has a strategy set Θ_i, strategy-proofness implies that, for all fixed $\alpha \in \mathcal{E}$, almost every individual has truthfulness as at least a weakly dominant strategy; almost strategy-proofness only implies that almost every individual has truthfulness as an "almost" weakly dominant strategy, in some obvious sense.

Next, a *potential finite coalition* is defined as a finite set C of pairs $(i, \theta) \in A$ with the property that each individual $i \in I$ features at most once in C. Thus, a potential finite coalition C is equivalent to a particular finite subset $I_C \subset I$ of individuals i whose respective characteristics $\theta_i \in \Theta_i$ are such that $C = \{ (i, \theta_i) \mid i \in I_C \}$. The idea is that the potential finite coalition could actually form whenever the individuals $i \in I_C$ all happen to have exactly the right characteristics.

Given the joint distribution $\alpha \in \mathcal{E}$, say that the potential finite coalition $C \subset A$ can *manipulate* the allocation mechanism $(\mathbf{x}, \mathbf{z})$ provided that, when all individuals $i \in I_C$ have respective characteristics $\theta_i \in \Theta_i$ satisfying $(i, \theta_i) \in C$, they can jointly find "manipulative" characteristics $\eta_i \in \Theta_i$ and net trade vectors $t_i \in \Re^T$ of tradeable goods satisfying $\sum_{i \in I_C} t_i = 0$ with the property that both $(x_i(\alpha, \eta_i) + t_i, z_i(\alpha, \eta_i)) \in F(\theta_i)$ and

$$(x_i(\alpha, \eta_i) + t_i, z_i(\alpha, \eta_i)) \succ_{\theta_i} (x_i(\alpha, \theta_i), z_i(\alpha, \theta_i)). \tag{3.2}$$

Note that in the special case when $\#C = 1$, or when there are no tradeable goods $(T = \emptyset)$ — or even only one $(\#T = 1)$ provided that individuals prefer more of it to less — then (3.2) holds iff the individual incentive constraint (3.1) is violated. As Guesnerie in [**22, 23**] and I have both shown, outside these special cases, multilateral strategy-proofness is considerably more demanding than ordinary individual strategy-proofness. Gale in [**19, 20**] also has similar but less general results.

Of course one is only interested in manipulation by potential finite coalitions who have a positive probability of being able to form. In particular, potential manipulation by finite coalitions does not matter if one could eliminate any possibility of it by removing a null set of agents. Accordingly, the allocation mechanism $(\mathbf{x}, \mathbf{z})$ is said to be *multilaterally strategy-proof* provided that, for any fixed $\alpha \in \mathcal{E}$, there exists a full measurable set $A^1_\alpha \subset A$ with $\alpha(A^1_\alpha) = 1$ such that no potential finite coalition $C \subset A^1_\alpha$ can manipulate the mechanism. Hence, if a mechanism is not multilaterally strategy-proof, it is because any full set $A^1_\alpha \subset A$ contains at least one potential finite coalition with the power to manipulate; excluding any null set from A cannot remove all such manipulative potential coalitions. Evidently, a mechanism is multilaterally strategy-proof only if it is (individually) strategy-proof.

If $\Theta_i = \{\theta_i\}$ for all $i \in I$ and so the economy has no private information, then the above definition reduces to a condition called "f-Pareto efficiency" in [**32**]. When in addition $G = \Re^T$, in effect, because there are only traded goods, the definition in [**34**] and in [**32**] of an f-core allocation $\mathbf{x}$ for the special case of an economy without widespread externalities can be obtained by replacing (3.2) with $t_i \succ_{\theta_i} x_i(\alpha, \theta_i)$ for all $i \in C$.

A rather weaker notion of multilateral strategy-proofness arises if one insists that the set of agents who belong to manipulative potential coalitions is more evidently of positive measure. Specifically, say that a finite family A_k $(k \in K)$ of measurable non-null subsets of A, not necessarily pairwise disjoint, can *discernibly manipulate* the allocation mechanism $(\mathbf{x}, \mathbf{z})$ provided that there exist corresponding natural numbers n_k $(k \in K)$ such that manipulation is possible by every potential finite coalition C which, for each $k \in K$, contains exactly n_k different individuals $i \in I$ with appropriate characteristics θ_i such that $(i, \theta_i) \in A_k$. And say that the allocation mechanism $(\mathbf{x}, \mathbf{z})$ is *weakly multilaterally strategy-proof* if there is no such finite family that can discernibly manipulate.

4 Budget decentralizations with linear prices for tradeable goods

A *budget correspondence* $B : I \times \mathcal{E} \twoheadrightarrow G$ specifies the budget set $B_i(\alpha)$ of each agent $i \in I$ as a function of the distribution α. But $B_i(\alpha)$ must be independent of agent i's own characteristic $\theta_i \in \Theta_i$. Such a correspondence *almost decentralizes* the allocation mechanism $(\mathbf{x}, \mathbf{z})$ provided that for all $\alpha \in \mathcal{E}$ and α-almost all $(i, \theta) \in A$, one has $(x_i(\alpha, \theta), z_i(\alpha, \theta)) \in B_i(\alpha)$ and

$$(x, z) \in B_i(\alpha) \cap F(\theta) \implies (x_i(\alpha, \theta), z_i(\alpha, \theta)) \succsim_\theta (x, z). \tag{4.1}$$

In other words, for all (i, θ) in some full set A^1_α with $\alpha(A^1_\alpha) = 1$, the pair of vectors $(x_i(\alpha, \theta), z_i(\alpha, \theta))$ must together maximize the preference ordering $\succsim_\theta$ w.r.t. (x, z) subject to the budget constraint $(x, z) \in B_i(\alpha)$ combined with the individual feasibility constraint $(x, z) \in F(\theta)$.

Second, the budget correspondence B *decentralizes* the mechanism $(\mathbf{x}, \mathbf{z})$ provided that it almost decentralizes $(\mathbf{x}, \mathbf{z})$, but in addition $(x_i(\alpha, \theta), z_i(\alpha, \theta)) \in B_i(\alpha)$ for all $\alpha \in \mathcal{E}$ and for *all* $(i, \theta) \in A$ without exception; it is still possible to have a null subset of pairs $(i, \theta) \in A$ for which $(x_i(\alpha, \theta), z_i(\alpha, \theta))$ is not the best point of $B_i(\alpha) \cap F(\theta)$.

Following **[26]**, it is obvious that if a mechanism is almost decentralizable then it must be almost strategy-proof, and if it is decentralizable then it must be strategy-proof. Conversely, by choosing the decentralization

$$B_i(\alpha) := \{\, (x, z) \in G \mid \exists \theta \in \Theta : (x, z) = (x_i(\alpha, \theta), z_i(\alpha, \theta)) \,\}$$

it is easy to see that a mechanism is almost decentralizable if it is almost strategy-proof, and decentralizable if it is strategy-proof.

Next, the budget correspondence $B : I \times \mathcal{E} \twoheadrightarrow G$ is said to have *linear prices for tradeable goods* provided that there exists a price mapping $p : \mathcal{E} \to \Re^T \setminus \{0\}$ and a (generally non-linear) *allowable expenditure function* $m : I \times \mathcal{E} \times Z \to \Re$ such that

$$B_i(\alpha) = \{\, (x, z) \in G \mid p(\alpha)\, x \leq m_i(\alpha, z) \,\}. \tag{4.2}$$

Notice that $m_i(\alpha, z)$ specifies how much agent i is allowed to spend on tradeable goods as a function of the distribution α and the non-tradeable good vector z chosen by the agent. Thus, $m_i(\alpha, z)$ reflects any non-linear pricing schedule that applies to collections of non-tradeable goods in the space Z.

If the allocation mechanism $(\mathbf{x}, \mathbf{z})$ is decentralizable by a budget correspondence $B_i(\alpha)$ satisfying (4.2) everywhere, then for almost all $(i, \theta) \in A$ and for all $(x, z) \in F(\theta)$, one has

$$(x, z) \succ_\theta (x_i(\alpha, \theta), z_i(\alpha, \theta)) \implies p(\alpha)\, x > m_i(\alpha, z). \tag{4.3}$$

On the other hand, if the allocation mechanism $(\mathbf{x}, \mathbf{z})$ is decentralizable or almost decentralizable by such a budget correspondence, then (4.3) is true for all (i, θ) belonging to some set A^1_α of full measure.

For the second lemma below, say that the budget set $B_i(\alpha)$ given by (4.2) is *almost a compensated linear decentralization* provided that for all $\alpha \in \mathcal{E}$ and α-almost all $(i, \theta) \in A$, one has $(x_i(\alpha, \theta), z_i(\alpha, \theta)) \in B_i(\alpha)$ and

$$(x, z) \succsim_\theta (x_i(\alpha, \theta), z_i(\alpha, \theta)) \implies p(\alpha)\, x \geq m_i(\alpha, z)$$

whenever $(x, z) \in F(\theta)$. Also, say that $B_i(\alpha)$ is a *compensated linear decentralization* provided that it is almost a compensated linear decentralization with $(x_i(\alpha, \theta), z_i(\alpha, \theta)) \in B_i(\alpha)$ for all $\alpha \in \mathcal{E}$ and all $(i, \theta) \in A$.

Theorem 4.1 First Decentralization Theorem. *An allocation mechanism* $(\mathbf{x}, \mathbf{z})$ *that is decentralizable by a linear budget constraint* $p(\alpha)\, x \leq m_i(\alpha, z)$ *is multilaterally strategy-proof.*

Proof Fix any $\alpha \in \mathcal{E}$. Suppose there is such a decentralization, with (4.1) holding for all $(i, \theta) \in A^1_\alpha$ where $\alpha(A^1_\alpha) = 1$. Suppose also that (3.2) is true for some pair $(i, \theta_i) \in A^1_\alpha$ and characteristic $\eta_i \in \Theta_i$. Then, faced with the budget set $B_i(\alpha)$ given by (4.2), individual i with characteristic η_i must be able to afford $(x_i(\alpha, \eta_i), z_i(\alpha, \eta_i))$, whereas i with characteristic θ_i must be unable to afford $(x_i(\alpha, \eta_i) + t_i, z_i(\alpha, \eta_i))$. Therefore

$$p(\alpha)\,[x_i(\alpha, \eta_i) + t_i] > m_i(\alpha, z_i(\alpha, \eta_i)) \geq p(\alpha)\, x_i(\alpha, \eta_i)$$

implying that $p(\alpha)\, t_i > 0$. Hence, if there exists a potential finite coalition C of pairs $(i, \theta_i) \in A^1_\alpha$ for individuals $i \in I_C$, together with characteristics $\eta_i \in \Theta_i$ and net trade vectors $t_i \in \Re^T$ satisfying (3.2) for each $i \in I_C$, then $\sum_{i \in I_C} p(\alpha)\, t_i > 0$. This excludes the possibility that $\sum_{i \in I_C} t_i = 0$, so no finite coalition $C \subset A^1_\alpha$ can manipulate. Hence, $(\mathbf{x}, \mathbf{z})$ must be multilaterally strategy-proof. □

This first result is similar to Arrow's first efficiency theorem of welfare economics in [**5**], saying that any Walrasian equilibrium allocation is (weakly) Pareto efficient. The theorem even has a similar proof. Its converse would be a second decentralization theorem, saying that any (weakly) multilaterally strategy-proof mechanism is decentralizable by a linear budget constraint. Similarly, Arrow's second efficiency theorem of welfare economics says that any (weakly) Pareto efficient allocation can be decentralized as a Walrasian equilibrium with suitable lump-sum redistribution. In fact the result is not true without several extra assumptions. Moreover, at least for an economy with a continuum of agents, the extra assumptions needed are somewhat similar, as is the proof.

Let $Q \subset \Re$ denote the set of rational numbers. Let Q^T denote the subset of vectors in $\Re^T$ whose coordinates are all rational.

Assumption 1 For each $\theta \in \Theta$, the feasible set $F(\theta)$ and strict preference relation $\succ_\theta$ satisfy *rational local non-satiation in tradeable goods* in the following sense: given any $(x, t, z) \in \Re^T \times \Re^T \times Z$ with $(x + t, z) \in F(\theta)$ and any neighbourhood V of t in $\Re^T$, there exists $t' \in V \cap Q^T$ such that $(x + t', z) \in F(\theta)$ with $(x + t', z) \succ_\theta (x + t, z)$.

Lemma 4.2 Budget Exhaustion. *Suppose that Assumption 1 is satisfied, and that* $B_i(\alpha)$ *given by (4.2) is almost a compensated linear decentralization. Then* $p(\alpha)\, x_i(\alpha, \theta) = m_i(\alpha, z_i(\alpha, \theta))$ *for all* $\alpha \in \mathcal{E}$ *and* α*-almost all* $(i, \theta) \in A$.

Proof Suppose that $\alpha \in \mathcal{E}$ and $p(\alpha)\, x_i(\alpha, \theta) < m_i(\alpha, z_i(\alpha, \theta))$ for some $(i, \theta) \in A$. By Assumption 1, there must then exist $t \in \Re^T$ such that $(x_i(\alpha, \theta) + t, z_i(\alpha, \theta)) \in F(\theta)$ with $(x_i(\alpha, \theta) + t, z_i(\alpha, \theta)) \succ_\theta (x_i(\alpha, \theta), z_i(\alpha, \theta))$ and $p(\alpha)\,[x_i(\alpha, \theta) + t] < m_i(\alpha, z_i(\alpha, \theta))$. But because $B_i(\alpha)$ is almost a compensated linear decentralization, this can be true for at most an α-null set of potential agents $(i, \theta) \in A$. □

Consider a given mechanism $(\mathbf{x}, \mathbf{z})$ and any fixed $\alpha \in \mathcal{E}$. Then, given any $a = (i, \theta) \in A$, define the two sets

$$T_a := \{\, t \in \Re^T \mid \exists \eta \in \Theta_i : (x_i(\alpha, \eta) + t, z_i(\alpha, \eta)) \succ_\theta (x_i(\alpha, \theta), z_i(\alpha, \theta))\} \quad (4.4)$$

$$U_a := \{\, t \in \Re^T \mid \exists \eta \in \Theta_i : (x_i(\alpha, \eta) + t, z_i(\alpha, \eta)) \succsim_\theta (x_i(\alpha, \theta), z_i(\alpha, \theta)\} \quad (4.5)$$

of trade vectors t which can be used by i with characteristic θ to manipulate (or manipulate weakly) the mechanism by behaving like an agent with some suitable characteristic $\eta \in \Theta_i$.

Lemma 4.3 COMPENSATED LINEAR DECENTRALIZATION. *Under the hypothesis that preferences are rationally local non-satiated in tradeable goods, a feasible allocation mechanism* $(\mathbf{x}, \mathbf{z})$ *is weakly multilaterally strategy-proof only if, for each* $\alpha \in \mathcal{E}$, *there exist* $p(\alpha) \in \Re^T \setminus \{0\}$ *and a mapping* $m_i(\alpha, z)$ *such that the budget constraint* $p(\alpha)\, x \leq m_i(\alpha, z)$ *is almost a compensated linear decentralization.*

Proof Because $(\mathbf{x}, \mathbf{z})$ is feasible, it loses no generality to replace A with the measurable subset $\{\, (i, \theta) \in A \mid (x_i(\alpha, \theta), z_i(\alpha, \theta)) \in F(\theta) \,\}$ whose measure is 1. Let $\alpha \in \mathcal{E}$ be any fixed continuum economy. Following the beginning of the proof of core equivalence in ([**7**], p. 45) and ([**41**], pp. 843–4), define for every $t \in Q^T$ the set

$$A(t) := \{\, a \in A \mid t \in T_a \,\}. \tag{4.6}$$

Claim The set $A(t)$ is measurable, for all $t \in Q^T$.

Proof of claim First, because preferences are continuous and the space Θ has been given the topology of closed convergence, the two sets

$$\begin{aligned} S_1 &:= \{\, (\theta, x, z, x', z') \in \Theta \times G \times G \mid (x, z), (x', z') \in F(\theta) \,\} \\ S_2 &:= \{\, (\theta, x, z, x', z') \in \Theta \times G \times G \mid (x, z) \succsim_\theta (x', z') \,\} \end{aligned}$$

are both closed in the relevant product topology. So the set

$$S_3 := S_1 \setminus S_2 = \{\, (\theta, x, z, x', z') \in \Theta \times G \times G \mid (x', z') \succ_\theta (x, z) \,\}$$

is measurable in the corresponding Borel σ-field. Next, define the two sets

$$\begin{aligned} H_1(t) &:= \{\, (h, \eta^1, x^1, z^1) \in A \times G \mid x^1 = x_h(\alpha, \eta^1) + t,\ z^1 = z_h(\alpha, \eta^1) \,\} \\ H_2 &:= \{\, (i, \eta^2, x^2, z^2) \in A \times G \mid x^2 = x_i(\alpha, \eta^2),\ z^2 = z_i(\alpha, \eta^2) \,\}. \end{aligned}$$

Both are clearly measurable in the relevant product σ-field because the allocation mechanism $(\mathbf{x}, \mathbf{z})$ involves measurable mappings $(i, \theta) \mapsto x_i(\alpha, \theta)$ and $(i, \theta) \mapsto z_i(\alpha, \theta)$ whose graphs must therefore be measurable. On the other hand, the set

$$\begin{aligned} E := \{\, (\theta, x, z, x', z', h, \eta^1, x^1, z^1, i, \eta^2, x^2, z^2) \in \Theta \times G \times G \times A \times G \times A \times G \\ \mid \theta = \eta^2,\ x' = x^1,\ z' = z^1,\ x = x^2,\ z = z^2,\ h = i \,\} \end{aligned} \tag{4.7}$$

is evidently closed. Now let $S(t) := S_3 \times H_1(t) \times H_2$, which is also measurable in the relevant product σ-field, as is $S(t) \cap E$. But then

$$\begin{aligned} A(t) = \{\, (i, \theta) \in A \mid \exists (x, z, x', z', h, \eta^1, x^1, z^1, \eta^2, x^2, z^2) \in G \times G \times A \times G \times \Theta \times G : \\ (\theta, x, z, x', z', h, \eta^1, x^1, z^1, i, \eta^2, x^2, z^2) \in S(t) \cap E \,\} \end{aligned} \tag{4.8}$$

which is the projection of $S(t) \cap E$ onto A. Therefore, $A(t)$ must be measurable. □

Having proved the claim, we now prove the lemma. To do so, first define

$$A^* := A \setminus \left[\bigcup \{\, A(t) \mid t \in Q^T,\ \alpha(A(t)) = 0 \,\} \right]. \tag{4.9}$$

Because Q^T is countable and each set $A(t)$ is measurable, the set A^* is also measurable and satisfies $\alpha(A^*) = \alpha(A) = 1$. Finally, define the convex hull

$$K := \mathrm{co} \left[\bigcup \{\, T_a \cap Q^T \mid a \in A^* \,\} \right]. \tag{4.10}$$

Suppose that $0 \in K$. Then there exists a finite collection of m individual–characteristic pairs $a_k = (i_k, \theta_{i_k}) \in A^*$, together with associated rational net trade vectors $\tilde{t}_k \in T_{a_k} \cap Q^T$ $(k = 1, \ldots, m)$ such that, for some positive convex weights $r_k \in \Re_+$, one has

$$\sum_{k=1}^{m} r_k = 1 \quad \text{and} \quad \sum_{k=1}^{m} r_k \, \tilde{t}_k = 0. \tag{4.11}$$

Equations (4.11) can be combined into the single matrix equation $\mathbf{T}\,\mathbf{r} = \mathbf{e}_1$, where $\mathbf{T}$ is a $(\#T + 1) \times m$ matrix, $\mathbf{r}$ is an m-vector, and $\mathbf{e}_1$ is the first unit vector in $\Re \times \Re^T$. This equation has a solution $\mathbf{r} \gg 0$. Because each $\tilde{t}_k \in Q^T$ has rational co-ordinates, the matrix $\mathbf{T}$ has rational elements. Now, if a positive solution exists, it can be found by Gaussian elimination or some other sequence of elementary row operations involving addition, subtraction, multiplication and division. Hence, there is at least one strictly positive solution $\mathbf{r}$ with only rational co-ordinates, which therefore involves *rational* positive convex weights $r_k \in Q_+$ $(k = 1, \ldots, m)$. After multiplying by a large enough common multiple of the denominators in the rational numbers r_k, it follows that there exist natural numbers n_k $(k = 1, \ldots, m)$ for which $\sum_{k=1}^{m} n_k \, \tilde{t}_k = 0$.

Because $a_k \in A^*$, there is no $t \in Q^T$ such that $a_k \in A(t)$ and $\alpha(A(t)) = 0$. Now, $\tilde{t}_k \in T_{a_k}$ and so $a_k \in A(\tilde{t}_k)$ $(k = 1, \ldots, m)$. But $\tilde{t}_k \in Q^T$, from which it follows that $\alpha(A(\tilde{t}_k)) > 0$ $(k = 1, \ldots, m)$. In particular, each set $A(\tilde{t}_k)$ has infinitely many members.

Now let $C \subset \cup_{k=1}^{m} A(\tilde{t}_k)$ be any potential finite coalition consisting of $\sum_{k=1}^{m} n_k$ different individuals $i \in I_C \subset I$ with characteristics $\theta_i \in \Theta_i$ such that, for each $k = 1, \ldots, m$, the set $A(\tilde{t}_k) \cap C$ has exactly n_k members. Also, for all $i \in I_C$, define $t_i := \tilde{t}_k$ for the unique k such that $(i, \theta_i) \in A(\tilde{t}_k) \cap C$. Then $\tilde{t}_k \in T_a$ whenever $a \in A(\tilde{t}_k) \cap C$. Also, $\sum_{i \in I_C} t_i = \sum_{k=1}^{m} n_k \, \tilde{t}_k = 0$. It follows that any such coalition can manipulate the mechanism $(\mathbf{x}, \mathbf{z})$. This shows that the finite family $A(\tilde{t}_k)$ $(k = 1$ to $m)$ can discernibly manipulate the mechanism $(\mathbf{x}, \mathbf{z})$. So $0 \in K$ implies that the mechanism $(\mathbf{x}, \mathbf{z})$ cannot be weakly multilaterally strategy-proof.

On the other hand, if $(\mathbf{x}, \mathbf{z})$ is weakly multilaterally strategy-proof, then $0 \notin K$. By construction, K is convex, so the point 0 can be separated from K by a hyperplane. Moreover, this hyperplane can be chosen so it passes through 0. Hence, there must exist $p(\alpha) \in \Re^T \setminus \{0\}$ for which $p(\alpha)\, t \geq 0$ whenever $t \in K$.

Now, given any $a \in A^*$ and any $t \in U_a$, rational local non-satiation in tradeable goods and transitivity of $\succsim_\theta$ together imply that t is the limit of a sequence t^q $(q = 1, 2, \ldots)$ of net trade vectors in $T_a \cap Q^T$. Then $t^q \in K$ and so $p(\alpha)\, t^q \geq 0$ $(q = 1, 2, \ldots)$, implying that $p(\alpha)\, t \geq 0$ in the limit as $q \to \infty$.

To complete the proof, first introduce the additional notation

$$\Theta_i^* := \{\, \theta \in \Theta_i \mid (i, \theta) \in A^* \,\}; \quad Z_i^* \;:=\; \{\, z \in Z \mid \exists \theta \in \Theta_i^* : z = z_i(\alpha, \theta) \,\};$$
$$Z_i \;:=\; \{\, z \in Z \mid \exists \theta \in \Theta_i : z = z_i(\alpha, \theta) \,\}.$$

Suppose now that the combination of $z \in Z_i^*$ with $a = (i, \theta) \in A^*$ and $\eta \in \Theta_i$ together satisfy $z_i(\alpha, \theta) = z_i(\alpha, \eta) = z$. Then (4.5) implies that $x_i(\alpha, \theta) - x_i(\alpha, \eta) \in U_a$ and so $p(\alpha)\, x_i(\alpha, \theta) \geq p(\alpha)\, x_i(\alpha, \eta)$. If $\eta \in \Theta_i^*$ also, then reversing θ and η in the above argument shows that $p(\alpha)\, x_i(\alpha, \theta) \leq p(\alpha)\, x_i(\alpha, \eta)$, and so $p(\alpha)\, x_i(\alpha, \theta) = p(\alpha)\, x_i(\alpha, \eta)$. Hence, one can define $m_i(\alpha, z)$ for all $z \in Z_i^*$ in order to satisfy $m_i(\alpha, z) = p(\alpha)\, x_i(\alpha, \theta)$ for all $\theta \in \Theta_i^*$ such that $z = z_i(\alpha, \theta)$. This already implies that $p(\alpha)\, x_i(\alpha, \theta) \leq m_i(\alpha, z_i(\alpha, \theta))$ for all $(i, \theta) \in A^*$.

To complete the definition of the function $m_i(\alpha,\cdot)$, for $z \in Z_i \setminus Z_i^*$ it is enough to define

$$m_i(\alpha, z) := \inf_{x,\theta} \{ p(\alpha)\, x \mid z = z_i(\alpha,\theta);\ (x,z) \in F(\theta);\ (x,z) \succsim_\theta (x_i(\alpha,\theta), z) \}.$$

For $z \in Z \setminus Z_i$, one can define

$$m_i(\alpha, z) := \inf_x \left\{ p(\alpha)\, x \mid (x,z) \in \bigcup_{\theta \in \Theta_i} F(\theta) \right\} - 1.$$

Suppose that $(i,\theta) \in A^*$ and $(x,z) \in F(\theta)$ satisfy $(x,z) \succsim_\theta (x_i(\alpha,\theta), z_i(\alpha,\theta))$. Clearly, in the case when $z \in Z \setminus Z_i$, it must be true that $p(\alpha)\, x \geq m_i(\alpha, z)$ by construction; the same is also true when $z \in Z_i \setminus Z_i^*$ because $z = z_i(\alpha,\eta)$ for some $\eta \in \Theta_i \setminus \Theta_i^*$. The other case is when $z \in Z_i^*$ because $z = z_i(\alpha,\eta)$ for some $\eta \in \Theta_i^*$. But then $p(\alpha)\, x_i(\alpha,\eta) = m_i(\alpha, z)$. Also, (4.5) implies that $x - x_i(\alpha,\eta) \in U_a$, so $p(\alpha)\, x \geq p(\alpha)\, x_i(\alpha,\eta) = m_i(\alpha,z)$ in this case as well. This completes the proof that the budget constraint $p(\alpha)\, x \leq m_i(\alpha,z)$ is almost a compensated linear decentralization. □

5 Second decentralization theorem

To go from a compensated linear decentralization to an ordinary linear decentralization requires at least two additional assumptions. The two presented below are alternatives to the three that were used in [**27**]. The extra conditions will be stated for each separate economy $\alpha \in \mathcal{E}$. Before doing so, first define, for each $\theta \in \Theta$ and $z \in Z$, the *conditional* feasible set given z as

$$X(z;\theta) := \{ x \in \Re^T \mid (x,z) \in F(\theta) \}$$

Assumption 2 For all $\theta \in \Theta$ and $z \in Z$, the set $X_i(z;\theta)$ is equal to the union $\cup_{k \in K(z;\theta)} X^k(z;\theta)$ of a finite or countably infinite family of closed convex subsets of $\Re^T$.

Note that the different sets $X^k(z;\theta)$ need not be disjoint. Assumption 2 is trivially satisfied when each set $X(z;\theta)$ is convex. It certainly excludes some forms of non-convexity in $X(z;\theta)$, but it does allow indivisible goods or other particular kinds of non-convexity. Assumption 2 leads to the following generalization of a well known result:

Lemma 5.1 Cheaper Point. *Suppose $(i,\theta) \in A$ is such that $p(\alpha)\, x \geq m_i(\alpha, z)$ whenever $(x,z) \in F(\theta)$ satisfies $(x,z) \succsim_\theta (x_i(\alpha,\theta), z_i(\alpha,\theta))$. Then for any $z \in Z$ and $k \in K(z;\theta)$ such that there exists some "cheaper point" $\underline{x} \in X^k(z;\theta)$ with $p(\alpha)\,\underline{x} < m_i(\alpha,z)$, one has $p(\alpha)\, x > m_i(\alpha,z)$ whenever $x \in X^k(z;\theta)$ satisfies $(x,z) \succ_\theta (x_i(\alpha,\theta), z_i(\alpha,\theta))$.*

Proof Fix any $(i,\theta) \in A$, $z \in Z$ and $k \in K(z;\theta)$. Suppose that $x \in X^k(z;\theta)$ satisfies $(x,z) \succ_\theta (x_i(\alpha,\theta), z_i(\alpha,\theta))$, and that there is a cheaper point $\underline{x} \in X^k(z;\theta)$. By continuity of preferences and Assumption 2, there must exist a scalar $\lambda \in (0,1)$ and a convex combination $\lambda\, x + (1-\lambda)\,\underline{x} \in X^k(z;\theta)$ such that $\lambda\, x + (1-\lambda)\,\underline{x} \succsim_\theta (x_i(\alpha,\theta), z_i(\alpha,\theta))$. By the hypotheses of the lemma, it follows that $p(\alpha)\,[\lambda\, x + (1-\lambda)\,\underline{x}] \geq m_i(\alpha,z)$. Therefore $\lambda\,[p(\alpha)\, x - m_i(\alpha,z)] \geq (1-\lambda)\,[m_i(\alpha,z) - p(\alpha)\,\underline{x}]$. Because $p(\alpha)\,\underline{x} < m_i(\alpha,z)$ and $0 < \lambda < 1$, it follows that $p(\alpha)\, x > m_i(\alpha,z)$. □

For each $\alpha \in \mathcal{E}$, $(i, \theta) \in A$, $z \in Z$ and $k \in K(z; \theta)$, define the minimum wealth level

$$\underline{m}_i^k(\alpha, z; \theta) := \min_x \{ p(\alpha)\, x \mid x \in X^k(z; \theta) \}.$$

Also, for each $\alpha \in \mathcal{E}$ and $z \in Z$, define

$$A_\alpha(z) := \{ (i, \theta) \in A \mid \exists k \in K(z; \theta) : m_i(\alpha, z) = \underline{m}_i^k(\alpha, z; \theta) \}.$$

Assumption 3* The distribution $\alpha \in \mathcal{E}$ satisfies $\alpha\left(\cup_{z \in Z} A_\alpha(z)\right) = 0$.

At least when Z is a finite or countably infinite set, assumption 3* seems a fairly obvious generalization of the "dispersion" condition due to [**61, 62**]. When Z is a continuum, it is not nearly so weak. Nevertheless, it is sufficient to use the following obviously weaker version of assumption 3*:

Assumption 3 There is at most a null set A^0 of potential agents $(i, \theta) \in A$ having points $z \in Z_i$ with the property that, for some $k \in K(z; \theta)$, the set $X^k(z; \theta)$ has a cheapest net trade vector $\underline{x}$ satisfying $(\underline{x}, z) \succ_\theta (x_i(\alpha, \theta), z_i(\alpha, \theta))$, as well as $p(\alpha)\, x \geq p(\alpha)\, \underline{x} = m_i(\alpha, z)$ for all $x \in X^k(z; \theta)$.

Theorem 5.2 Second Decentralization Theorem. *In any economy $\alpha \in \mathcal{E}$ for which Assumptions 1–3 all hold, a feasible allocation mechanism $(\mathbf{x}, \mathbf{z})$ is weakly multilaterally strategy-proof only if it is almost decentralizable by the linear budget constraint $p(\alpha)\, x \leq m_i(\alpha, z)$ that was constructed in proving the Compensated Decentralization Lemma.*

Proof By the Compensated Linear Decentralization Lemma 4.3, there must exist $p(\alpha) \in \Re^T \setminus \{0\}$ and a mapping $m_i(\alpha, z)$ such that, for all $(i, \theta) \in A^*$, whenever $(x, z) \in F(\theta)$ satisfies $(x, z) \succsim_\theta (x_i(\alpha, \theta), z_i(\alpha, \theta))$, then $p(\alpha)\, x \geq m_i(\alpha, z)$.

Suppose that $(i, \theta) \in A^*$ and $(\underline{x}, z) \in F(\theta)$ satisfy $(\underline{x}, z) \succ_\theta (x_i(\alpha, \theta), z_i(\alpha, \theta))$ with $p(\alpha)\, \underline{x} = m_i(\alpha, z)$. By construction of the function $m_i(\alpha, z)$, it must be true that $z \in Z_i$. Also, by the Cheaper Point Lemma 5.1, there must exist $k \in K(z; \theta)$ such that $\underline{x}$ is the cheapest net trade vector of $X^k(z; \theta)$ at prices $p(\alpha)$. So $(i, \theta) \in A^0$. By Assumption 3, it follows that $(\mathbf{x}, \mathbf{z})$ is almost decentralizable, as claimed. □

Acknowledgements

It is a great pleasure to recall the extensive fruitful interaction that I have enjoyed with Bob Aumann during the many summer workshops on economic theory we have both attended at Stanford over the years since 1975, as well as on other occasions elsewhere, especially during my visit to Jerusalem in 1983. It has been a great privilege also to observe how strongly Bob upholds the highest standards of both scientific and personal integrity.

References

[1] Aghion, P. and Howitt, P. [1995], *Technical Progress in the Theory of Economic Growth*, in Economics in a Changing World, Vol. V: Economic Growth and Capital and Labour Markets (J.-P. Fitoussi, ed.), Macmillan, London, 101–122.

[2] Al-Najjar, N. I. [1995], *Decomposition and Characterization of Risk with a Continuum of Random Variables*, Econometrica **63**, 1195–1224.

[3] Anderson, R. M. [1991], *Non-Standard Analysis with Applications to Economics*, in Handbook of Mathematical Economics, Vol. IV (W. Hildenbrand and H. Sonnenschein, eds.), North-Holland, Amsterdam, 2145–2208.

[4] Anderson, R. M. [1994], *The Core in Perfectly Competitive Economies*, in Handbook of Game Theory with Economic Applications, Vol. I (R. J. Aumann and S. Hart, eds.), North-Holland, Amsterdam, 413–457.

[5] Arrow, K. J. [1951], *An Extension of the Basic Theorems of Classical Welfare Economics*, in Proceedings of the Second Berkeley Symposium on Mathematical Statistics and Probability (J. Neyman, ed.), University of California Press, Berkeley, 507–532.

[6] Artstein, Z. [1980], *Discrete and Continuous Bang-Bang and Facial Spaces, or: Look for the Extreme Points*, SIAM Review **22**, 172–185.

[7] Aumann, R. J. [1964], *Markets with a Continuum of Traders*, Econometrica **32**, 39–50.

[8] Aumann, R. J. [1964], *Mixed vs. Behavior Strategies in Infinite Extensive Games*, Annals of Mathematics Studies **52**, 627–630.

[9] Aumann, R. J. [1965], *Integrals of Set-Valued Functions*, Journal of Mathematical Analysis and Applications **12**, 1–12.

[10] Aumann, R. J. [1966], *Existence of Competitive Equilibria in Markets with a Continuum of Traders* Econometrica, **34**, 1–17.

[11] Aumann, R. J. and Shapley, L. S. [1974], *Values of Non-Atomic Games*, Princeton University Press, Princeton.

[12] Barro, R. J. and Sala-i-Martin, X. [1995], *Economic Growth*, McGraw Hill, New York.

[13] Bhattacharya, R. N. and Majumdar, M. [1973], *Random Exchange Economies*, Journal of Economic Theory **6**, 37–67

[14] Blackorby, C. and Donaldson, D. [1988], *Cash versus Kind, Self-Selection, and Efficient Transfers*, American Economic Review **78**, 691–700.

[15] Brown, D. J. and Robinson, A. [1975], *The Cores of Large Standard Exchange Economies*, Journal of Economic Theory **9**, 245–254.

[16] Brown, D. J. and Robinson, A. [1975], *Nonstandard Exchange Economies*, Econometrica **43**, 41–55.

[17] Clarke, F. H. [1983], *Optimization and Nonsmooth Analysis*, John Wiley, New York.

[18] Feldman, M. and Gilles, C. [1985], *An Expository Note on Individual Risk without Aggregate Uncertainty*, Journal of Economic Theory **35**, 26–32.

[19] Gale, D. M. [1980], *Money, Information and Equilibrium in Large Economies*, Journal of Economic Theory **23**, 28–65.

[20] Gale, D. M. [1982], *Money: In Equilibrium*, James Nisbet, Welwyn.

[21] Grodal, B. [1972], *A Second Remark on the Core of an Atomless Economy*, Econometrica **40**, 581–583.

[22] Guesnerie, R. [1981], *On Taxation and Incentives: Further Reflections on the Limits of Redistribution*, Discussion Paper No. 89, Sonderforschungsbereich 21, University of Bonn; revised as ch. 1 of [**23**].

[23] Guesnerie, R. [1995], *A Contribution to the Theory of Taxation*, Cambridge University Press, Cambridge.

[24] Hahn, F. H. [1995], *On Growth Theory*, in Economics in a Changing World, Vol. V: Economic Growth and Capital and Labour Markets (J.-P. Fitoussi, ed.), Macmillan, London, 85–100.

[25] Halmos, P. R. [1950], *Measure Theory*, Van Nostrand Reinhold, New York.

[26] Hammond, P. J. [1979], *Straightforward Individual Incentive Compatibility in Large Economies*, Review of Economic Studies **46**, 263–282.

[27] Hammond, P. J. [1987], *Markets as Constraints: Multilateral Incentive Compatibility in a Continuum Economy*, Review of Economic Studies **54**, 399–412.

[28] Hammond, P. J. [1989], *The Power of Small Coalitions in Large Economies*, Stanford University, Institute of Mathematical Studies in the Social Sciences, Economics Technical Report No. 501; to appear in M.O.L. Bacharach, M.A.H. Dempster and J.L. Enos (eds.) *Mathematical Models in Economics: Oxford University Mathematical Economic Seminar 25th Anniversary Volume*, Oxford University Press, Oxford:.

[29] Hammond, P. J. [1992], *On the Impossibility of Perfect Capital Markets*, in Economic Analysis of Markets and Games: Essays in Honor of Frank Hahn (P. Dasgupta, D. Gale, O. Hart, and E. Maskin, eds.), M. I. T. Press, Cambridge, Mass., 527–560.

[30] Hammond, P. J. [1993], *Irreducibility, Resource Relatedness, and Survival in Equilibrium with Individual Non-Convexities*, in General Equilibrium, Growth, and Trade II: The Legacy of Lionel W. McKenzie (R. Becker, M. Boldrin, R. Jones, and W. Thomson, eds.), Academic Press, San Diego, 73–115.

[31] Hammond, P. J. [1993], *History as a Widespread Externality in Some Arrow–Debreu Market Games*, University of Munich, Center for Economic Studies Working Paper No. 47; for inclusion in The Formulation of Economic Theory and Other Essays in Honor of Kenneth Arrow (G. Chichilnisky, ed.), Cambridge University Press, Cambridge (in preparation).

[32] Hammond, P. J. [1995], *Four Characterizations of Constrained Pareto Efficiency in Continuum Economies with Widespread Externalities*, Japanese Economic Review **46**, 103–124.

[33] Hammond, P. J. [1998], *On f-Core Equivalence with General Widespread Externalities*, Working Paper 95-004, Department of Economics, Stanford University; to appear in the Journal of Mathematical Economics.

[34] Hammond, P. J., Kaneko, M. and Wooders, M. H. [1989], *Continuum Economies with Finite Coalitions: Core, Equilibrium and Widespread Externalities*, Journal of Economic Theory **49**, 113–134.

[35] Hammond, P.J. and Rodríguez-Clare, A. [1993], *On Endogenizing Long-Run Growth*, Scandinavian Journal of Economics, **95**, 391–425; also in Endogenous Growth, (T. M. Andersen and K. O. Moene, eds.), Blackwell, Oxford, 1–35.

[36] Haubrich, J. G. [1988], *Optimal Financial Structure in Exchange Economies*, International Economic Review **29**, 217–235.

[37] Haubrich, J. G. and King, R. G. [1990], *Banking and Insurance*, Journal of Monetary Economics **26**, 361–386.

[38] Helpman, E. [1992], *Endogenous Macroeconomic Growth Theory*, European Economic Review **36**, 237–267.

[39] Hildenbrand, W. [1971], *Random Preferences and General Economic Equilibrium*, Journal of Economic Theory **3**, 414–429.

[40] Hildenbrand, W. [1974], *Core and Equilibria of a Large Economy*, Princeton University Press, Princeton.

[41] Hildenbrand, W. [1982], *Core of an Economy*, in Handbook of Mathematical Economics, Vol. II (K. J. Arrow and M. D. Intriligator, eds.), North-Holland, Amsterdam, 831–877.

[42] Jacklin, C. J. [1987], *Demand Deposits, Trading Restrictions, and Risk Sharing*, in Contractual Arrangements for Intertemporal Trade (E. C. Prescott and N. Wallace, eds.), University of Minnesota Press, Minneapolis, 26–47.

[43] Judd, K. [1985], *The Law of Large Numbers with a Continuum of Random Variables*, Journal of Economic Theory **35**, 19–25.

[44] Kaneko, M. and Wooders, M. H. [1986], *The Core of a Game with a Continuum of Players and Finite Coalitions: The Model and Some Results*, Mathematical Social Sciences **12**, 105–137.

[45] Kaneko, M. and Wooders, M. H. [1989], *The Core of a Continuum Economy with Widespread Externalities and Finite Coalitions: From Finite to Continuum Economies*, Journal of Economic Theory **49**, 135–168.

[46] Kaneko, M. and Wooders, M. H. [1994], *Widespread Externalities and Perfectly Competitive Markets: Examples*, in Imperfections and Behavior in Economic Organizations, (R. P. Gilles and P. H. M. Ruys, eds.), Kluwer Academic Publishers, Boston, 71–87.

[47] Lucas, R. E. [1993], *Making a Miracle*, Econometrica **61**, 251–272.

[48] Lyapunov, A. A. [1940], *On Completely Additive Vector-Functions*, Proceedings of the Academy of Sciences of the USSR **4**, 465–478.

[49] Mas-Colell, A. [1989], *An Equivalence Theorem for a Bargaining Set*, Journal of Mathematical Economics **18**, 129–139.

[50] Mertens, J.-F. [1970], *Upper Semi-Continuity of the Walras–Correspondence: A Probabilistic Proof*, CORE Discussion Paper No. 7003.

[51] Mirrlees, J. A. [1971], *An Exploration in the Theory of Optimum Income Taxation*, Review of Economic Studies **38**, 175–208.

[52] Parthasarathy, K. R. [1967], *Probability Measures on Metric Spaces*, Academic Press, New York.

[53] Richter, H. [1963], *Verallgemeinerung eines in der Statistik benötigten Satzes der Masstheorie*, Mathematische Annalen **150**, 85–90; correction 440–441.

[54] Romer, P. M. [1991], *Increasing returns and new developments in the theory of growth*, in Equilibrium Theory and Applications: Proceedings of the 6th International Symposium in Economic Theory and Econometrics (W. Barnett, ed.), Cambridge University Press, Cambridge, 83–110.

[55] Royden, H. L. [1968], *Real Analysis*, 2nd ed., Macmillan, New York.

[56] Schmeidler, D. [1972], *A Remark on the Core of an Atomless Economy*, Econometrica **40**, 579–580.

[57] Stern, N. [1991], *The Determinants of Growth*, Economic Journal **101**, 122–133.

[58] Vickrey, W. S. [1945], *Measuring Marginal Utility by the Reactions to Risk*, Econometrica **13**, 319–333.

[59] Vind, K. [1972], *A Third Remark on the Core of an Atomless Economy*, Econometrica **40**, 585–586.

[60] Weiss, E.-A. [1981], *Finitely Additive Exchange Economies*, Journal of Mathematical Economics **8**, 221–240.

[61] Yamazaki, A. [1978], *An Equilibrium Existence Theorem without Convexity Assumptions*, Econometrica **46**, 541–555.

[62] Yamazaki, A. [1981], *Diversified Consumption Characteristics and Conditionally Dispersed Endowment Distribution: Regularizing Effect and Existence of Equilibria*, Econometrica **49**, 639–654.

Fields Institute Communications
Volume **23**, 1999

Arbitrage with Price-Dependent Preferences: Equilibrium and Market Stability

Frank H. Page, Jr.
Department of Finance
University of Alabama
Tuscaloosa, Alabama 35487 USA
fpage@cba.ua.edu

Myrna H. Wooders
Department of Economics
University of Toronto
Toronto, Ontario, Canada M5S 1A1
mwooders@chass.utoronto.ca

Abstract. We introduce price-dependent preferences, a form of externalities, into a general equilibrium model of an economy with short sales. The condition of "no unbounded arbitrage," a similarity assumption on preferences and choice sets, is defined and shown to be sufficient for existence of an equilibrium, for balancedness of the economy, and for nonemptiness of the core. These conclusions rest on the result that no unbounded arbitrage is necessary and sufficient for compactness of the set of individually rational and feasible allocations. We also show that with additional assumptions on the model no unbounded arbitrage is necessary and sufficient for these properties to hold. Finally, we show that there is duality relationship between net trade conditions limiting arbitrage and price conditions limiting arbitrage. Since all conditions found in the literature fall into one of these two categories, our duality result allows us to clarify how the various conditions are related.

1 Introduction: arbitrage in models of economies

In an economy with short sales, unlike the standard Arrow-Debreu-MacKenzie general equilibrium model, there is no exogenous lower bound on consumption sets. Even when standard assumptions of convexity of preferences are satisfied, the potential for unbounded short sales poses problems for the existence of economic equilibrium. If agents have sufficiently dis-similar expectations about future

1991 *Mathematics Subject Classification.* Primary 90A09; Secondary 90A09, 90D12.
The authors are grateful to the Social Sciences and Humanities Research Council of Canada for support. We thank Erik Balder, Jonathan Borwein, Jean-Michel Grandmont, Birgit Grodal, Yakar Kannai, Donald Saari, and Norman Schofield for helpful comments. The authors are also grateful to two anonymous referees for helpful comments on a shortened version of this paper.

prices they may engage in arbitrarily large purchases of securities on one market for immediate resale on another market in order to profit from a price discrepancy; in other words, agents may engage in unbounded arbitrage. In a general equilibrium model, it is intuitvely clear that if the preferences for any two agents in the economy are sufficiently dissimilar, then there may be no set of prices at which the agents' willingnesses to trade can be equated. At any set of prices, one agent may be willing to sell an arbitrarily large amount of one commodity and another agent may be willing to buy an arbitrarily large amount. As discussed even by the classical economists, in order for **any** trade to take place there must be a mutual coincidence of wants. For the existence of an equilibrium, this coincidence of wants must be reconcilable by the trading process.

Using arbitrage conditions based on recession cones, a number of recent papers have addressed the question of existence of equilibrium in models of economies with short sales. Recession cones provide a very precise way of describing net trades that are arbitrarily large and utility non-decreasing. The recession cone of the preferred set of an agent describes the directions in which the agent will seek to make unbounded trades. If two agents have diametrically opposed preferences, then neither agent will *ever* cease to want to make yet further trades with the other agent; the potential for mutually advantageous trades cannot be exhausted. Hence there is no equilibrium. Since the possibility of unbounded utility increasing trades creates the possibly for nonexistence of equilibrium, the literature on models of economies with short sales has focused on conditions limiting arbitrage opportunities stated in terms of recession cones.

Arbitrage conditions characterizing competitive equilibrium have appeared in several economic contexts. These contexts include (a) temporary equilibrium models,[1] (b) general equilibrium models of asset markets,[2] and (c) general equilibrium models more broadly.[3] Many of these authors consider both necessary and sufficient conditions. Several papers also consider the core.[4].

In the context of asset market models, it is natural to have preferences depend on prices. Relative prices may well be taken as indicators of the desirability of assets and may convey information about future asset returns. In fact, in their models of asset markets, [**18**], [**16**], and [**35**] all allow price dependent preferences. In the case of the general equilibrium models noted above, however, with the exception of [**36**] and some brief remarks in [**51**] price dependent preferences are not permitted.

In this paper we extend the general equilibrium model of an economy to incorporate price-dependent preferences and price-dependent, convex, and possibly unbounded choice sets (or consumption sets). The fact that choice sets are arbitrary and price dependent introduces new complexities that do not appear in [**51**]. We establish that no unbounded arbitrage, an arbitrage condition introduced in [**35**], implies existence of equilibrium, compactness of the set of utility possibilities, "balancedness" of the economy, and nonemptiness of the core.[5] Since mutually compatible desires to trade can be equated by prices, we call economies satisfying

[1]For example, Green in [**13**] and Grandmont in [**14, 15**].

[2]For example, [**18**], [**16**], [**33**], [**34**], [**35**], [**37**], [**7**], [**25**], and [**24**].

[3]For example, [**9**], [**26**], [**36**], [**11**], [**51**], [**32**], [**6**], [**38, 39, 41**], [**8**], [**29, 30**], and [**42**].

[4]For example, [**25**] and [**41**].

[5]In fact, in the appendix, we show that no unbounded arbitrage is necessary and sufficient for the compactness of the set of individually rational and feasible *allocations*. Boundedness of the set of individually rational and feasible allocations underlies the existence of equilibrium in models of convex economies with unbounded short sales. Kousougeras in [**25**] shows that Page's condition in [**35**] of no unbounded arbitrage implies this boundedness.

our assumptions *reconcilable*. In "strictly reconcilable" economies,[6] including, for example, those in which agents have price dependent utility functions with indifference surfaces containing no half lines, no unbounded arbitrage is necessary and sufficient for these properties to hold.

In this paper, to limit the extent of arbitrage opportunities, we extend the condition of no unbounded arbitrage to a general equilibrium setting with price dependent preferences.[7] No unbounded arbitrage is a similarity assumption, ruling out situations where any two groups of agents have diametrically opposed preferences and choice sets. The condition implies that all opportunities for utility-increasing trades can be exhausted by trades bounded in size.

In addition to presenting results on the connections between no unbounded arbitrage, existence of equilibrium, compactness of utility possibilities, and nonemptiness of the core, we also present a careful discussion of the relationship between no unbounded arbitrage and other conditions limiting arbitrage found in the literature. Our discussion is not exhaustive. For another treatment, focused on general equilibrium models where preferences do not depend on prices, Dana, Le Van and Magnien in [**8**] provide an excellent discussion (see also [**29, 30, 31**]).

Conditions limiting arbitrage in the literature generally fall into two broad categories:

- conditions on net trades (e.g., [**9**], [**18**], [**35**], [**33**], [**32**], [**41, 38**], [**24**]);
- conditions on prices (e.g., [**13**], [**14, 15**], [**26**], [**16**], [**37, 36**], [**51**], [**6**]).

Here, using an elementary result due to ([**47**], Corollary 16.2.2), we show that there is duality relationship between net trade conditions limiting arbitrage and price conditions limiting arbitrage. Our duality result allows us to clarify how the various conditions are related. For example, we show that, in strictly reconcilable economies, Werner's condition [**51**] limiting arbitrage is a price condition dual to our condition of no unbounded arbitrage - a condition on net trades.

Necessary and sufficient conditions for nonemptiness of the core of a game have appeared in a number of papers (cf., [**5**], [**50**], [**19**], and [**23**]). The results of Bondareva and Shapley are immediately applicable to economies with quasi-linear utilities (where all agents have utility functions of the form $u_i(x, \xi) = \widehat{u}_i(x) + \xi$) while those of [**19**] apply to economies with coalition structures.[8] In the context of economies without short sales, it follows immediately from [**10**] that existence of equilibrium and nonemptiness of the core are equivalent. The study of the core in models with unbounded short sales was initiated in [**25**]. Page and Wooders in [**41, 38, 39**] demonstrate that no unbounded arbitrage is necessary and sufficient for nonemptiness of the core. In this paper we define the core relative to given prices; thus, in cooperative activities, individuals take prices as given.[9] Our main result for the core is that in a strictly reconcilable economy with price dependent preferences, no unbounded arbitrage is necessary and sufficient for nonemptiness.

---

[6] An economy is said to be strictly reconcilable if it satisfies two conditions. Roughly, these conditions are: the recession cones used to represent potential unbounded arbitrage opportunities must be invariant with respect to the endowments and prices, and indifference surfaces must contain no half lines.

[7] Page and Wooders in [**38, 39, 41**] use the same condition of no unbounded arbitrage as [**35**] but for the case where preferences do not depend on prices.

[8] Classes of economies corresponding to the class of games studied in [**23**] have not yet been characterized.

[9] This is reasonable in "large" economies where small coalitions are effective but cannot significantly influence aggregate outcomes, as in the literature on the f-core (cf., [**17**] and [**20**]).

But in fact we show more: no unbounded arbitrage is necessary and sufficient for the balancedness of the game derived from the economy. This implies nonemptiness of the "partnered core," introduced in [**43**] and [**44, 45**].[10]

Finally, before proceeding, we note that there have been some recent developments in the study of economies without price dependent preferences. These include [**8**], who introduce an arbitrage condition which implies existence of equilibrium but is weaker than compactness of the set of feasible and individually rational allocations. Their condition, however, implies compactness of the set of utility possibilities and existence of equilibrium. Another condition limiting arbitrage opportunities, called *inconsequential arbitrage*, was introduced in [**42**]. Inconsequential arbitrage is implied by compactness of the set of utility possibilities but it is unknown if the opposite implication holds. More recently, Le Van and Magnien [**27**] introduced a no arbitrage condition for economies with a continuum of agents. The arbitrage conditions of these papers have not been studied in situations with price-dependent preferences.

2 An economy with price-dependent preferences

Let $E = (X_i(\cdot), \omega_i, u_i(\cdot,\cdot))_{i=1}^n$ denote an unbounded exchange *economy*. For each agent i there is given a set-valued mapping $p \to X_i(p)$ specifying, for each vector p of prices, the ith agent's *choice set* $X_i(p) \subset \mathbf{R}^L$. The ith agent's *endowment* is given by $\omega_i \in \mathbf{R}^L$. The ith agent's preferences are specified via a *utility function*

$$u_i(\cdot,\cdot) : \mathbf{R}^L \times \mathbf{R}^L \longrightarrow \mathbf{R}.$$

We will maintain the following standard assumptions on the economy $E = (X_i(\cdot), \omega_i, u_i(\cdot,\cdot))_{i=1}^n$ throughout the paper.

[A-1] For each agent $i = 1, ..., n$, the choice correspondence $p \to X_i(p)$ is zero-homogeneous (depends only on relative prices), convex-valued, lower hemicontinuous with a closed graph and the endowment is in the interior of the choice set $X_i(p)$ for each p.

- **[A-1.1]** $p \to X_i(p)$ is convex-valued and lower hemicontinuous with a closed graph,[11]
- **[A-1.2]** for each $p \in \mathbf{R}^L$ and $\lambda > 0$, $X_i(\lambda p) = X_i(p)$, and
- **[A-1.3]** for each $p \in \mathbf{R}^L$, $\omega_i \in intX_i(p)$, where "*int*" denotes interior.

[A-2] For each agent $i = 1, ..., n$, the utility function $u_i(\cdot,\cdot)$ is continuous, preferred sets are zero-homogeneous in prices, and preferences are locally non-satiated.

- **[A-2.1]** $u_i(\cdot,\cdot)$ is continuous on $\mathbf{R}^L \times \mathbf{R}^L$,
- **[A-2.2]** for each $(x,p) \in \mathbf{R}^L \times \mathbf{R}^L$, the preference set

$$P_i(x,p) = \{x' \in \mathbf{R}^L : u_i(x',p) > u_i(x,p)\}$$

  is convex, and for $\lambda > 0$, $P_i(x,p) = P(x, \lambda p)$, and
- **[A-2.3]** for each $(x,p) \in \mathbf{R}^L \times \mathbf{R}^L$,

$$clP_i(x,p) = \{x' \in \mathbf{R}^L : u_i(x',p) \geq u_i(x,p)\}.$$

Remark 1. *On the consumption set and preferences.* Note that preferences are defined over all of $\mathbf{R}^L \times \mathbf{R}^L$ instead of over the consumption set $X_i(p)$ for each price p. While it is standard to define preferences only over consumption sets, in

---

[10] See also [**1**], [**2**], [**3**], [**46**], and [**22**] for related concepts, results and further discussion of partnership.

[11] $X_i(\cdot)$ is *lower hemicontinuous* if, for every open set $E \subset \mathbf{R}^L$, the set $\{p \in \mathbf{R}^L : X_i(p) \cap E \neq \emptyset\}$ is open. The graph of $p \to X_i(p)$ is given by $\Gamma(X_i(\cdot)) = \{(x,p) \in \mathbf{R}^L \times \mathbf{R}^L : x \in X_i(p)\}$.

our context there is no reason to do so. In fact, there may be legal or tax restrictions on consumption sets that make them different from the domain of preferences.

It follows from [A-1.2] and [A-2.2] that agents' preferences and choice sets are sensitive only to relative prices. Thus, without loss of generality we can restrict prices to be in the unit ball

$$\mathcal{B} := \{p \in \mathbf{R}^L : \|p\| \leq 1\}.$$

The local nonsatiation assumption [A-2.3] can be replaced by a weaker global nonsatiation assumption without changing the results in the paper. This can be accomplished via a technique introduced by Gale and Mas-Colell in [**12**] - see [**42**].

2.1 Individually rational and feasible allocations. The "at least as good as" set of the *ith* agent, $\widehat{P}_i(x_i, p)$, is the set of choices that are at least as desirable as his endowment. Formally,

$$\widehat{P}_i(x_i, p) = \{x' \in \mathbf{R}^L : u_i(x', p) \geq u_i(x_i, p)\}.$$

Given prices p and endowments $\omega = (\omega_1, ..., \omega_n)$, the set of all individually rational and feasible *allocations* for the economy $(X_i(\cdot), \omega_i, u_i(\cdot, \cdot))_{i=1}^n$ is given by

$$F(\omega, p) := \{(x_1, ..., x_n) \in \Pi_{i=1}^n X_i(p) : \sum_{i=1}^n x_i = \sum_{i=1}^n \omega_i \text{ and for each } i,\ x_i \in \widehat{P}_i(\omega_i, p)\}. \tag{1}$$

Note that given [A-1.2] and [A-2.2] for all $p \in \mathbf{R}^L$ and for all $\lambda > 0$, $F(\omega, p) = F(\omega, \lambda p)$.

Let $\Gamma(F(\omega, \cdot))$ denote the graph of the individually rational allocation mapping $p \to F(\omega, p)$ where $p \in \partial\mathcal{B}$ and ∂ denotes the boundary, that is, $\partial\mathcal{B} := \{p' \in \mathbf{R}^L : \|p'\| = 1\}$. Thus,

$$\Gamma(F(\omega, \cdot)) = \{(x_1, ..., x_n, p) \in \mathbf{R}^L \times \cdots \times \mathbf{R}^L \times \partial B : (x_1, ..., x_n) \in F(\omega, p)\ \} \tag{2}$$

The corresponding set of *utility possibilities* is given by

$$U(F(\omega, \cdot)) = \{(u_1, ..., u_n) \in \mathbf{R}^n : \text{for some } (x_1, ..., x_n, p) \in \Gamma(F(\omega, \cdot)),\ u_i = u_i(x_i, p) \text{ for each } i = 1, ..., n\}.$$

2.2 Equilibrium. Given prices $p \in \mathbf{R}^L$, the cost of a consumption vector $x = (x_1, ..., x_L)$ is

$$\langle x, p \rangle = \sum_{\ell=1}^L p_\ell \cdot x_\ell.$$

Given prices p and endowment ω_i, the *budget set of the* i^{th} agent is defined by

$$B_i(\omega_i, p) = \{x \in \mathbf{R}^L : \langle x, p \rangle \leq \langle \omega_i, p \rangle\}.$$

[D-1] *Equilibrium.* An *equilibrium* for the economy $E = (X_i(\cdot), \omega_i, u_i(\cdot, \cdot))_{i=1}^n$ is an $(n+1)$-tuple of vectors $(x_1^*, ..., x_n^*, p^*)$ such that
$(x_1^*, ..., x_n^*) \in F(\omega, p^*)$ (the allocation is feasible);
$p^* \in \mathcal{B}\backslash\{0\}$ (prices are in the unit ball and not all prices are zero); and
for each agent i,

$\langle x_i^*, p^* \rangle = \langle \omega_i, p^* \rangle$ (budget constraints are satisfied), and $P_i(x_i^*, p^*) \cap X_i(p^*) \cap B_i(\omega_i, p^*) = \emptyset$ (there are no affordable preferred net trades).

2.3 Balanced economies and the core. Let C be a nonempty subset of $N = \{1, ..., n\}$, called a *coalition*. A collection of choice vectors $(z_i : i \in C)$ is a $C-$*allocation* given prices $p \in \mathbf{R}^L$ if, for each $i \in C$, it holds that $z_i \in \widehat{P}_i(\omega_i, p) \cap X_i(p)$ and $\sum_{i \in C} z_i = \sum_{i \in C} \omega_i$.

Let $F_C(\omega, p)$ denote the set of all $C-$allocations given prices p. Also, given prices p, for each nonempty coalition C define the set

$$V(C;p) = \{(u_1, ..., u_n) \in \mathbf{R}^n : \text{for some } C\text{-allocation} \\ (x_i : i \in C) \in F_C(\omega, p),\ u_i(x_i, p) \geq u_i \ \text{ for each } i \in C\ \}. \tag{3}$$

Expression (3) defines a set-valued mapping

$$V(\cdot; p) : 2^N \backslash \emptyset \rightarrow \text{ subsets of } \mathbf{R}^n.$$

If for each price vector $p \in \mathbf{R}^L$ and coalition C, $V(C;p)$ is nonempty and closed, and if $V(C;p)$ is bounded from above, then the pair $(N, V(\cdot;p))$ defines a game in characteristic function form (in the sense of [**48**]) corresponding to the economy $E = (X_i(\cdot), \omega_i, u_i(\cdot, \cdot))_{i=1}^n$ and given prices $p \in \mathbf{R}^L$.

[D-2] *Balanced economies.* Let β denote a balanced collection of subsets of N with balancing weights δ_C for $C \in \beta$ and let $\beta(i) = \{C \in \beta : i \text{ is contained in } C\}$. Thus, $\delta_C \geq 0$ for all $C \in \beta$ and for each $i \in M$

$$\sum_{C \in \beta(i)} \delta_C = 1 \ .$$

Given a price vector p the economy E is *well-defined and balanced relative to* p if $(N, V(\cdot, p))$ is a well-defined game and for every balanced collection β of subsets of N it holds that

$$\bigcap\nolimits_{C \in \beta} V(C;p) \subseteq V(N;p).$$

The economy is *well-defined and balanced* if for each $p \in \partial\mathcal{B}$ the economy is well-defined and balanced relative to p.

[D-3] *The core.* An allocation $(x_1^*, ..., x_n^*)$ is in the *core of the economy E given prices p* if there does not exist a coalition $C \subset \{1, ..., n\}$ of agents and an C-allocation $(z_i : i \in C) \in F_C(w, p)$ at p such that for all $i \in C$

$$u_i(z_i, p) > u_i(x_i, p).$$

We say that an economy *has a nonempty core* if for some $p \in \partial\mathcal{B}$ there exists an allocation $(x_1, ..., x_n) \in F(\omega, p)$ in the core of the economy given prices p.

Remark 2. Note that in the above definition of the core coalitions take prices as given. In large economies, if coalitions are relatively small and take prices as given, this assumption may be natural. (See [**17**] and [**20**] for more motivation for our approach to the core.) It would be interesting to consider a concept of the core where coalitions take into account the effect of their behaviour on economic aggregates, including prices. This is beyond the scope of the present paper.

Remark 3. It is immediate from Scarf's Theorem [**48**] that a balanced economy has a nonempty core. From [**44**] and [**38**] it follows that a balanced economy has a nonempty partnered core.

Remark 4. Note that the definition of the core above, standard in game theory, is different from the standard economic defintion, where it is only required that

$$\begin{array}{c} u_i(z_i, p) \geq u_i(x_i, p) \text{ for all } i \in C \text{ and} \\ \text{for at least one } i \text{ in } C,\ u_i(z_i, p) > u_i(x_i, p). \end{array} \qquad (*)$$

In general these two definitions of the core are not equivalent. It is obvious that if a subset of agents can all be better off in an improving coalition then $(*)$ is satisfied. Thus, the core defined by $(*)$ contains the core defined by [D-3].

3 No unbounded arbitrage

In order to describe unbounded arbitrage for individual agents and for the economy, we will utilize the notions of recession cone and increasing cone.[12]

3.1 Recession cones, lineality spaces, and positive dual cones.

[D-4] *Directions of recession and recession cones.*
Given any convex subset S of $\mathbf{R}^L$, we say that $y \in \mathbf{R}^L$ is a *direction of recession* for S if $x + \lambda y \in S$ for all $\lambda \geq 0$ and for all $x \in S$. We shall denote by $\mathcal{R}(S)$ the set of all recession directions of S, and we shall refer to $\mathcal{R}(S)$ as the *recession cone* corresponding to S.

It is easy to show that $\mathcal{R}(S)$ is a convex cone containing the origin. Moreover, if S is closed then $\mathcal{R}(S)$ is closed and $x + \lambda y \in S$ for all $\lambda \geq 0$ and *some* $x \in S$ implies that $y \in \mathcal{R}(S)$.

[D-5] *Lineality spaces.*

Given any convex subset S of $\mathbf{R}^L$, we say that $y \in \mathbf{R}^L$ is a *direction of lineality* for S if $x + \lambda y \in S$ for all $\lambda \in \mathbf{R}$. We shall denote by $L(S)$ the set of all lineality directions of S. Given the definition of the recession cone, it is easy to see that for S convex, the *lineality space* $L(S)$ of S is given by

$$L(S) = -\mathcal{R}(S) \bigcap \mathcal{R}(S).$$

[D-6] *Positive dual cones.*

Given any convex subset S of $\mathbf{R}^L$, the positive dual cone corresponding to S is given by

$$\mathcal{D}(S) = \{p \in \mathbf{R}^L : \langle y, p \rangle > 0 \text{ for all } y \in S \backslash \{0\}\}.$$

If we think of the set S as a set of net trades and the corresponding positive dual cone $\mathcal{D}(S)$ as a set of prices, then any price vector $p \in \mathcal{D}(S)$ assigns a *positive* value to any nonzero net trade vector $y \in S$.

---

[12] Here we use the extended definition of the increasing cone, introduced in [**40**]. The increasing cone was first used in an asset market model to show that Page's [**37**] price conditon limiting arbitrage is necessary and sufficient for the existence of equilibrium (see also [**33**]). In [**41, 39**], the increasing cone is used to show that no unbounded arbitrage is necessary and sufficient for existence, compactness of utility possibilities, and nonemptiness of the core.

3.2 Arbitrage cones and strict increasing cones. Define

$$\mathcal{A}_i(x,p) := \mathcal{R}\left(X_i(p) \bigcap \widehat{P}_i(x,p)\right) \tag{4}$$

We shall refer to the recession cone $\mathcal{A}_i(x,p)$ as the ith agent's *arbitrage cone* given prices p and choice vector x. Note that a vector of net trades y is contained in $\mathcal{A}_i(x,p)$ if and only if

1. $x + \lambda y \in X_i(p)$ for all $\lambda \geq 0$ and
2. $u_i(x + \lambda y, p) \geq u_i(x,p)$ for all $\lambda \geq 0$.

A cone closely related to the arbitrage cone is the increasing cone, $\mathcal{I}_i(x,p)$, defined as follows:

$$\mathcal{I}_i(x,p) := \left\{ \begin{array}{c} y \in \mathcal{A}_i(x,p) : \forall\lambda \geq 0, \exists\lambda' > \lambda \text{ such that} \\ \\ u_i(x + \lambda' y, p) > u_i(x + \lambda y, p) \end{array} \right\} \tag{5}$$

We shall refer to the positive dual cone $\mathcal{D}(\mathcal{A}_i(x,p))$ as the ith agent's *no arbitrage price cone.*

3.3 Individual arbitrage opportunities. Define

$$H(p) = \{y \in \mathbf{R}^L : \langle y, p \rangle \leq 0\}.$$

Given prices p, $H(p)$ is the ith agent's set of net trades that are at most costless. If $\langle y, p \rangle = 0$ then net trades y are costless to make and if $\langle y, p \rangle < 0$ then the net trades y generate a positive return.

[D-7] *Individual arbitrage opportunities.*

[D-7.1] From the perspective of the ith agent, a nonzero vector of net trades $y \in \mathbf{R}^L$ represents *an opportunity for unbounded arbitrage given prices* $p \in \mathbf{R}^L$ *and choice vector* $x_i \in X_i(p)$ if and only if

$$y \in H(p) \cap \mathcal{A}_i(x_i, p).$$

[D-7.2] From the perspective of the ith agent, a price vector $p \in \mathbf{R}^L$ *does not admit unbounded arbitrage at choice vector* x if and only if

$$p \in \mathcal{D}(\mathcal{A}_i(x_i, p)).$$

Thus, a vector of net trades y is an unbounded arbitrage for the ith agent given prices p and choice vector $x_i \in X_i(p)$ if and only if the net trades y are at most costless to make and y is contained in the ith agent's arbitrage cone $\mathcal{A}_i(x_i, p)$. Given a choice vector x_i, a price vector p admits no unbounded arbitrage if and only if p is a fixed point of the mapping $p \to \mathcal{D}(\mathcal{A}_i(x_i, p))$.

Note that if $y \in H(p) \cap \mathcal{A}_i(x,p)$ then $y \in H(\lambda p) \cap \mathcal{A}_i(x, \lambda p)$ for all $\lambda > 0$. Thus, arbitrages are determined by *relative* prices. Moreover, if $p \in \mathcal{D}(\mathcal{A}_i(x,p))$ then $\lambda p \in \mathcal{D}(\mathcal{A}_i(x, \lambda p))$ for all $\lambda > 0$.

3.4 No unbounded arbitrage. Given endowments and given any nonzero vector of prices, no unbounded arbitrage rules out the possibility that any agent can find a *mutually compatible* trading partner (or group of trading partners) with whom to engage in unbounded and utility nondecreasing trade.

[D-8] *No unbounded arbitrage.* Given endowments $(\omega_1, ..., \omega_n)$, the economy satisfies *no unbounded arbitrage* if and only if

$$\begin{array}{c} \text{for each } p \in \mathcal{B}\backslash\{0\} \text{ if } \sum_{i=1}^n y_i = 0 \text{ and } y_i \in \mathcal{A}_i(\omega_i, p) \text{ for all } i, \\ \text{then } y_i = 0 \text{ for all } i. \end{array} \tag{6}$$

In words, if for any price vector $p \in \mathcal{B}\backslash\{0\}$, there is a set of mutually compatible net trades (that is, if the y_i, $i = 1, ..., n$ are such that $\sum_{i=1}^{n} y_i = 0$) each of which is feasible and possibly utility increasing on any scale (that is, each y_i, $i = 1, ..., n$ is such that $y_i \in \mathcal{A}_i(\omega_i, p)$), then these mutually compatible net trades must be trivial (that is, $y_i = 0$ for all i).

It is important to note that no unbounded arbitrage does not rule out mutually compatible *bounded* arbitrages. Thus, even if no unbounded arbitrage is satisfied there may still be present in the economy arbitrages which can be exhausted via bounded trades. (See [**34**] and [**33**] for a detailed discussion of bounded and unbounded arbitrage in asset markets.)

In an investigation of the irreversibility assumption in general equilibrium models with production and with choice sets and preferences *independent* of prices, Debreu in [**9**] introduces a recession direction condition similar to our condition of no unbounded arbitrage, but applied to choice sets only. Translating Debreu's condition to our model, we have,

$$\text{for each } p \in \mathcal{B}\backslash\{0\} \text{ if } \sum_{i=1}^{n} y_i = 0 \text{ and } y_i \in \mathcal{R}(X_i(p)) \text{ for all } i, \\ \text{then } y_i = 0 \text{ for all } i.$$

Debreu's condition is essentially an assumption concerning the degree of similarity of agents' choice sets, $X_i(p)$. Note that, within the context of our model, Debreu's condition implies our condition.

In the analysis to follow, we will show that no unbounded arbitrage (6) is sufficient for the existence of equilibrium, compactness of the set of utility possibilities and nonemptiness of the core. Moreover, since we prove our results within the context of a general equilibrium model in which agents' preferences are price-dependent and consumption sets are price dependent and possibly unbounded, our model subsumes many of the models in the existing literature. Under additional conditions on the economy, we will also show that no unbounded arbitrage is necessary and sufficient for the existence of equilibrium, the compactness of the utility possibilities, and nonemptiness of the core. The additional conditions we will need are:

[A-3] *Extreme desirability.* Given any price vector $p \in \mathcal{B}\backslash\{0\}$ and choice vector $x \in \mathbf{R}^L$, each nonzero unbounded arbitrage opportunity $y \in \mathcal{A}_i(x, p)$ is eventually utility increasing. Thus,

$$\text{for all } p \in \mathcal{B}\backslash\{0\} \text{ and } x \in \mathbf{R}^L, \\ \mathcal{A}_i(x, p)\backslash\{0\} = \mathcal{I}_i(x, p)$$

[A-4] *Uniformity of arbitrage cones.* Each agent's utility function and choice correspondence is such that the set of unbounded arbitrage opportunities is invariant with respect to the price vector $p \in \mathcal{B}\backslash\{0\}$ and the choice vector $x \in \mathbf{R}^L$. Thus,

$$\mathcal{A}_i(x, p) = \mathcal{A}_i(x', p') \\ \text{for all } p \text{ and } p' \text{ in } \mathcal{B}\backslash\{0\} \text{ and for all } x \text{ and } x' \text{ in } \mathbf{R}^L.$$

We shall refer to an economy in which agents' utility functions satisfy assumptions [A-2]-[A-4] as a *strictly reconcilable economy.* (See [**41**].)

In Section 4 below, we construct an example of a strictly reconcilable asset exchange economy. Besides providing an example, this allows us to more easily discuss the conditions limiting arbitrage introduced by [**18**] and [**16**].

3.5 Characterizing no unbounded arbitrage: the duality of conditions on trades and conditions on prices. We begin by stating a general result on the connection between conditions on net trades limiting arbitrage (such as no unbounded arbitrage (6)) and conditions on prices limiting arbitrage. Theorem 1 below is obtained by specializing Corollary 16.2.2 in [**47**] to cover the exchange economy developed here.

From the perspective of unbounded arbitrage, the lineality space corresponding to the arbitrage cone $\mathcal{A}_i(x_i, p)$ is particularly noteworthy. Consider a vector of net trades $y \in L(\mathcal{A}_i(x_i, p))$. For this vector of net trades, $x_i + \lambda y \in X_i(p)$ for all $\lambda \in \mathbf{R}$ and more importantly, $u_i(x_i + \lambda y, p) = u_i(x_i, p)$ for all $\lambda \in \mathbf{R}$. Thus, given any unbounded arbitrage opportunity $y \in L(\mathcal{A}_i(x_i, p))$, trading in y direction or the $-y$ direction is feasible and utility constant. To save writing we will sometimes let

$$L(\mathcal{A}_i(x_i, p)) := \mathcal{L}_i(x_i, p) \tag{7}$$

Theorem ([**47**]) *The duality between conditions on trades and conditions on prices.* Let $E = (X_i(\cdot), \omega_i, u_i(\cdot, \cdot))_{i=1}^n$ be an economy satisfying [A-1]-[A-2]. For any nonzero price vector $p \in \mathbf{R}^L$ and any $n - tuple$ of choice vectors $(x_1, \ldots, x_n) \in \mathbf{R}^L \times \cdots \times \mathbf{R}^L$ such that for all i, $\mathcal{A}_i(x_i, p) \setminus \mathcal{L}_i(x_i, p) \neq \emptyset$, the following statements are equivalent:

1. (*a condition on net trades*)

$$\sum_{i=1}^{n} y_i = 0 \text{ and } y_i \in \mathcal{A}_i(x_i, p) \text{ for all } i = 1, \ldots, n,$$
$$\text{then } y_i \in \mathcal{L}_i(x_i, p) \text{ for all } i = 1, \ldots, n.$$

2. (a condition on prices)

$$\cap_i \mathcal{D}(\mathcal{A}_i(x_i, p) \setminus \mathcal{L}_i(x_i, p)) \neq \emptyset.$$

Corollary to Rockafellar's Theorem *Characterization of no unbounded arbitrage as the nonempty intersection of the no arbitrage price cones.* Let $E = (X_i(\cdot), \omega_i, u_i(\cdot, \cdot))_{i=1}^n$ be an economy satisfying [A-1]-[A-2]. For any nonzero price vector $p \in \mathbf{R}^L$ and any $n - tuple$ of choice vectors $(x_1, \ldots, x_n) \in \mathbf{R}^L \times \cdots \times \mathbf{R}^L$ such that for all i, $\mathcal{L}_i(x_i, p) = \{0\}$ and $\mathcal{A}_i(x_i, p) \setminus \mathcal{L}_i(x_i, p) \neq \emptyset$, the following statements are equivalent:

1. If $\sum_{i=1}^n y_i = 0$ and $y_i \in \mathcal{A}_i(x_i, p)$ for all i,then $y_i = 0$ for all i.
2. $\cap_i \mathcal{D}(\mathcal{A}_i(x_i, p)) \neq \emptyset$.

If in economy E agents utility functions are such that indifference surfaces contain no half lines, then for any nonzero price vector $p \in \mathbf{R}^L$ and any n-*tuple* of choice vectors $(x_1, \ldots, x_n) \in \mathbf{R}^L \times \cdots \times \mathbf{R}^L$ it will be true that for all i $\mathcal{L}_i(x_i, p) = \{0\}$. It follows from Theorem 1 in [**30**], that the no half lines condition is equivalent to [A-3] (extreme desirability). Thus, in view of the Corollary, in economies satisfying the no half lines condition, no unbounded arbitrage (6) can be stated equivalently as follows:

$$\text{for each } p \in \mathcal{B} \setminus \{0\}, \ \bigcap_i \mathcal{D}(\mathcal{A}_i(\omega_i, p)) \neq \emptyset. \tag{8}$$

Geometrically, $\cap_i \mathcal{D}(\mathcal{A}_i(\omega_i, p)) \neq \emptyset$ if and only if all of the arbitrage cones $\mathcal{A}_i(\omega_i, p)$, $i = 1, 2..., n$ lie *strictly* on one side of a hyperplane through the origin.

4 The main theorems

In this subsection we state our main results. We note that parts of some of the following Theorems are already proven in the literature for economies with unbounded short sales and without price-dependent preferences. We discuss the related results in a later section.

Theorem 1 *No unbounded arbitrage is sufficient for existence of equilibrium, compactness of the set of utility possibilities, balancedness of the economy, and nonemptiness of the core.* Let $E = (X_i(\cdot), \omega_i, u_i(\cdot,\cdot))_{i=1}^n$ be an economy satisfying [A-1]-[A-2]. No unbounded arbitrage (6) implies the following:

1. E has an equilibrium.
2. The set of utility possibilities $U(F(\omega,\cdot))$ is compact.
3. E is well-defined and balanced.
4. E has a nonempty core.

In the context of a strictly reconcilable economy, existence of a competitive equilibrium, compactness of the set of utility possibilities, balancedness of the economy, and nonemptiness of the core are each necessary and sufficient for no unbounded arbitrage.[13]

Theorem 2 *In a strictly reconcilable economy no unbounded arbitrage is necessary and sufficient for the existence of equilibrium, compactness of the set of utility possibilities, balancedness of the economy, and nonemptiness of the core.* Let $E = (X_i(\cdot), \omega_i, u_i(\cdot,\cdot))_{i=1}^n$ be an economy satisfying [A-1]-[A-4]. Then the following statements are equivalent:

1. E satisfies no unbounded arbitrage.
2. For each $p \in \mathcal{B}\backslash\{0\}$, $\cap_i \mathcal{D}(\mathcal{A}_i(\omega_i, p)) \neq \emptyset$.
3. E has an equilibrium.
4. The set of utility possibilities $U(F(\omega,\cdot))$ is compact.
5. E is well-defined and balanced.
6. E has a nonempty core.

Our next result shows that with fewer assumptions on the economy E no unbounded arbitrage is necessary for the existence of equilibrium.

Theorem 3 *A necessary conditions for the existence of equilibrium.* Let $E = (X_i(\cdot), \omega_i, u_i(\cdot,\cdot))_{i=1}^n$ be an economy satisfying [A-1]-[A-3]. If $(x_1^*, ..., x_n^*, p^*)$ is an equilibrium for the economy E then the following conditions hold:

1. If $\sum_{i=1}^n y_i = 0$ and $y_i \in \mathcal{A}_i(x_i^*, p^*)$ for all i, then $y_i = 0$ for all i.
2. $p^* \in \cap_i \mathcal{D}(\mathcal{A}_i(x_i^*, p^*))$.

Consider now the following weakening of assumption [A-4] concerning uniformity of arbitrage cones.

[A-5] *Uniformity of arbitrage cones with respect to choice.* Each agent's utility function and choice correspondence are such that for each price vector $p \in$

[13]In fact all the results for strictly reconcilable economies hold if only $n-1$ agents have strictly reconcilable preferences - see [**38**] .

$\mathcal{B}\backslash\{0\}$, the set of potential unbounded arbitrage opportunities is invariant with respect to the choice vector $x \in \mathbf{R}^L$. Thus,

$$\text{for } p \in \mathcal{B}\backslash\{0\}$$
$$\mathcal{A}_i(x,p) = \mathcal{A}_i(x',p)$$
$$\text{and for all } x \text{ and } x' \text{ in } \mathbf{R}^L.$$

If, in addition to assuming [A-1]-[A-3], we assume [A-5], then we have the following Corollary to Theorem 3.

Corollary 1 *A stronger necessary condition for the existence of equilibrium.* Let $E = (X_i(\cdot), \omega_i, u_i(\cdot,\cdot))_{i=1}^n$ be an economy satisfying [A-1]-[A-3] and [A-5]. If $(x_1^*, ..., x_n^*, p^*)$ is an equilibrium for the economy E then the following conditions hold:

1. If $\sum_{i=1}^n y_i = 0$ and $y_i \in \mathcal{A}_i(\omega_i, p^*)$ for all i,then $y_i = 0$ for all i.
2. $p^* \in \cap_i \mathcal{D}(\mathcal{A}_i(\omega_i, p^*))$.

Note that in the Corollary we have replaced $\mathcal{A}_i(x_i^*, p^*)$ with $\mathcal{A}_i(\omega_i, p^*)$. In a temporary equilibrium setting, the condition $p^* \in \cap_i \mathcal{D}(\mathcal{A}_i(\omega_i, p^*))$ is equivalent to the necessary condition given by [**13**]. (See also [**15**] and, for further discussion, [**16**], [**33**] and [**34**]).

5 Conditions limiting arbitrage in the literature

In this section we describe some of the other conditions limiting arbitrage found in the literature, and we relate these conditions to the condition of no unbounded arbitrage presented here. Our coverage is not exhaustive. We focus on the conditions of [**18**], [**16**], and [**51**]. For a more complete treatment of conditions in general equilibrium models, see [**8**].

In order to sensibly discuss the conditions limiting arbitrage introduced by [**18**] and [**16**], we begin by constructing an asset markets example.

5.1 An asset markets example of a strictly reconcilable economy. Consider an asset exchange economy $E = (X_i(\cdot), \omega_i, u_i(\cdot,\cdot))_{i=1}^n$ in which n agents have expected utility over portfolios given by

$$u_i(x,p) = \int_{\mathbf{R}_+^L} v_i(\langle x, r\rangle) d\mu_i(dr \mid p)$$

and portfolio choice sets $X_i(p)$ for each price vector $p \in \mathbf{R}^L$. Here

- $\mathbf{R}_+^L$ denotes the nonnegative orthant,
- $v_i(\cdot) : \mathbf{R} \to \mathbf{R}$ is the ith agent's utility function,
- $\mu_i(\cdot \mid p)$ is a probability measure defined on the Borel subsets $B(\mathbf{R}_+^L)$ of $\mathbf{R}_+^L$ and represents the ith agent's probability beliefs concerning asset returns given prices p,
- $\langle x, r\rangle := \sum_{i=1}^L x_\ell \cdot r_\ell$ where, for $\ell = 1, ..., L$,
  x_ℓ is the number of shares of asset ℓ held in portfolio $x \in \mathbf{R}^L$, and
  r_ℓ is the gross return on one share of asset ℓ.

Let $S[\mu_i(\cdot \mid p)]$ denote the support of the probability measure $\mu_i(\cdot \mid p)$ and let

$$s_i(+) := \lim_{c\to\infty} \tfrac{dv_i(c)}{dc} \text{ and}$$

$$s_i(-) := \lim_{c\to-\infty} \tfrac{dv_i(c)}{dc}.$$

Thus, $s_i(+)$ and $s_i(-)$ are the asymptotic derivatives of the ith agent's utility function.

Now consider the following list of assumptions concerning each agent's utility function and probability beliefs:

[a-1]: the choice correspondence $p \to X_i(p)$ satisfies [A-1],
[a-2]: $v_i(\cdot)$ is concave, increasing, and everywhere differentiable with $s_i(+) < s_i(-)$,
[a-3]: the mapping $p \to \mu_i(\cdot \mid p)$ is weakly continuous and positively homogeneous of degree zero,[14]
[a-4]: $S[\mu_i(\cdot \mid p)]$ is bounded,
[a-5]: if, given asset prices p, net trades y_i generate a nonnegative gross return with probability 1 (i.e., if y_i is such that $\mu_i(\{r \in \mathbf{R}^L_+ : \langle y_i, r\rangle \geq 0\} \mid p) = 1$), then $y_i \in \mathcal{R}\,(X_i(p))$,
[a-6]: assets are desirable, that is, $\mathbf{R}^L_{++} \cap kc\{S[\mu_i(\cdot \mid p)]\} \neq \emptyset$ (here, $\mathbf{R}^L_{++}$ denotes the positive orthant of $\mathbf{R}^L$and $kc\{S[\mu_i(\cdot \mid p)]\}$ denotes the convex cone generated by the support of the ith agent's probability beliefs concerning asset returns),
[a-7]: there are no redundant assets, that is,

$$int\, kc\{S[\mu_i(\cdot \mid p)]\} \neq \emptyset,$$

("*int*" denotes interior),
[a-8]: the ith agent is sufficiently risk averse. That is, either

$$\text{(a) for all } c \in \mathbf{R},\ 0 \leq s_i(+) < \frac{dv_i(c)}{dc} < s_i(-) \leq \infty,$$

or

$$\text{(b) } s_i(+) = 0 \text{ and/or } s_i(-) = +\infty,$$

[a-9]: the closed, convex cone $kc\{S[\mu_i(\cdot \mid p)]\}$ generated by the ith agent's probability beliefs concerning asset returns is invariant with respect to $p \in \mathcal{B}\backslash\{0\}$.

If assumptions [a-1]-[a-6] hold, then the asset exchange economy E satisfies [A-1] and [A-2]. If assumptions [a-1]-[a-5] and assumptions [a-7] - [a-9], hold then the economy is strictly reconcilable (i.e., satisfies [A-1]-[A-4] - see [**33, 34, 35, 37**]). Note that since $v_i(\cdot)$ is concave, assumption [A-5] - uniformity of arbitrage cones with respect to (portfolio) choice - holds automatically.

5.2 Asset market models: the no unbounded arbitrage conditions of [18] and [16]. Having introduced the basic ingredients used in general equilibrium models of asset markets, we are now in a position to state the conditions limiting arbitrage introduced by [**18**] and [**16**].

Hart's asset exchange model is essentially identical to the model given in Section 4 under assumptions [a-1]- [a-6]. Hart in fact assumes that each agent's choice set is a closed, convex subset of $\mathbf{R}^L$ independent of asset prices. Thus, in Hart, $\mathcal{A}_i(\omega_i, p) = \mathcal{R}\left(X_i \cap \widehat{P}_i(\omega_i, p)\right)$. As shown in [**18**], the following no unbounded

[14] That is, for any bounded continuous function $f(\cdot) : \mathbf{R}^L_+ \to \mathbf{R}$, the function $p \to \int_{\mathbf{R}^L_+} f(r)d\mu_i(dr \mid p)$ is continuous, and for each Borel subset $E \in B(\mathbf{R}^L_+)$, $\mu_i(E \mid p) = \mu_i(E \mid \lambda p)$ for all $\lambda > 0$.

arbitrage condition, suffices for existence of equilibrium:

$$\begin{array}{c}\text{for each } p' \in \mathcal{B}\backslash\{0\}, \text{ if } \sum_{i=1}^{n} y_i = 0 \text{ and } y_i \in \mathcal{A}_i(\omega_i, p') \text{ for all } i, \\ \text{then, for all } i \text{ and } p \text{ in some neighborhood of } p', \\ \mu_i(\{r \in \mathbf{R}^L_+ : \langle y_i, r\rangle = 0\} \mid p) = 1.\end{array} \tag{9}$$

Note that because agents' utility functions $v_i(\cdot)$ are concave, [A-5] holds automatically. Thus,

$$\mathcal{A}_i(\omega_i, p) = \mathcal{A}_i(x_i, p) \text{ for all } x_i \in \mathbf{R}^L.$$

Note also that [a-5] and [a-6] together imply that for any nonzero price vector $p \in \mathbf{R}^L$ and any $n - tuple$ of choice vectors $(x_1, \ldots, x_n) \in \mathbf{R}^L \times \cdots \times \mathbf{R}^L$, $\mathcal{A}_i(x_i, p) \backslash \mathcal{L}_i(x_i, p) \neq \emptyset$.

Finally, note that if, given asset prices p, net trades y_i generate zero gross return with probability 1 (i.e., if y_i is such that $\mu_i(\{r \in \mathbf{R}^L_+ : \langle y_i, r\rangle = 0\} \mid p) = 1$), then $y_i \in \mathcal{L}_i(x_i, p)$ for all $x_i \in \mathbf{R}^L$.

By Proposition 5.1 in [**35**], if there are no redundant assets (i.e., if [a-7] holds), then Hart's condition reduces to

$$\begin{array}{c}\text{for each } p \in \mathcal{B}\backslash\{0\}, \text{ if } \sum_{i=1}^{n} y_i = 0 \text{ and } y_i \in \mathcal{A}_i(\omega_i, p) \text{ for all } i, \\ \text{then } y_i = 0 \text{ for all } i.\end{array}$$

Thus, under [a-7] Hart's condition reduces to our condition of no unbounded arbitrage (6).

If in addition to satisfying [a-1]-[a-5] and [a-7], the model also satisfies and [a-8] (sufficient risk aversion), then by Corollary 2 (see also [**33**], Proposition 5), Hart's condition is equivalent to

$$\text{for each } p \in \mathcal{B}\backslash\{0\}, \bigcap\nolimits_i \mathcal{D}(\mathcal{A}_i(x_i, p)) \neq \emptyset. \tag{10}$$

Moreover, under [a-1]-[a-5] and [a-7]-[a-8], it follows from Lemma 5 in [**33**] that $\mathcal{A}_i(x_i, p) \backslash \{0\} = \mathcal{I}_i(x_i, p)$ - i.e., [A-3] holds. If [a-1]-[a-5] and [a-7] and [a-8](b) hold, then (10) reduces to Hammond's overlapping expectations condition given by

$$\text{for each } p \in \mathcal{B}\backslash\{0\} \bigcap\nolimits_i int\, kc\{S[\mu_i(\cdot \mid p)]\} \neq \emptyset. \tag{11}$$

Thus, under assumptions [a-1]-[a-5] and [a-7] and [a-8](b), the conditions of Hart and Hammond are equivalent to our condition of no unbounded arbitrage. Moreover, if in addition assumption [a-9] holds, then it follows from our Theorem 2 that the conditions of Hart and Hammond are necessary and sufficient for the existence of equilibrium, compactness of utility possibilities, and nonemptiness of the core for the asset exchange economy E.

5.3 Abstract general equilibrium models: the no unbounded arbitrage conditions of Page [36], Werner [51], and Nielsen [32]. Werner in [**51**] considers a general equilibrium model, $E = (X_i, \omega_i, u_i(\cdot))_{1=1}^n$, without price dependent preferences in which agents' choice sets are arbitrary closed, convex sets. Thus, in Werner arbitrage cones are given by $\mathcal{A}_i(x, p) = \mathcal{R}\left(X_i \cap \widehat{P}_i(x)\right)$. With the exception that [**51**] requires *uniformity of arbitrage cones* ([A-4]), the conditions on his economic model are weaker than ours.[15] Let $\mathcal{A}_i$ denote the *ith*

[15]We do not require uniformity to show that no unbounded arbitrage is sufficient for existence. Uniformity means that the set of potential arbitrage opportunities available to the agent is the same no matter what the agent's starting point x and no matter what the prices p. See also Remark 5.

agent's arbitrage cone and let $\mathcal{L}_i$ denote the corresponding lineality space (under [A-4], $\mathcal{A}_i(x,p) = \mathcal{A}_i$ for all $(x,p) \in \mathbf{R}^L \times \mathbf{R}^L$). Werner's nonsatiation assumption is

$$\text{[W-1] for all } i,\ \mathcal{A}_i \backslash \mathcal{L}_i \neq \emptyset$$

(see assumption A.4 in [**51**]).[16] By our Theorem 4, Werner's price condition limiting arbitrage, given by

$$\bigcap{}_i \mathcal{D}(\mathcal{A}_i \backslash \mathcal{L}_i) \neq \emptyset, \tag{12}$$

is weaker than no unbounded arbitrage. Finally, Werner's assumptions on endowments is given by

$$\text{[W-2] for all } i,\ \langle \omega_i, p\rangle \quad > \quad \inf\{\langle x, p\rangle : x \in X_i\} \text{ for all } p \in cl\left\{\bigcap{}_i \mathcal{D}(\mathcal{A}_i \backslash \mathcal{L}_i)\right\},$$
$$\text{where "}cl\text{" denotes closure.}$$

To summarize, within the context of an exchange economy, $(X_i, \omega_i, u_i(\cdot))_{1=1}^n$, without price dependencies satisfying assumption [A-1.1], [A-2.1], [A-4] (uniformity), [W-1] ($\mathcal{A}_i \backslash \mathcal{L}_i \neq \emptyset$), and [W-2] (endowments), Werner shows that his price condition limiting arbitrage, $\cap_i \mathcal{D}(\mathcal{A}_i \backslash \mathcal{L}_i) \neq \emptyset$, is sufficient for the existence of an equilibrium. Werner also notes that his condition (12) is necessary for existence provided indifference curves contain no half lines. Since in the no half lines case $\mathcal{L}_i = \{0\}$, it follows from our Corollary 2 that for this case, Werner's condition (12) is equivalent to our condition of no unbounded arbitrage (6). Moreover, the no half lines assumption, along with Werner's assumption of uniformity implies the economy, $(X_i, \omega_i, u_i(\cdot))_{1=1}^n$, is strictly reconcilable.

Remark 5. In the conclusion of his paper, Werner in [**51**] remarks that his results can be generalized to treat price dependent preferences so long as the recession cones of the "at least as good as" sets do not depend on prices and utility functions depend continuously on prices. As he also notes, uniformity assumptions have a long tradition in the theory of concavifiable preference orderings (see [**21**]). As we have noted, our sufficiency results do not require uniformity – arbitrage cones may vary as prices vary.

Page in [**36**] considers the question of existence of equilibrium within the context of an exchange economy with price dependent preferences in which agents' choice sets are all of $\mathbf{R}^L$. In [**36**], the economic model is given by a triple, $(\mathbf{R}^L, \omega_i, u_i(\cdot,\cdot))_{i=1}^n$, satisfying [A-1] and [A-2]. Thus, in [**36**] arbitrage cones are given by $\mathcal{A}_i(\omega_i, p) = \mathcal{R}\left(\widehat{P}_i(\omega_i, p)\right)$. Page in [**36**] shows that the price condition

$$\text{for each } p \in \mathcal{B}\backslash\{0\},\ \bigcap{}_i \mathcal{D}\left(\mathcal{R}\left(\widehat{P}_i(\omega_i, p)\right)\right) \neq \emptyset. \tag{13}$$

is sufficient for the existence of equilibrium.

For the sake of comparisons, Werner's condition (12) can be rewritten as

$$\bigcap{}_i \mathcal{D}\left(\mathcal{R}\left(X_i \bigcap \widehat{P}_i(\omega_i)\right) \backslash L\left(\mathcal{R}\left(X_i \bigcap \widehat{P}_i(\omega_i)\right)\right)\right) \neq \emptyset. \tag{14}$$

[16]Recall that our results continue to hold even if we replace our local nonsatiation assumption [A-2.3] with a weaker global nonsatiation assumption. However, the relationship between the nonsatiation condition $\mathcal{A}_i \backslash \mathcal{L}_i \neq \emptyset$ and global nonsatiation in economic models satisfying uniformity is an open question.

If in [**36**] utility functions are *not* price dependent and if in [**51**], $X_i = \mathbf{R}^L$ and $L\left(\mathcal{R}\left(X_i \cap \widehat{P}_i(\omega_i)\right)\right) = \{0\}$ (e.g., if indifference surfaces contain no half lines), then the price conditions of [**36**] and [**51**] are equivalent.

In a general equilibrium model without price dependences in which agents' choice sets are arbitrary closed, convex sets, [**32**] shows that the condition

$$\text{if } \sum_{i=1}^{n} y_i = 0 \text{ and } y_i \in \mathcal{R}(X_i \bigcap \widehat{P}_i(\omega_i)) \text{ for all } i \qquad \text{then } y_i = 0 \text{ for all } i, \tag{15}$$

is sufficient for the existence of equilibrium. Nielsen's economic model is similar to the model presented here, but without price dependencies. As Nielsen notes, the condition in [**32**] can be viewed as a variant of Page's condition in [**35**].

We close this section by noting that it follows from Corollary 2 that in a strictly reconcilable economy without price dependencies and choice sets given by all of $\mathbf{R}^L$, the conditions limiting arbitrage given by [**36**], [**51**], and [**32**] are all equivalent.

5.4 Comments on Green [13], and Grandmont [14, 15]. It follows from Lemma 2.5 in [**13**] that, in a temporary equilibrium model, overlapping expectations is necessary for the existence of equilibrium.[17] Also in a temporary equilibrium setting [**14, 15**] shows that if the economy satisfies conditions analogous to those needed to guarantee strict reconcilability in an asset market model, then the overlapping expectations condition is necessary and sufficient for the existence of a temporary equilibrium. Grandmont in [**14, 15**] makes an assumption similar to that of [**51**] for the case of price dependent preferences. It may be that our approach, using the result in [**49**], could be applied to temporary equilibrium models to treat a broader class of economies.

5.5 Comments on Kousougeras [25]. Kousougeras in [**25**] was the first to study the core in models with unbounded short sales. Kousougeras uses the no unbounded arbitrage condition of [**35**] to show that the set of individually rational and feasible allocations is bounded and then appeals to Scarf's Theorem [**48**] to prove that no unbounded arbitrage is sufficient for the nonemptiness of the core. This nonemptiness also follows from existence of a competitive equilibrium and is shown for the partnered core in [**39**] by appealing to the [**46**] result on the partnered core of a game without side payments.

6 Conclusions

We conclude this paper with some remarks on the core and the partnered core. Even in situations without externalities, the competitive equilibrium may be subject to asymmetric dependencies between groups of players.. If an allocation is in the core, it still may be the case that one player is dependent on another player – one player may need the cooperation of another player but the second player may not need the cooperation of the first. Similarly, a group of players may need the cooperation of another group. The partnered core, introduced in [**43**] and [**46**], is not subject to such asymmetric dependencies. Using the Reny and Wooders result, Page and Wooders in [**39**], and [**38**] show that the partnered core of a balanced economy is nonempty. From our result showing balancedness of an economy with

[17] The Green condition is sometimes referred to as common expectations (see [**13**], p. 1114).

price dependent preferences and satisfying no unbounded arbitrage it follows that the partnered core of the economy is nonempty.

In general, the competitive equilibrium payoff is not partnered. In strictly reconcilable economies, however, the partnership property of the competitive payoff was shown by [**3**]. It is immediate from our results in this paper and [**38, 39**] that in strictly reconcilable economies no unbounded arbitrage is necessary and sufficient for the nonemptiness of the partnered core and for existence of partnered competitive equilibrium.

7 Appendix: proofs

First, note that it follows from [A-2] that for each agent i the preference correspondences

$$(x,p) \to \widehat{P}_i(x,p) := \{x' \in \mathbf{R}^L : u_i(x',p) \geq u_i(x,p)\}$$
$$(x,p) \to P_i(x,p) := \{x' \in \mathbf{R}^L : u_i(x',p) > u_i(x,p)\}$$

exhibits the following continuity properties:

1. $(x,p) \to \widehat{P}_i(x,p)$ is continuous and, in particular, for each $x \in \mathbf{R}^L$, $p \to \widehat{P}_i(x,p)$ is continuous,
2. the graph of $P_i(\cdot,\cdot)$ is open relative to $(\mathbf{R}^L \times \mathcal{B}) \times \mathbf{R}^L$.

7.1 Preliminary lemmas. We begin with some preliminary lemmas. Lemma 1 is from [**35**].

Lemma 1 Let $\Lambda(\cdot)$ be a set-valued mapping defined on $\mathcal{B} \subset \mathbf{R}^L$ with nonempty convex values. Suppose $\Lambda(\ )$ is lower hemicontinuous (lhc) with a closed graph. Let $\{(p_k, x_k)\}_k \subset \mathcal{B} \times \mathbf{R}^L$ be any sequence with $p_k \to p^* \in \mathcal{B}$ and $x_k \in \Lambda(p_k)$ for all k. Further, let $\{\lambda_k\}_k$ be any sequence of positive real numbers such that $\lambda_k \downarrow 0$. Then any cluster point of the sequence $\{\lambda_k x_k\}_k$ is contained in the recession cone $\mathcal{R}(\Lambda(p^*))$.

The next Lemma can be proven using elementary facts concerning sequences.

Lemma 2 Let $\{x_{1k}, ..., x_{nk}\}_k$ be a sequence such that $\sum_{i=1}^n \|x_{ik}\| \to \infty$ as $k \to \infty$. Then for any cluster point $(y_1, ..., y_n)$ of the sequence $\{\lambda_k x_{1k}, ..., \lambda_k x_{nk}\}_k$ where $\lambda_k = \left(\sum_{i=1}^n \|x_{ik}\|\right)^{-1}$ it holds that $\sum_{i=1}^n y_i = 0$ and $\sum_{i=1}^n \|y_i\| = 1$.

7.2 Bounded economies. Let $\{r_k\}$ be a sequence of positive numbers such that $r_k \to \infty$ and let $S(k) \subset \mathbf{R}^L$ be a closed ball of radius r_k centered at the origin. The sequence $\{S(k)\}_k$ is an increasing sequence of closed balls. Suppose now that for each agent i, $\omega_i \in intS(k)$ for all k. Consider the k-bounded economy $E^k := (X_{ik}(\cdot), \omega_i, u_i(\cdot,\cdot))_{i=1}^n$ where, for each price vector p,

$$X_{ik}(p) := S(k) \bigcap X_i(p).$$

Given prices p and endowments $\omega = (\omega_1, ..., \omega_n)$, the set of all rational allocations for the economy E^k is given by

$$F_k(\omega,p) := \{(x_1, ..., x_n) \in \prod_{i=1}^n X_{ik}(p) : \sum_{i=1}^n x_i = \sum_{i=1}^n \omega_i$$
$$\text{and for each } i,\ x_i \in \widehat{P}_i(\omega_i, p)\}.$$

Let $\Gamma(F_k(\omega,\cdot))$ denote the graph of the rational allocation mapping $p \to F_k(\omega,p)$ for prices $p \in \partial\mathcal{B}$. Thus,

$$\Gamma(F_k(\omega,\cdot)) = \{((x_1,...,x_n,p) \in \mathbf{R}^L \times \cdots \times \mathbf{R}^L \times \partial\mathcal{B} : \\ (x_1,...,x_n) \in F_k(\omega,p)\}.$$

Proposition 1 *No unbounded arbitrage is necessary and sufficient for compactness of the graph of the rational allocation mapping.* Let $E = (X_i(\cdot),\omega_i,u_i(\cdot,\cdot))_{i=1}^n$ be an economy satisfying [A-1] and [A-2]. The following statements are equivalent:

1. For each $p \in \mathcal{B}\backslash\{0\}$ if $\sum_{i=1}^n y_i = 0$ and $y_i \in \mathcal{A}_i(\omega_i,p)$ for all i, then $y_i = 0$ for all i (there is no unbounded arbitrage).
2. $\Gamma(F(\omega,\cdot))$ is compact.

Proof The graph $\Gamma(F(\omega,\cdot))$ of the rational allocation mapping $p \to F(\omega,p)$ is contained in the space

$$\mathbf{R}^L \times \cdots \times \mathbf{R}^L = \mathbf{R}^{L(n+1)}.$$

Equip $\mathbf{R}^{L(n+1)}$ with the norm $\|\cdot\|_E$ defined as follows: given

$$(x',p') = (x'_1,...,x'_n,p') \in \mathbf{R}^{L(n+1)} \text{ and } (x,p) = (x_1,...,x_n,p) \in \mathbf{R}^{L(n+1)}, \\ \|(x',p') - (x,p)\|_E = \|p' - p\| + \textstyle\sum_{i=1}^n \|x'_i - x_i\|,$$

where $\|\cdot\|$ is the usual norm in $\mathbf{R}^L$. It is easy to check that $\Gamma(F(\omega,\cdot))$ is a closed subset of $\mathbf{R}^{L(n+1)}$ under the $\|\cdot\|_E$ norm. Moreover, for each k, $\Gamma(F_k(\omega,\cdot))$ is closed in $\mathbf{R}^{L(n+1)}$ in the topology induced by $\|\cdot\|_E$.

We will in fact prove the following: If the economy $E = (X_i(\cdot),\omega_i,u_i(\cdot,\cdot))_{i=1}^n$ satisfies [A-1] and [A-2], then the following are equivalent:

(i) For each $p \in \partial\mathcal{B}$, if $\sum_{i=1}^n y_i = 0$ and $y_i \in \mathcal{A}_i(\omega_i,p)$ for all i, then $y_i = 0$ for all i.

(ii) There exists an integer k_0 such that $\Gamma(F(\omega,\cdot)) = \Gamma(F_k(\omega,\cdot))$ for all $k \geq k_0$.

If (ii) holds then $\Gamma(F(\omega,\cdot))$ is $\|\cdot\|_E$-bounded and thus (i) holds. To see that (i) implies (ii), consider the following: Let $\{(\overline{x}_{1k},...,\overline{x}_{nk},\overline{p}_k)\}_k$ be a sequence contained in $\Gamma(F(\omega,\cdot))$ and satisfying the property that for each k, $(\overline{x}_{1k},...,\overline{x}_{nk},\overline{p}_k) \notin \Gamma(F_k(\omega,\cdot))$. Thus, for each k, there is some i such that $\overline{x}_{ik} \notin S(k)$ so that $\sum_{i=1}^n \|\overline{x}_{ik}\| \to \infty$ as $k \to \infty$.

Let $(\overline{y}_1,...,\overline{y}_n,\overline{p}) \in \mathbf{R}^L \times \cdots \times \mathbf{R}^L \times \partial\mathcal{B}$ be a cluster point of the sequence $\{(\lambda_k\overline{x}_{1k},...,\lambda_k\overline{x}_{nk},\overline{p}_k)\}_k$ where $\lambda_k = (\sum_{i=1}^n \|\overline{x}_{ik}\|)^{-1.}$ Note that the set-valued mapping $p \to X_i(p) \cap \widehat{P}_i(\omega_i,p)$ is convex-valued, lower hemicontinuous, and has a closed graph. Thus, since for each k, $\overline{x}_{ik} \in X_i(\overline{p}_k) \cap \widehat{P}_i(\omega_i,\overline{p}_k)$, it follows from Lemma 1 that $\overline{y}_i \in \mathcal{R}\left(X_i(\overline{p}) \cap \widehat{P}_i(\omega_i,\overline{p})\right) = \mathcal{A}_i(\omega_i,p)$ for each agent i. By Lemma 2, $\sum_{i=1}^n \overline{y}_i = 0$ and $\sum_{i=1}^n \|\overline{y}_i\| = 1$. Thus, for some agent i, $\overline{y}_i \neq 0$ and thus the no unbounded arbitrage condition fails to hold, a contradiction. □

Let $\Gamma(F_C(\omega,\cdot))$ denote the graph of the rational, C-allocation mapping $p \to F_C(\omega,p)$ for $p \in \partial\mathcal{B}$. The following is an easy corollary of Proposition 1.

Corollary Let $E = (X_i(\cdot),\omega_i,u_i(\cdot,\cdot))_{i=1}^n$ be an economy satisfying [A-1] and [A-2]. The following statements are equivalent:

1. For each $p \in \mathcal{B}\backslash\{0\}$ if $\sum_{i=1}^n y_i = 0$ and $y_i \in \mathcal{A}_i(\omega_i,p)$ for all i, then $y_i = 0$ for all i (there is no unbounded arbitrage).
2. For any coalition $C \in 2^N$, $\Gamma(F_C(\omega,\cdot))$ is compact.

An *equilibrium* for the $k-$bounded economy $(X_{ik}(\cdot),\omega_i,u_i(\cdot,\cdot))_{i=1}^n$ is an $(n+1)$-tuple of vectors $(x_1^*,...,x_n^*,p^*)$ such that

1. $(x_1^*, ..., x_n^*) \in F_k(\omega, p^*)$ (the allocation is feasible);
2. $p^* \in \mathcal{B}\backslash\{0\}$ (prices are in the unit ball and not all prices are zero); and
3. for each agent i,
   (a) $\langle x_i^*, p^* \rangle = \langle \omega_i, p^* \rangle$ (budget constraints are satisfied), and
   (b) $P_i(x_i^*, p^*) \cap X_{ik}(p^*) \cap B_i(\omega_i, p^*) = \emptyset$ (there are no affordable preferred net trades).

For each k and $(x, p) \in S(k) \times \mathcal{B}$ define the constraint correspondences

$$H_i(k, \omega_i, p) := X_i(p) \cap B_i(\omega_i, p) \cap S(k), \ i = 1, 2, ..., n.$$

and the preference correspondences

$$P_i(k, x, p) := P_i(x, p) \cap S(k), \ i = 1, ..., n \text{ and}$$
$$\widehat{P}_i(k, x, p) := \widehat{P}_i(x, p) \cap S(k), \ i = 1, ..., n.$$

Note that

$$H_i(\infty, \omega_i, p) := H_i(\omega_i, p) := X_i(p) \cap B_i(\omega_i, p),$$
$$\widehat{P}_i(\infty, x, p) := \widehat{P}_i(x, p)$$

and

$$P_i(\infty, x, p) := P_i(x, p).$$

The correspondences $H_i(k, \omega_i, \cdot)$, $P_i(k, \cdot, \cdot)$ and $\widehat{P}_i(k, \cdot, \cdot)$ have the following properties:

1. Given [A-1], for each $i = 1, ..., n$ and $k = 1, 2, ...$

$$p \to H_i(k, \omega_i, p)$$

   is continuous with nonempty, closed convex values.
2. Given [A-2], for each $i = 1, 2, ..., n$ and $k = 1, 2, ...$

$$(x, p) \to P_i(k, x, p)$$

   has nonempty, convex values in $S(k)$ and an open graph relative to $[S(k) \times \mathcal{B}] \times S(k)$. Moreover, for each $i = 1, 2, ..., n$ and $k = 1, 2, ...$

$$p \to \widehat{P}_i(k, \omega_i, p)$$

   is continuous.

Proposition 2 *Equilibrium for the k-bounded economies.*
Let $E = (X_i(\cdot), \omega_i, u_i(\cdot, \cdot))_{i=1}^n$ be an economy satisfying [A-1] and [A-2] and consider the sequence of k-bounded economies $E^k = (X_{ik}(\cdot), \omega_i, u_i(\cdot, \cdot))_{i=1}^n$, $k = 1, 2....$ Each $k-$bounded economy has an equilibrium $(x_{1k}^*, ..., x_{nk}^*, p_k^*)$ with $p_k^* \in \partial\mathcal{B}$.

Given observations 1. and 2. above, the proof of Proposition 2 follows directly from [**49**].

Now we have two important observations concerning the $k-$ bounded equilibria:

1. If $(x_{1k}^*, ..., x_{nk}^*, p_k^*)$ is an equilibrium for the k-bounded economy E^k, then for each agent $i = 1, 2, ..., n$, $x_{ik}^* \in X_i(p_k^*) \cap \widehat{P}_i(\omega_i, p_k^*)$.
2. If $(x_{1k}^*, ..., x_{nk}^*, p_k^*)$ is an equilibrium for the k-bounded economy E^k such that for each agent $i = 1, ..., n$, $x_{ik}^* \in intS(k)$, then $(x_{1k}^*, ..., x_{nk}^*, p_k^*)$ is an equilibrium for the economy $E = (X_i(\cdot), \omega_i, u_i(\cdot, \cdot))_{i=1}^n$.

7.3 Proof of Theorem 1. Consider an economy $E = (X_i(\cdot), \omega_i, u_i(\cdot,\cdot))_{i=1}^n$ satisfying [A-1] and [A-2].

1. *No unbounded arbitrage implies the existence of equilibrium.* Consider a sequence $\{(x_{1k}^*, ..., x_{nk}^*, p_k^*)\}_{k=1}^{\infty}$ of equilibria for the bounded economies E^k, with $p_k^* \in \partial\mathcal{B}$. Suppose that for each k, $x_{ik}^* \notin intS(k)$ for some agent i. Thus, $\sum_{i=1}^n \|x_{ik}^*\| \to \infty$ as $k \to \infty$. Let $(y_1^*, ..., y_n^*, p^*) \in \mathbf{R}^L \times \cdots \mathbf{R}^L \times \partial\mathcal{B}$ be a cluster point of the sequence $\{(\lambda_k x_{1k}^*, ..., \lambda_k x_{nk}^*, p_k^*)\}_k$ where $\lambda_k = \left(\sum_{i=1}^n \|x_{ik}^*\|\right)^{-1}$. Again, note that the set-valued mapping $p \to X_i(p) \cap \widehat{P}_i(\omega_i, p_k^*)$ is convex-valued, lower hemicontinuous, and has a closed graph. Thus, since for each k and i, $x_{ik}^* \in X_i(p_k^*) \cap \widehat{P}_i(\omega_i, p_k^*)$, it follows from Lemma 1 that for each agent i, $y_i^* \in \mathcal{R}\left(X_i(p^*) \cap \widehat{P}_i(\omega_i, p^*)\right) = \mathcal{A}_i(\omega_i, p^*)$. By Lemma 2, $\sum_{i=1}^n y_i = 0$ and $\sum_{i=1}^n \|y_i^*\| = 1$. Thus, for some agent i, $y_i^* \neq 0$, contradicting the condition of no unbounded arbitrage.
2. *No unbounded arbitrage implies that $U(F(\omega,\cdot))$ is compact.* By Proposition 1, no unbounded arbitrage implies that $\Gamma(F(\omega,\cdot))$ is compact. The compactness of $U(F(\omega,\cdot))$ then follows directly from assumption [A-2.1] that the utility functions $u_i(\cdot,\cdot)$ are continuous.
3. *No unbounded arbitrage implies that E is well-defined and balanced.* It follows from the definition (see expression 3) that $V(C,p)$ is nonempty for any price vector $p \in \partial\mathcal{B}$ and any coalition $C \in 2^N$. By [A-2.2] (the quasi-concavity of preferences), $V(C,p)$ is convex for any $C \in 2^N$ and $p \in \partial\mathcal{B}$. Finally, since no unbounded arbitrage implies that $\Gamma(F_C(\cdot,\omega))$ is compact for any coalition C, $V(C,p)$ is closed and bounded from above for any coalition C and $p \in \partial\mathcal{B}$. Thus, for each $p \in \partial\mathcal{B}$ the characteristic form game $(N, V(\cdot,p))$ corresponding to the economy E is well-defined.

   To see that the game is balanced, consider the following: Let β be a balanced collection of nonempty subsets of N with balancing weights δ_C for $C \in \beta$. Thus, $\delta_C \geq 0$ for all $C \in \beta$ and, for each $i \in N$,

$$\sum_{C \in \beta(i)} \delta_C = 1.$$

   Given $p \in \partial\mathcal{B}$, let $(u_1, ..., u_n) \in \cap_{C\in\beta} V(C,p)$. It follows that for each $C \in \beta$ there is a individually rational C-allocation $(x_i^C : i \in C) \in F_C(\omega, p)$ such that

$$u_i(x_i^C, p) \geq u_i \text{ for each } i \in C. \quad (*)$$

   For each $i \in N$, let

$$x_i = \sum_{C\in\beta(i)} \delta_C x_i^C,$$

   and observe that

$$\sum_{i\in N} x_i = \sum_{i\in N}\sum_{C\in\beta(i)} \delta_C x_i^C = \sum_{i\in N} \omega_i.$$

   By inequality $(*)$ and the quasi-concavity of $u_i(\cdot,p)$ we have for each $i \in N$

$$u_i = \sum_{C\in\beta(i)} \delta_C u_i \leq \sum_{C\in\beta(i)} \delta_C u_i(x_i^C, p) \leq u_i(x_i, p).$$

Moreover, since $(x_i^C : i \in C) \in F_C(\omega, p)$, we have for each $i \in N$,

$$u_i(\omega_i, p) = \sum_{C \in \beta(i)} \delta_C u_i(\omega_i, p) \leq \sum_{C \in \beta(i)} \delta_C u_i(x_i^C, p) \leq u_i(x_i, p).$$

Thus, $(x_1, ..., x_n) \in F(w, p)$ and we can conclude that $(u_1, ..., u_n) \in V(N, p)$. This implies that the game $(N, V(\cdot, p))$ corresponding to the economy is well-defined and balanced for each $p \in \partial\mathcal{B}$.

4. *No unbounded arbitrage implies that E has a nonempty core.* By part 1, no unbounded arbitrage implies that E has an equilibrium, say $(x_1^*, ..., x_n^*, p^*)$, with $p^* \in \partial\mathcal{B}$. We will show that $(x_1^*, ..., x_n^*)$ is in the core given p^*. Suppose not. Then for some nonempty coalition C there is a rational C-allocation $(x_i : i \in C) \in F_C(\omega, p^*)$ such that

$$u_i(x_i, p^*) > u_i(x_i^*, p^*) \text{ for all agents } i \in C. \ (**)$$

Since $\sum_{i \in C} x_i = \sum_{i \in C} \omega_i$, it holds that $\langle \sum_{i \in C} x_i, p^* \rangle = \langle \sum_{i \in C} \omega_i, p^* \rangle$. However, given $(**)$, $\langle x_i^*, p^* \rangle > \langle \omega_i, p^* \rangle$ for all $i \in C$. Therefore

$$\left\langle \sum_{i \in C} x_i, p^* \right\rangle > \left\langle \sum_{i \in C} \omega_i, p^* \right\rangle,$$

a contradiction. □

7.4 Proof of Theorem 2. Consider a strictly reconcilable economy $E = (X_i(\cdot), \omega_i, u_i(\cdot, \cdot))_{i=1}^n$. Thus, in addition to satisfying [A-1] and [A-2] the economy satisfies [A-3] and [A-4]. First, note that the equivalence of (1) and (2) follows immediately from the Corollary to Rockafellar's Theorem.

1. *The existence of an equilibrium implies no unbounded arbitrage.*

   Let $(x_1^*, ..., x_n^*, p^*)$ be an equilibrium for the economy E. If no unbounded arbitrage fails to hold then by the Corollary to Rockafellar's Theorem we have for some $p' \in \partial\mathcal{B}$, $\cap_i \mathcal{D}(\mathcal{A}_i(\omega_i, p')) = \emptyset$. This implies via assumption [A-4] (the uniformity of arbitrage cones) that $\cap_i \mathcal{D}(\mathcal{A}_i(x_i^*, p^*)) = \emptyset$. Thus, for some agent i', there is a nonzero vector of net trades $y_{i'} \in \mathcal{A}_i(x_{i'}^*, p^*)$ such that $\langle y_{i'}, p^* \rangle \leq 0$. By assumption [A-3] (extreme desirability), for some $\lambda' > 0$ $u_{i'}(x_{i'}^* + \lambda' y_{i'}, p^*) > u_{i'}(x_{i'}^*, p^*)$. Thus for λ',

$$x_{i'}^* + \lambda' y_{i'} \in P_{i'}(x_i^*, p^*) \cap X_{i'}(p^*) \cap B_{i'}(\omega_{i'}, p^*),$$

   contradicting the fact that $(x_1^*, ..., x_n^*, p^*)$ is an equilibrium.

2. *Compactness of the utility possibilities set $U(F(\omega, \cdot))$ implies no unbounded arbitrage.* Suppose that $U(F(\omega, \cdot))$ is compact and no unbounded arbitrage fails to hold. Then for some $p' \in \partial\mathcal{B}$, there exists $(y_1', ..., y_n') \neq (0, ..., 0)$ such that

$$\sum_{i=1}^n y_i' = 0 \text{ and } y_i' \in \mathcal{A}_i(\omega_i, p') \text{ for all } i.$$

By the uniformity assumption [A-4], for any $(x_1, ..., x_n, p) \in \Gamma(F(\omega, \cdot))$, $y_i' \in \mathcal{A}_i(x_i, p)$ for all i and $(x_1 + \lambda y_1', ..., x_n + \lambda y_n', p) \in \Gamma(F(\omega, \cdot))$ for all $\lambda > 0$. Now let $\delta = (\delta_1, ..., \delta_n)$ be a vector in $\mathbf{R}^n$ with strictly positive components and let $\overline{u} = (\overline{u}_1, ..., \overline{u}_n) \in U(F(\omega, \cdot))$ satisfy

$$\langle \delta, \overline{u} \rangle = \max\{\langle \delta, u \rangle : u \in U(F(\omega, \cdot))\}.$$

Since $U(F(\omega,\cdot)$ is compact such a $\overline{u}$ exists. Suppose $(\overline{x}_1,...,\overline{x}_n,\overline{p})\in\Gamma(F(\omega,\cdot))$ is such that $\overline{u}_i = u_i(\overline{x}_i,\overline{p})$ for each i. By the extreme desirability assumption [A-3], we have for some $\lambda' > 0$,

$$\sum_{i=1}^{n}\delta_i u_i(\overline{x}_i + \lambda' y_i', \overline{p}) > \sum_{i=1}^{n}\delta_i u_i(\overline{x}_i,\overline{p}) = \sum_{i=1}^{n}\delta_i\overline{u}_i = \langle\delta,\overline{u}\rangle$$
$$= \max\{\langle\delta,u\rangle : u\in U(F(\omega,\cdot))\}.$$

But since $(u_1(\overline{x}_1 + \lambda y_1',\cdot),...,u_n(\overline{x}_n + \lambda y_n',\cdot)) \in U(F(\omega,\cdot)$ for all $\lambda > 0$, we have a contradiction. □

3. *Well-definedness and balancedness imply no unbounded arbitrage.* If the economy E is well-defined and balanced then E has a nonempty core. Let $(\overline{x}_1,...,\overline{x}_n,\overline{p}) \in \Gamma(F(\omega,\cdot))$ be such that $(\overline{x}_1,...,\overline{x}_n)$ is in the core. Suppose no unbounded arbitrage fails to hold. Then for some $p' \in \partial\mathcal{B}$, there exists $(y_1',...,y_n') \neq (0,...,0)$ such that

$$\sum_{i=1}^{n} y_i' = 0 \text{ and } y_i' \in \mathcal{A}_i(\omega_i,p') \quad \text{for all } i.$$

By the uniformity assumption (i.e., [A-4]), $y_i' \in \mathbf{A}_i(\overline{x}_i,\overline{p}_i)$ for all i and $(\overline{x}_1 + \lambda y_1',...,\overline{x}_n + \lambda y_1',\overline{p}) \in \Gamma(F(\omega,\cdot))$ for all $\lambda > 0$. By the extreme desirability assumption (i.e., [A-3]), there exists $\lambda' > 0$ such that $u_i(\overline{x}_i+\lambda' y_i',\overline{p}) > u_i(\overline{x}_i,\overline{p})$ for any agent i such that $y_i' \neq 0$. Thus, the individually rational allocation $(\overline{x}_1 + \lambda' y_1',...,\overline{x}_n + \lambda' y_1') \in F(\omega,\overline{p})$ Pareto dominates the individually rational allocation $(\overline{x}_1,...,\overline{x}_n) \in F(\omega,\overline{p})$, contradicting the supposition that $(\overline{x}_1,...,\overline{x}_n)$ is in the core.
4. *Nonemptiness of the core implies no unbounded arbitrage.* The proof follows from the arguments given in (3) above.

7.5 Proof of Theorem 3. Let $(X_i(\cdot),\omega_i,u_i(\cdot,\cdot))_{i=1}^n$ be an economy satisfying [A-1]- [A-3] and let $(x_1^*,...,x_n^*,p^*)$ be an equilibrium for E. Suppose $p^* \notin \cap_i\mathcal{D}(\mathcal{A}_i(x_i^*,p^*))$. Thus, for some agent i', there is a nonzero vector of net trades $y_{i'} \in \mathcal{A}_{i'}(x_{i'}^*,p^*)$ such that $\langle y_{i'},p^*\rangle \leq 0$. By assumption [A-3] (extreme desirability), $u_{i'}(x_{i'}^* + \lambda' y_{i'},p^*) > u_{i'}(x_{i'}^*,p^*)$ for some $\lambda' > 0$. Thus, for any λ',

$$x_{i'}^* + \lambda' y_{i'} \in P_{i'}(x_{i'}^*,p^*) \cap X_{i'}(p^*) \cap B_{i'}(\omega_{i'},p^*),$$

contradicting the fact that $(x_1^*,...,x_n^*,p^*)$ is an equilibrium. The fact that the condition

$$\text{if } \sum_{i=1}^{n} y_i = 0 \text{ and } y_i \in \mathcal{A}_i(x_i^*,p^*) \text{ for all } i, \text{then } y_i = 0 \text{ forall } i,$$

holds if $(x_1^*,...,x_n^*,p^*)$ is an equilibrium now follows from the Corollary to Rockafellar's Theorem. □

References

[1] Albers, W. [1979], *Core and kernel variants based on imputations and demand profiles*, in Game Theory and Related Topics (O. Moeschlin and D. Pallaschke, eds.), North Holland, Amsterdam.

[2] Bennett, E. [1983], *The aspiration approach to predicting coalition formation and payoff distribution in sidepayment games*, International Journal of Game Theory **12**, 1-28.

[3] Bennett, E. and Zame, W. R. [1988], *Bargaining in cooperative games*, International Journal of Game Theory **17**, 279-300.

[4] Berge, C. [1963], *Topological Spaces*, Macmillan, New York.

[5] Bondareva, O. [1962], *Theory of the core in an n -Person Game*, Vestnik, LGU13, 141-142 (in Russian), (Leningrad State University, Leningrad).

[6] Brown, J. D. and Werner, J. [1995], *Arbitrage and existence of equilibrium in infinite asset markets*, Review of Economic Studies **62**, 101-114.

[7] Cheng, H. C. C. [1991], *Asset Market Equilibrium in Infinite Dimensional Complete Markes*, Journal of Mathematical Economics **20**, 137-152.

[8] Dana, R. A., Le Van, C. and Magnien, F. [1996], *On different notions of arbitrage and existence of equilibrium*, Journal of Economic Theory (to appear).

[9] Debreu, G. [1962], *New concepts and techniques for equilibrium analysis*, International Economic Review **3**, 257-273.

[10] Debreu G. and Scarf, H. [1963], *A limit theorem on the core of an economy*, International Economic Review **4**, 235-246.

[11] Duffie, D. [1986], *Competitive equilibrium in general choice spaces*, Journal of Mathematical Economics **14**, 9-16.

[12] Gale, D. and Mas-Colell, A. [1975], *An equilibrium existence theorem for a general model without ordered preferences*, Journal of Mathematical Economics **2**, 9-16.

[13] Green, J. R. [1973], *Temporary general equilibrium in a sequential trading model with spot and futures transactions*, Econometrica **41**, 1103-1123.

[14] Grandmont, J. M. [1977], *Temporary general equilibrium theory*, Econometrica **45**, 535-572.

[15] Grandmont, J. M. [1982], *Temporary general equilibrium theory*, Handbook of Mathematical Economics, Volume II, North Holland.

[16] Hammond, P. J. [1983], *Overlapping expectations and Hart's condition for equilibrium in a securities model*, Journal of Economic Theory **31**, 170-175.

[17] Hammond, P., Kaneko, M. and Wooders M. H. [1989], *Continuum economies with finite coalitions: Core, equilibria, and widespread externalities*, Journal of Economic Theory **49**, 113-134.

[18] Hart, O. D. [1974], *On the Existence of equilibrium in a securities model*, Journal of Economic Theory **9**, 293-311.

[19] Kaneko, M. and Wooders, M. H. [1982], *Cores of partitioning games*, Mathematical Social Sciences **3**, 313-327.

[20] Kaneko, M. and Wooders, M. H. [1989], *The core of a continuum economy with widespread externalities and finite coalitions: From finite to continuum economics*, Journal of Economic Theory **49**, 135-168.

[21] Kannai, Y. [1977], *Concavifability and constructions of concave utility functions*, Journal of Mathematical Economics **4**, 1-56.

[22] Kannai, Y. and Wooders, M. H. [1996], *A further extension of the KKMS Theorem*, University of Bielefeld IMW Working Paper 251.

[23] Keiding, H. and Thorlund-Petersen, L. [1987], *The core of a cooperative game without side payments*, Journal of Optimization Theory **54**, 273-288.

[24] Kim, C. [1998], *Stochastic dominance, Pareto optimality, and equilibrium asset pricing*, Review of Economic Studies **65**, 341-356.

[25] Kousougeras, L. [1992], *A two-stage core with applications to asset market and differential information economies*, Revised as Economic Theory **11(3)**, 563-584, 4/1998.

[26] Kreps, D. [1981], *Arbitrage and equilibrium in economies with infinitely many commodities*, Journal of Mathematical Economics **8**, 15-35.

[27] Le Van, C. and Magnien, F. [1998], *No arbitrage condition and existence of equilibria in asset markets with a continuum of traders*, presented at the 1998 Association for Public Economic Theory Conference, University of Alabama May 1998.

[28] Monteiro, P., Page, F. and Wooders, M. H. [1998], *Increasing cones, global cones and recession cones*, Optimization (to appear).

[29] Monteiro, P., Page, F. and Wooders, M. H. [1999], *Arbitrage and global cones; another counterexample*, Social Choice and Welfare **16**, 337-346.

[30] Monteiro, P., Page, F. and Wooders, M. H. [1997], *Arbitrage, equilibrium and gains from trade; A counterexample*, Journal of Mathematical Economics **28**, 481-501.

[31] Monteiro, P., Page, F. H. Jr. and Wooders, M. H. [1995], *Arbitrage and global cones; Global cones are not open*, Department of Finance University of Alabama Working Paper 254.

[32] Nielsen, L. [1989], *Asset Market Equilibrium with Short Selling*, Review of Economic Studies **56**, 467-474.

[33] Page, F. H., Jr. [1996], *Arbitrage and Asset Prices*, Mathematical Social Sciences **31**, 183-208.

[34] Page, F. H., Jr. [1989], *Utility arbitrage and asset prices*, Graduate School of Business, Indiana University Working Paper No. 401.

[35] Page, F. H., Jr. [1987], *On equilibrium in Hart's securities exchange model*, Journal of Economic Theory **41**, 392-404.

[36] Page, F. H., Jr. [1984], *Equilibria in Unbounded Economies*, Department of Finance, The University of Texas at Austin, Working Paper 83/84-2-19.

[37] Page, F. H., Jr. [1982], *Information, arbitrage, and equilibrium*, Department of Finance, The University of Texas at Austin, Working paper 81/82-2-51.

[38] Page, F. H., Jr. and Wooders, M. H. [1996a], *The partnered core and the partnered competitive equilibrium*, Economics Letters **52**, (1996) 143-152.

[39] Page, F. H., Jr. and Wooders, M. H. [1996b], *A necessary and sufficient condition for compactness of individually rational and feasible outcomes and existence of an equilibrium*, Economics Letters **52**, (1996) 153-162.

[40] Page, F. H., Jr. and Wooders, M. H. [1995], *The partnered core and the partnered competitive equilibrium*, University of Alabama Department of Finance Working Paper No. 250 (Revised November 1995).

[41] Page, F. H., Jr. and Wooders, M. H. [1993], *Arbitrage in Markets with unbounded short sales: Necessary and sufficient conditions for nonemptiness of the core and existence of equilibrium*, University of Toronto Working Paper No. 9409.

[42] Page, F. H., Jr., Wooders, M. H. and Monteiro, P. [1997], *Inconsequential arbitrage*, Presented at the Fall 1997 Meetings of the Midwest Math Econ Association and at the 1998 Winter Meetings of the Econometric Society.

[43] Reny, P. J., Winter, E. and Wooders, M. H. [1993], *The partnered core of a game with side payments*, Hebrew University Center for Rationality and Interactive Decision Theory, Discussion Paper 33.

[44] Reny, P. J. and Wooders, M. H. [1996], *The partnered core of a game without side payments*, Journal of Economic Theory **70**, 298-311.

[45] Reny, P. J. and Wooders, M. H. [1996], *Credible threats of secesion, partnership, and commonwealths*, in Understanding Strategic Interaction; Essays in Honor of Reinhard Selten, (W. Gueth, P. Hammerstein, B. Moldovanu, E. van Damme, eds.), Springer Verlag Berlin/Heidelberg/New York/Tokyo 305-312.

[46] Reny, P. J. and Wooders, M. H. [1998], *An extension of the KKMS Theorem*, Journal of Mathematical Economics **29**, 125-134.

[47] Rockafellar, R. T. [1970], *Convex Analysis,* Princeton University Press.

[48] Scarf, H. E. [1967], *The Core of an n-person Game*, Econometrica **35**, 50-67.

[49] Shafer, W. J. and Sonnenschein, H. [1975], *Equilibrium in abstract economies without ordered preferences*, Journal of Mathematical Economics **2**, 345-348.

[50] Shapley, L. S. [1967], *On balanced sets and cores*, Naval Research Logistics Quarterly **14**, 453-460.

[51] Werner, J. [1987], *Arbitrage and existence of competitive equilibrium*, Econometrica **55**, 1403-1418.

Fields Institute Communications
Volume **23**, 1999

A Smooth Social Choice Method of Preference Aggregation

Norman Schofield
Center in Political Economy
Washington University
St. Louis, Missouri 63130-4899 USA
schofld@wuecon.wustl.edu

Abstract. A process h, parametrized by preference, assigns to each point x in the state space, W, a cone of feasible vectors at x. For a typical voting process, the equilibrium $E(h)$ of h will be empty and the voting trajectories induced by h can wander everywhere. To solve this social choice problem, a particular solution set in W, termed the *heart* $\mathcal{H}(h)$ as defined, is shown to exist and to admit a "Paretian" selection, g, which is continuous with respect to a C^1-topology on preferences. For representative democracy, g induces a game form whose Nash equilibria are proposed as the outcome of the electoral process.

1 Introduction

Unlike economic theory, where equilibrium can be expected, social choice and voting theory have to accommodate the possibility of cycles and non-existence of equilibrium. In the case of smooth preferences on a topological space, W, an extension of Fan's theorem can be used to obtain a duality condition which is sufficient to guarantee existence of equilibrium. Using this result, a cooperative solution concept, called the *heart*, is constructed. It is shown that the heart is always non-empty, and is a subset of the "critical Pareto set". Viewed as a correspondence from the space of smooth convex preference profiles, the heart is lower hemi-continuous, and admits a continuous selection.

The outcome of bargaining between parties in a Parliament is seen as a probability measure over the set of outcomes belonging to a political heart. The possibility of a continuous selection from the set of such probability measures implies that political choice is a continuous game form, which under certain conditions will exhibit Nash equilibria. The paper concludes with a brief interpretation of bargaining in the Israeli Knesset.

1991 *Mathematics Subject Classification.* Primary 90A06; Secondary 90A14.

This is the fifth in a series of papers (following [**31, 32, 33, 34**]) on the aggregation of preferences by voting or by a system of electoral representation. The research has been supported by NSF grants SBR-94-22048, 96-17708, and 97-30275.

I am very grateful to Jeff Banks, Marcus Berliant, John Duggan and Myrna Wooders for helpful comments.

2 Equilibrium of a dynamical process and a theorem of fan

In economic theory and social choice it is usual to consider general equilibrium, or rest points, of some process like exchange or voting. Such processes are defined or parametrized in terms of the preference correspondences of the individuals in the economy or society. To seek global equilibria however, requires that individuals search over the entire economic or political landscape. Given the information requirements of such a global search, it makes more sense to look for local equilibria. In situations where all preferences are convex (that is, all preferred sets are convex objects), then local and global equilibria will generally coincide. However the local theory allows one to consider non-convex preferences, and to then consider additional constraints sufficient to guarantee that the equilibria are stable. This is obviously the procedure in differential calculus, where the critical points are characerized by the index, and this can then be used to determine the nature of the critical point (a maximum, minimum, saddle, etc.). A further advantage of this differential approach is that the analysis can be done on a smooth state space (namely a manifold), and the process interpreted as a vector field. The Euler characteristic of the state space can then be seen as an obstruction to the existence of equilibria of the process. For example, any even-dimensional sphere has Euler characteristic equal to two. Thus any vector field on such a sphere must have a zero. The odd dimensional sphere has zero Euler characteristics and exhibits vector fields with no zeros. (For example clockwise rotation on the 1-sphere, S^1, has no zeros.) Since the surface of the earth is S^2, this is probably the underlying reason why there are singularities in plate tectonics, weather systems, etc.

In economics, game theory and social choice, it is usual to use fixed point analysis [**4, 12**] on the simplex, to show existence of equilibrium. However, the proof of Brouwer's Theorem was based on an argument using the Euler characteristic (in fact the "hairy ball" theorem, see [**20**]), so a more general proof strategy for existence of equilibria is to use differential techniques on a smooth manifold, and then to seek whether the "first order" conditions are sufficient to guarantee existence of equilibrium. Since the Euler characteristic is the most important obstruction, it is convenient to use the term "Fan manifold" to denote a *finite dimensional* smooth compact manifold W, of dimension w with non-zero Euler characteristic. A manifold admits smooth "charts" (homeomorphisms) from any neighborhood V in W to $\Re^w$. At any interior point $x \in W$, the tangent space T_xW at x is a copy of $\Re^w$ comprising all tangent vectors at x. For a neighborhood V, there is a smooth homeomorphism between $TV = \cup_{x \in V} T_xW$ and $V \times \Re^w$. Use TW to denote the tangent space. There is a natural projection $\pi : TW \to W$, with $\pi^{-1}(x) \simeq T_xW$. A *vector field* on W is a differentiable function $v : W \to TW$ where $v(x) \in \pi^{-1}(x)$ for each x. In some cases we need to restrict W to be a compact convex affine subspace of finite dimensional Euclidean space. In this case we shall call W a *Fan space* (of dimension w). Note that a Fan space has non-zero Euler characteristic. For $W \equiv \Re^w$, Euclidean space, a vector $v(x)$ at x may be interpreted simply as $(y - x)$ for some $y \in \Re^w$.

Here we shall be interested in a *process*, of some kind, whether economic or choice theoretic. A process can be thought of as a generalization of a vector field.

A (dynamical) *process* is an operator $h : W \to TW$ where for each $x \in W$, $h(x)$ is a cone in T_xW (thus if $v \in h(x)$, then $\lambda v \in h(x)$ for all $\lambda > 0$). The *equilibrium* of h is $E(h) = \{x \in W : h(x) = \phi\}$.

Since we assume $h(x)$ is a cone in T_xW, it does no harm to normalize $h(x)$. That is, if we pick a coordinate chart of x, then assuming $x \notin E(h)$, we can normalize $h(x)$ by identifying it with a set of vectors of unit length. In this case we can write $h : W \to SW$, where we use SW to denote the normalized tangent space. The normalization depends, of course, on the chart. But topologically speaking we can identify $h(x)$, for $x \notin E(h)$, with some subset of the sphere $S^{w-1} \simeq S_xW$, the spherical tangent space above x. For example, suppose $u_i : W \to \Re$ is a smooth utility function. At any point x the gradient $du_i(x)$ is a linear operator $du_i(x) : T_xW \to \Re$, but $du_i(x)(v)$ can be identified under a chart with a real number. Whether or not this number is positive, or not, is independent of the chart. Thus we can define a preference *process*, associated with u_i, to be:

$$h(u_i) : W \to SW$$

where $h(u_i)(x) = \{v \in S_xW : du_i(x)(v) > 0\}$.

The critical points of the utility, u_i, coincide (on the interior of W) with the equilibria $E(h(u_i))$ of the process $h(u_i)$. (Notice that at a point, x, on the boundary ∂W of W, the tangent space at x is a half space, and so $h(u_i)(x)$ may be null for $x \in \partial W$, yet $du_i(x)(v) > 0$ for some $v \in \Re^w \setminus T_xW$.) Suppose now that $u = (u_1, \ldots, u_n) : W \to \Re^W$ is a smooth utility *profile* for a society N. Then, for any coalition (a subset $M \subset N$) the *Pareto process for* M is given by

$$h_M(u) : W \to SW$$

where $v \in h_M(u)(x)$ iff $v \in S_xW$ and $du_i(x)(v) > 0$ for all $i \in M$.

Smale [**36**] originally studied the Pareto process and obtained a characterization of the critical Pareto set for M. This set is simply $E(h_M(u))$; obviously, if x is an interior point in $E(h_M(u))$, then the gradients $\{du_i(x)\}_M$ must be semi-positively dependent (there is a vector $\{\lambda_i\}_M$ where all $\lambda_i \geq 0$, but not zero, such that $\sum_M \lambda_i \, du_i(x) = 0$).

Smale showed, for almost all u (in an appropriate topology) that $E(h_M(u))$ is a stratified manifold in W, whose dimension is of order $\mid M \mid - 1$. Lower-dimensional singularity strata are nested inside this manifold. We shall introduce the C^1-topology on profiles below. Let us, for the moment, denote the set of all smooth utility profiles, for the society N on W, by $U(W)^N$. (Throughout we use superfix N for profiles. If $\mid N \mid = 1$, then we delete it from the notation.) Economics and social choice are concerned with the properties of equilibrium correspondences, E_h, from $U(W)^N$ to W (here $E_h(u) \equiv E(h(u))$ is the equilibrium of a process $h(u)$ parametrized by the profile u).

For example, the economic exchange process occurs on a subset W of $\Re^{ne}$ where e is the number of commodities. At a point $x = (x_1, \ldots, x_n) \in (\Re^e)^n$, the feasible trades $f_M(x)$ for coalition M are all the infinitesimal exchanges of commodities amongst the members of the coalition. Thus $f_M(x)$ can simply be viewed as a cone in T_xW generated by the linear constraints $\sum_{i \in M} v_{ij} = 0$, for $j = 1, \ldots, e$ (j indexes the e commodities).

The *exchange* process h_ϵ is then defined by

$$h_\epsilon(u) = \cup_{M \subset N} \, h_M(u) \cap f_M.$$

That is, at any $x \in W$, the feasible and preferred cone of exchanges of coalition M is $h_M(u)(x) \cap f_M(x) \subset S_xW$.

As before, we view this cone as a subset of the tangent sphere S_xW above the point x. The *equilibrium* of this exchange process is $E_\epsilon(u) = E(h_\epsilon(u)) \subset W$.

Existence of such a set can give insight into existence of what is known as the *core* of an economy. (The core of an economy is the set of allocations of commodities that exhausts all the possible advantages of trade. It depends on the initial set of commodity endowments of the individuals.)

In this paper we shall be concerned with the existence of an equilibrium for a voting process. We define a voting process formally.

Definition 1 A voting rule is a family $\mathcal{D}$ of subsets of the society N, (where $|N| = n$ is finite). The *Nakamura* [**22**] number of $\mathcal{D}$ is defined as follows.

The *collegium* of $\mathcal{D}$ is $K(\mathcal{D}) = \cap_{M\in\mathcal{D}} M$.

If $K(\mathcal{D}) \neq \phi$, then $\mathcal{D}$ is termed *collegial*, and each member of $K(\mathcal{D})$ is called a *vetoer*. In this case the *Nakamura* number of $\mathcal{D}$, denoted $k(\mathcal{D})$, is ∞. If $K(\mathcal{D})$ is empty, then $\mathcal{D}$ is termed *non-collegial*. In this case, the Nakamura number $(k(\mathcal{D}))$ of $\mathcal{D}$ is the cardinality of the smallest subfamily $\mathcal{D}'$ of $\mathcal{D}$ such that $K(\mathcal{D}') = \phi$. The *voting process* determined by a profile $u \in U(W)^N$ and $\mathcal{D}$ is defined to be the operator $h_{\mathcal{D}}(u) : W \to SW$, where $h_{\mathcal{D}}(u)(x) = \cup_{M\in\mathcal{D}} h_M(u)(x)$.

The equilibrium $E(h_{\mathcal{D}}(u))$ of the voting process $h_{\mathcal{D}}(u)$ is also denoted $E_{\mathcal{D}}(u)$.

Obviously a point in $E(h_{\mathcal{D}}(u))$ only satisfies the first order conditions of equilibrium. To determine whether such a point is in fact a global equilibrium we need to introduce the underlying preferences. Thus given $u_i \in U(W)$, there is a natural projection $proj(u_i)$ into preferences, where $p_i = proj(u_i) : W \to 2(W)$ is defined by $p_i(x) = \{y \in W : u_i(y) > u_i(x)\}$. (Here $2(W)$ is the set of all subsets of W, including the empty set, ϕ.) Write $proj : U(W)^N \to P(W)^N$, where $P(W)^N$ denotes the set of preference profiles. The *voting game* induced by $\mathcal{D}$ at the profile $p \in P(W)^N$ is represented by the preference correspondence

$$\sigma_{\mathcal{D}}(p) = \cup_{M\in\mathcal{D}} p_M : W \to 2(W)$$

where $p_M = \cap_{i\in M} p_i$ is the preference of coalition M.

The equilibrium or *choice* of the voting game is $E(\sigma_{\mathcal{D}}(p))$. By analogy with an exchange economy, this equilibrium is often termed the *core* of the polity. In general the equilibrium of the process $h_{\mathcal{D}}(u)$ contains the equilibrium of $\sigma_{\mathcal{D}}(proj(u))$, since it is possible that $h_{\mathcal{D}}(u)(x) = \phi$ yet $\sigma_{\mathcal{D}}(proj(u))(x) \neq \phi$. However, imposing convexity on the component preferences induces equality between these two equilibrium notions.

A standard way of proving existence of a choice of a *preference* is by a fixed point argument, using convexity and continuity in some fashion. The most elegant fixed point theorem is due to Fan [**12**]. Fan assumed the state space W was a compact, convex topological vector space. As indicated above, we shall call such a space a *Fan space*. (Although Fan's result was not restricted to the case when W is finite dimensional, we shall assume that W is indeed finite dimensional.) This theorem can also be adopted to the case of a dynamical process, but to do so we need to introduce the various continuity properties that will prove useful.

Definition 2

1. A correspondence $\Psi : X \to 2(Y)$ between the topological spaces X, Y is
   (a) *upper hemi-continuous* (uhc) iff for any V open in Y,
   $\{x : \Psi(x) \subset V\}$ is open in X,
   (b) *lower hemi-continuous* (lhc) iff, for any V open in Y,
   $\{x : \Psi(x) \cap V \neq \phi\}$ is open in X.
2. A preference $\Psi : W \to 2(W)$ is *lower demi-continuous* (ldc) iff $\Psi^{-1}(x) = \{y \in W : x \in \Psi(y)\}$ is open in W.

We can also consider continuity properties on dynamical processes.

Definition 3

1. A *preference field* $q : W \to SW$ is a differentiable process induced from a vector field $v : W \to TW$: that is there exists a vector field $v : W \to TW$ such that
$$E(q) = \{x \in W : q(x) = \phi\} = \{x \in W : v(x) = 0\}$$
and $q(x) = \frac{v(x)}{\|v(x)\|} \in S^{w-1}$, whenever $v(x) \neq 0$.
(Here $\| \ \|$ is an appropriate norm on T_xW.)
2. A process $h : W \to SW$ is $S-$*continuous* iff h admits a preference field as a selection: that is there exists a preference field $q : W \to SW$ such that $E(q) = E(h)$ and $q(x) \in h(x) \; \forall \, x \notin E(h)$.
3. A process h is uhc (lhc) iff the local representations $h_V : V \to SV$ are uhc (lhc).

In part (3), taking a coordinate chart gives a smooth homeomorphism $SV = V \times S^{w-1}$, so h can be viewed locally as a map $h_V : V \to 2(S^{w-1})$. Then h is uhc (say) iff $h_V : V \to SV$ is uhc, for any such chart on V.

Fan in [**12**] generalized the fixed point result of Tychonoff in the following way. Suppose $\Psi : W \to 2(W)$ is an ldc preference correspondence on a Fan space such that Ψ is semi-convex (that is the point $x \in W$ never belongs to the convex hull of $\Psi(x)$). Then the choice $E(\Psi) = \{x \in W : \Psi(x) = \phi\}$ must be non-empty. It is an easy matter [**13, 29, 37**] to use this result to show that if $p = (p_1, \dots, p_n)$ is a ldc, semi-convex profile (each p_i is ldc and semi-convex) and W is a Fan space of dimension less than $k(\mathcal{D}) - 2$, then $E(\sigma_{\mathcal{D}}(p)) \neq \phi$. To extend this result of Fan to the case of a manifold we need an analogue of "semi-convexity".

Definition 4

1. For any process $h : W \to SW$ define the dual process $h^* : W \to SW$ as follows:
$$h^*(x) = S^{w-1} \text{ whenever } h(x) = \phi,$$
and
$$h^*(x) = \{v \in S^{w-1} : h(x) \cdot v > 0\} \text{ if } h(x) \neq \phi.$$
(Here $\cdot$ is the scalar product induced from $\Re^w$ and $h(x) \cdot v > 0$ means $t \cdot v > 0$ for all $t \in h(x)$.)
2. The *cycle set* of h, denoted $E^*(h)$, is the equilibrium set $E(h^*)$ of h^*.

The term *cycle set* of h is used for $E(h^*)$ because it can be shown [**28**] that if $x \in E(h^*)$ then cyclic trajectories can be constructed in W by integrating the process. (The construction of cycles depends on showing that the process h^* is uhc. Using this, it can be shown that $E(h^*)$ is an open set. In contrast, the process h will usually be lhc, so that $E(h)$, if it exists, will be closed.) More importantly, if $E(h^*)$ is empty, then a local convexity property holds, and the equilibrium $E(h)$ must be non-empty (in the case W is a Euclidean space, or more generally, a Fan manifold). We can now present the principal "fixed point" theorem that we shall use. See [**30**].

Theorem 1 Suppose $h : W \to SW$ is a process on the Fan manifold, such that $E^*(h) = \phi$ and h^* is $S-$continuous. Then $E(h) \neq \phi$.

Outline of Proof Suppose in fact that $E(h) = \phi$. Since h^* is $S-$continuous there is a vector field which induces a selection g of the process $h^* \cap h$. Since both

$E(h)$ and $E(h^*)$ are null, the vector field has no zeros. But then the vector field induces a deformation $f_1 : W \to W$, which has no fixed point ($f_1(x) = x$ for no $x \in W$). Here f_1 is homotopic to the identity $Id : W \to W$ via a (continuous) homotopy $f_t : W \to W$. Since W has non-zero Euler characteristic, there cannot exist such a fixed point free deformation [**5**]. Thus $E(h) \neq \phi$. □

Observation 1 A key condition of Theorem 1 is that $E(h^*) \equiv E^*(h) = \phi$. If, in contrast, $x \in E(h^*)$ then the set $h(x)$, viewed as a cone in T_xW, does not belong to an open half space in the tangent space. In this case it is impossible to construct the deformation f_1 at x (see [**30**]). On the other hand if $E(h^*) = \phi$, then at every point $x \in W$ the "cone" $h(x)$ does indeed belong to a half space. In this case the dual $h^*(x)$ may be viewed as a generalized gradient for the process at x. To illustrate Theorem 1, consider a voting process $h_{\mathcal{D}}(u)$ at a profile u. Since $h_{\mathcal{D}}(u)(x)$ can be interpreted as the union of a family of cones in T_xW, the dual of $h_{\mathcal{D}}(u)$ at x is

$$[h_{\mathcal{D}}(u)]^*(x) = \cap_{M \in \mathcal{D}} [h_M(u)]^*(x).$$

For each coalition, M, the dual $[h_M(u)]^*(x)$ at x is a cone in S^{w-1} obtained essentially by taking the convex hull of $\{du_i(x)\}_{i \in M}$.

Thus the condition that $E(h^*) = \phi$ requires that this family of cones always intersects (in a non-empty set). In the case $dim(W) \leq k(\mathcal{D}) - 2$, an easy argument using Helly's Theorem [**2**] shows that this intersection is indeed always non-empty. Moreover, the voting process and its dual are both $S-$continuous. Thus we obtain the following corollary.

Corollary 1 Let $\mathcal{D}$ be a voting rule with Nakamura number $k(\mathcal{D})$, and let W be a Fan manifold of dimension $\leq k(\mathcal{D}) - 2$. Then for any $u \in U(W)^N$ the voting process $h_{\mathcal{D}}(u) : W \to SW$ satisfies the two properties that $E((h_{\mathcal{D}}(u))^*) = \phi$ and $E(h_{\mathcal{D}}(u)) \neq \phi$.

It immediately follows (with the dimension constraint on a Fan space) that if the component utilities induce convex preferences ($p_i(x)$ is a convex set or empty $\forall i$, $\forall x$), then the choice, $E(\sigma_{\mathcal{D}}(p))$, are non-empty [**13, 29, 37**]. Thus Corollary 1 generalizes standard results on voting equilibria to a more abstract context.

Observation 2 In principle Theorem 1 can be used to show the existence of a *core* in an exchange economy. Since the exchange process $h_\epsilon(u)$, at $x \in W$, is a union of cones, the dual $[h_\epsilon(x)]^*$ at x can be identified with the intersection of a family of dual cones

$$V(x) = \cap_{M \subset N} [h_M(u) \cap f_M(x)]^*.$$

If this is non-null at all x, so $E(h_\epsilon(u)^*) = \phi$, then Theorem 1 can be used to show $E(h_\epsilon(u)) \neq \phi$. Recent papers have explored conditions under which a duality condition of this kind is sufficient to guarantee existence of an equilibrium price vector and of the core of the exchange economy. (See, for example [**23, 24, 38**]. See also [**21**] for a counter-example to the conditions as presented in [**8**] and a discussion of the problems in [**9**].[1]) For example, if W is a Fan space, and the intersection denoted $V(x)$ above is non-null for all x, then V can be used to construct a vector field v on W whose zeros satisfy the first order conditions. In particular, if $v(x)$ when projected into $\Re^w$ is independent of x, then it can be regarded as an equilibrium price vector (at which supply and demand of all commodities are equated). Note

[1]Editors note added in proof: The discussion of problems in [**9**] appears in Monteiro, P., F. Page and M. H. Wooders (1999) "Arbitrage and equilibrium; another counterexample", *Social Choice and Welfare* **16**, 337-346

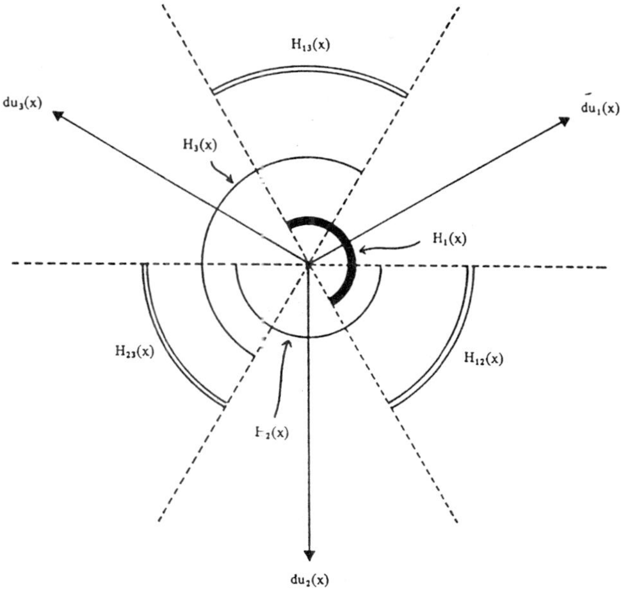

Figure 1 Preference cones at x for the majority voting rule with $n = 3$

however that the key property in this proof technique is that $V(x)$ be non-null, for all $x \in W$. It is possible (but not necessary) to prove this using some version of Helly's Theorem. (See also [**7**] for discussion of Helly's Theorem, [**23, 24, 38**] for results not directly using Helly's Theorem, and [**21**] for a counterexample to existence of equilibrium in [**7**].)

Observation 3 The dimension condition of Corollary 1 is necessary as well as sufficient. To see this, consider a voting rule $\mathcal{D}$ on $N = \{1,2,3\}$ where each two-person coalition belongs to $\mathcal{D}$. Clearly $k(\mathcal{D}) = 3$. In two dimensions the three gradients $\{du_i(x)\}_N$ can be strictly positively dependent (as in Figure 1). Then the process, h, at x cannot satisfy $E(h^*) = \phi$. To see this, observe that the cone $h_{\mathcal{D}}(u)(x)$ can be identified with the union

$$H_{12}(x) \cup H_{13}(x) \cup H_{23}(x).$$

But then the dual cone $[h_{\mathcal{D}}(u)]^*(x)$ at x is the intersection of the three cones, q_{12}, q_{23}, q_{13}, where q_{ij} is the cone in S^1 generated by $\{du_i(x), du_j(x)\}$. Figure 2 makes it clear that these three cones do not intersect. Thus $h_{\mathcal{D}}(u)^*(x) = \phi$ at all such points, so $E(h^*) \neq \phi$. It should be obvious that it is certainly possible for $E(h)$ to be null. The fact that $E(h^*) \neq \phi$ and $E(h) = \phi$ for the process h is related to Arrow's Impossibility Theorem [**1**].

It is clear that the duality condition $h(u)^*(x) \neq \phi$ is theoretically important in proving existence both of an economic core and of a voting equilibrium. The failure of this condition can be termed *social diversity* (A distinct notion of diversity occurs in [**9**] but this notion is unrelated to existence of equilibrium and nonemptiness of

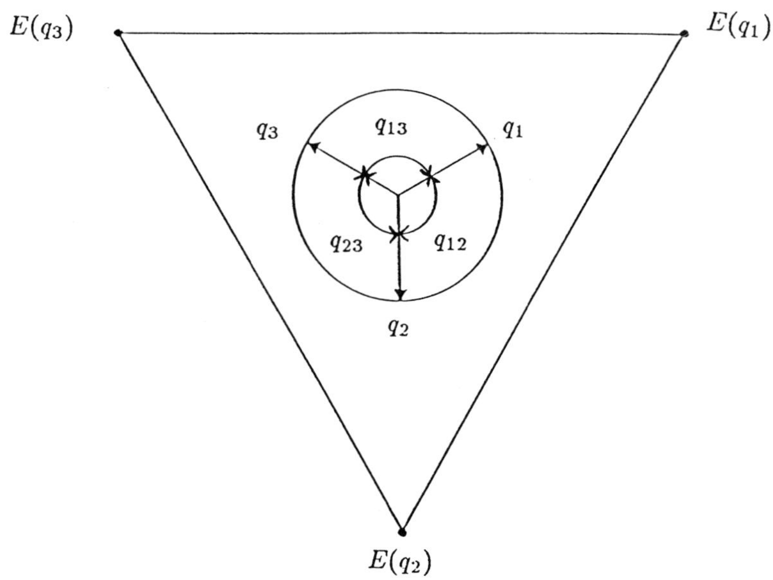

Figure 2 Dual cones at x for majority rule

the core – see [**21**].) If social diversity occurs at any point x, so $h(u)^*(x) = \phi$, then there is no guarantee of existence of an equilibrium $E(h(u))$. Indeed for a voting rule, social diversity can occur almost everywhere, for almost all profiles [**26, 28**]. However, Theorem 1 suggests that, for any process $h(u)$, we regard $E(h(u)) \cup E(h(u)^*)$ as a "solution", since this set will be non-empty on a Fan manifold. For a voting rule $E(h_{\mathcal{D}}(u)^*)$ can be open-dense. However, if we modify the notion of a voting process, then we can guarantee that $E(h(u)) \cup E(h(u)^*)$ is not only non-empty, but also a subset of the "critical Pareto set". This is the idea behind the "heart". Moreover, we shall show the heart exhibits a continuity property, with respect to a C^1–topology on preference profiles.

3 The C^1–topology and the "Heart"

We use the C^1–topology on utilities to generate a C^1–topology on preferences and preference fields. First the C^0– (or compact open) topology for $U(W)$ is generated by basic neighborhoods of the form

$$Z_0(u_i, \epsilon) = \{u_i' \in U(W) : \| u_i(x) - u_i'(x) \| < \epsilon \, \forall \, x \in W\}.$$

(We have assumed W is compact, and $\| \ \|$ is any norm compatible with the topology on W; since W is a manifold it is paracompact and thus metrizable.)

Now let

$$Z_1(u_i, \epsilon) = \{u_i' \in U(W) : \| du_i(x) - du_i'(x) \| < \epsilon, \ \forall \, x \in W\}$$

where again $\| \ \|$ is a norm on the dual space $T_x^* W$. A set V is open in the C^1–topology on $U(W)$ iff, for all $u_i \in V$ there exists $\epsilon > 0$ and basic neighborhoods $Z_0(u_i, \epsilon)$, $Z_1(u_i, \epsilon)$ such that $Z_0(u_i, \epsilon) \cap Z_1(u_i, \epsilon) \subset V$. Any set open in the C^0 topology is open in the C^1- topology, and indeed the latter is strictly finer than the former. Write $U_0(W), U_0(W)^N, U_1(W), U_1(W)^N$ for the obvious topological spaces (where suffix 0, 1 denote C^0 and C^1 respectively). For convenience we restrict attention to the subspaces $U_j^*(W), U_j^*(W)^N$, $j = 0, 1$. Here $u_i \in U^*(W)$ iff u_i is a *Morse function* [**19**] all of whose critical points are non-degenerate. Since W is compact, critical points of a Morse function are locally isolated and finite in number.

Now define operations

$$proj : U^*(W)^N \rightarrow P^*(W)^N \text{ and } d : P^*(W)^N \rightarrow S^*(W)^N.$$

As before $proj(u_i)$ is a preference in $P(W)$, and $P^*(W)$ denotes simply the set of preferences lying in the image of *proj*. Secondly, $d(p_i)$ is a preference field obtained from p_i in the following way: let u_i be any utility such that $p_i = proj(u_i)$, and let $du_i : W \rightarrow T^*W$ be the gradient field (as before, T^*W, is the dual tangent space). Locally du_i can be identified with the gradient vector field $grad(u_i) : W \rightarrow TW$. This field, when normalized gives a preference field, denoted $d(p_i) : W \rightarrow SW$, which we shall call the gradient preference field induced by p_i (or u_i). Let $S^*(W)$ be the image of $P^*(W)$ under d. There is a natural topology on $P^*(W)$ based on the Hausdorff topology on subsets of W. Given a preference p_i, the weak preference r_i is defined by $y \in r_i(x)$ iff $x \notin p_i(y)$. The *closed convergence topology* on $P^*(W)^N$ is the Hausdorff topology induced from the Hausdorff metric applied to the graphs of weak preferences. By [**15**] the map $proj : U_0^*(W) \rightarrow P^*(W)$ is open with respect to the C^0–topology on $U^*(W)$ and the closed convergence topology on $P^*(W)$. Now define the C^1–quotient topology on $S^*(W)$ to be the coarsest topology such that the map $(d \circ proj) : U_1^*(W) \rightarrow S^*(W)$ is open. Use $S_1^*(W)$ to denote this space. Also define the C^1–quotient topology on $P^*(W)$ to be the coarsest topology on $P^*(W)$ such that the map $d : P^*(W) \rightarrow S_1^*(W)$ is continuous. Recent work [**33**] shows that there is a C^1-topology on $P^*(W)$ which is strictly finer than both the closed convergence and C^1–quotient topologies (this topology is obtained by using basic neighborhoods of the form $Z_0(p_i, \epsilon) \cap Z_1(p_i, \epsilon)$, where Z_0 is open in the closed convergence topology and Z_1 is open in the C^1–quotient topology). Use $P_1^*(W)$ to denote this space. Then we can state as a Lemma the main result of [**33**].

Lemma 1 The C^1–topologies on utilities, preferences and fields are such that the maps

$$d \circ proj : U_1^*(W)^N \rightarrow P_1^*(W)^N \rightarrow S_1^*(W)^N$$

are compatible: that is, $d \circ proj : U_1^*(W)^N \rightarrow S_1^*(W)$ is open and $d : P_1^*(W)^N \rightarrow S_1^*(W)^N$ is continuous. In addition, *proj* maps a neighborhood $Z_0(u_i, \epsilon) \cap Z_1(u_i, \epsilon)$ to a neighborhood $Z_0(p_i, \epsilon') \cap Z_1(p_i, \epsilon'')$, and is thus open.

Having established the topologies on the "preference spaces", we may demonstrate continuity of the equilibrium process.

Definition 5 Let $q = (q_1, \ldots, q_n) \in S^*(W)^N$.

1. The *efficient Pareto process* for coalition M at q, denoted $\hat{h}_M(q) : W \rightarrow SW$ is defined by $\hat{h}_M(q)(x) = h_M(u)(x) \cap h_M(u)^*(x)$.
   Here $h_M(u)^*$ is the process dual to $h_M(u)$ and u is any utility profile satisfying $q = d \circ proj(u)$.
2. The *efficient process* for the rule, $\mathcal{D}$, at q is $\hat{h}_{\mathcal{D}}(q) : W \rightarrow SW$, where $\hat{h}_{\mathcal{D}}(q)(x) = \cup_{M \in \mathcal{D}} \hat{h}_M(q)(x)$.

3. The *heart* of $\mathcal{D}$ at q is the subset of W given by

$$\mathcal{H}_{\mathcal{D}}(q) = E(\hat{h}_{\mathcal{D}}(q)) \cup Clos\ E^*(\hat{h}_{\mathcal{D}}(q)).$$

4. The heart correspondence is $\mathcal{H}_{\mathcal{D}} : S_1^*(W)^N \to 2(W)$. This can be pulled back to $U_1^*(W)^N$ using the identity $\mathcal{H}_{\mathcal{D}}(u) \equiv \mathcal{H}_{\mathcal{D}}(proj \circ d(u))$.

Since $\hat{h}_M(q)$ is induced from smooth preferences it is easy to show it is $S-$continuous. Consequently $E(\hat{h}_{\mathcal{D}}(q))$ is closed. On the other hand the cycle set $E^*(\hat{h}_{\mathcal{D}}(q))$ is open: it is convenient to take the closure of this set (designated by *Clos*).

To modify the earlier notation, define the *critical Pareto set* of the society, N, at the profile, q, to be $E(h_N(q))$. We denote this by $E_N(q)$. In general $E_N(q)$ contains the *global Pareto set* $E_N(p) \equiv E(p_N)$ of any preference profile p, such that $d(p) = q$.

We now present the existence theorem on the heart (see also [**31**]).

Theorem 2 Let $\mathcal{D}$ be a voting rule and W a Fan manifold. Then, for all $q \in S^*(W)^N$, $\mathcal{H}_{\mathcal{D}}(q)$ is non-empty and is a subset of the critical Pareto set $E_N(q)$.

Proof Existence follows immediately from Theorem 1. The Pareto property can be shown as follows. If $x \notin E_N(q)$ then the vectors $\{q_i(x)\}_N$ all belong to a half space, in which case $\hat{h}_{\mathcal{D}}(q)(x)$ belongs to a half space, so its dual is non-empty. Thus $[\hat{h}_{\mathcal{D}}(q)]^*(x) \neq \phi$. But also $h_M(u)(x) \neq \phi$ for some coalition M, so $\hat{h}_{\mathcal{D}}(q)(x) \neq \phi$. Thus $x \notin \mathcal{H}_{\mathcal{D}}(q)$. □

We seek to show that the correspondence $\mathcal{H}_{\mathcal{D}}$ is lhc. This we can do for the case of "convex" preferences. In the case W is a Fan space, let $U^C(W)^N$ denote the set of strictly pseudo concave utility profiles (u_i is strictly pseudo-concave iff $u_i(y) \geq u_i(x)$ implies $du_i(x)(y-x) > 0$). Let $U_1^C(W)^N$ denote this set with the C^1-topology and let $P_1^C(W)$ and $S_1^C(W)^N$ denote the images of $U_1^C(W)^N$ under *proj* and *proj* $\circ\ d$ respectively, with the topologies induced from inclusions. Clearly d and *proj* are compatible on these subspaces. Notice that if u_i is strictly pseudo-concave, then there is a unique point at which u_i is maximized. Clearly $U^C(W)^N$ is a subset of $U^*(W)^N$.

Theorem 3 Let $\mathcal{D}$ be a voting rule, on a Fan space. Then $\mathcal{H}_{\mathcal{D}} : S_1^C(W)^N \to 2(W)$ is a lhc correspondence, and admits a continuous selection $g : S_1^C(W)^N \to W$, namely a function g such that $g(q) \in \mathcal{H}_{\mathcal{D}}(q)\ \forall\, q \in S_1^C(W)^N$.

Proof The proof of lhc is derived from [**31**]. There it is shown that, for a Fan manifold, $\mathcal{H}_{\mathcal{D}}$ is lhc, as a correspondence from a subset of $U_1^*(W)^N$ to $2(W)$. (This subset is defined by an index condition on the Hessian of the profile.) When W is a Fan space, this subset of $U_1^*(W)^N$ contains $U_1^C(W)^N$. Consequently $\mathcal{H}_{\mathcal{D}} : U_1^C(W)^N \to 2(W)$ is lhc. But by Lemma 1, the map $d \circ proj : U_1^C(W)^N \to S_1^C(W)^N$ is open. Thus $\mathcal{H}_{\mathcal{D}}$ is lhc on $S_1^C(W)^N$. To prove existence of a selection, consider first the subset $S^E(W)$ of $S^C(W)$ comprising Euclidean preference fields. Here $q \in S^E(W)$ iff $q \equiv d(p)$, where $p(x_i) = \phi$ at the "bliss" point x_i and $y \in p(z)$ iff $\| y - x_i \| < \| z - x_i \|$.

Let $S_1^E(W)^N$ denote the topological space of all Euclidean preference fields. Schofield in [**34**] shows that $S_1^C(W)^N$ is a deformation retract of $S_1^E(W)^N$. That is, there is a continuous homotopy $F : S_1^C(W)^N \times [0,1] \to S_1^C(W)^N$ such that $f_0 = F(\ ,0)$ is the identity on $S_1^C(W)^N$ and $f_1 = F(\ ,1)$ is a retraction of $S_1^C(W)^N$ onto $S_1^E(W)^N$ (so f_1 restricted to $S_1^E(W)^N$ is the identity).

Moreover, it is shown that $\mathcal{H}_{\mathcal{D}}(q)$ is "star-shaped" for each $q \in S_1^E(W)^N$. By lhc of $\mathcal{H}_{\mathcal{D}}$, a theorem of Mas-Colell [**16**] shows that $\mathcal{H}_{\mathcal{D}}$ admits a continuous selection, g, on the domain $S_1^E(W)^N$. To see that the selection can be extended to $S_1^C(W)^N$, note that the retraction, f_1, induces for each $q \in S_1^C(W)$, a homeomorphism, f_*, between $\mathcal{H}_{\mathcal{D}}(q)$ and $\mathcal{H}_{\mathcal{D}}(f_1(q))$.

Thus

$$\mathcal{H}_{\mathcal{D}}(q) = f_*^{-1}\mathcal{H}_{\mathcal{D}}(f_1(q)).$$

Consequently if $g(f_1(g)) \in \mathcal{H}_{\mathcal{D}}(f_1(q))$, then we can define $g' = g \circ f_1$ to show that $f_*^{-1} \circ g'(q) \in \mathcal{H}_{\mathcal{D}}(q) \;\forall\; q \in S_1^C(W)^N$. Since f_*^{-1}, g and f_1 are continuous, this gives a continuous selection of $\mathcal{H}_{\mathcal{D}}$ on the domain $S_1^C(W)^N$. □

Observation 4 The heart has been defined in terms of a profile, q, of preference fields. It is also possible to define the heart of a voting rule, $\mathcal{D}$, for a preference profile p. This can be done by localizing the notion of the covering relation induced by the voting game $\sigma_{\mathcal{D}}(p)$. Say a point x "covers" y iff $x \in \sigma_{\mathcal{D}}(p)(y)$ and $\sigma_{\mathcal{D}}(p)(x) \subset \sigma_{\mathcal{D}}(p)(y)$. It is easy to show that the set of "uncovered" points is non-empty [**3, 10, 17**]. Then the heart $\mathcal{H}_{\mathcal{D}}(p)$ at p is the set of "locally uncovered" points [**32**]. In general $\mathcal{H}_{\mathcal{D}}(p)$ and $\mathcal{H}_{\mathcal{D}}(q)$ will not coincide (when $q = d(p)$) because $\mathcal{H}_{\mathcal{D}}(q)$ is based on first order conditions alone. For non-convex preferences the equilibrium set $E(\hat{h}_{\mathcal{D}}(q))$ may be a superset of $E(\sigma_{\mathcal{D}}(p))$. However, for convex preferences, on a Fan space, the two sets $\mathcal{H}_{\mathcal{D}}(p)$ and $\mathcal{H}_{\mathcal{D}}(dp)$, do coincide. Thus for $p \in P^C(W)^N$ we may interpret the heart, $\mathcal{H}_{\mathcal{D}}(p)$, of $\mathcal{D}$ at p to be the set $\mathcal{H}_{\mathcal{D}}(dp)$. By Theorem 2, $\mathcal{H}_{\mathcal{D}}(p)$ will be non-empty. The advantage of using the heart is that it is always non-null under the stated conditions of convexity, and finite dimensionality. As we noted in Observation 3, the equilibrium $E(\sigma_{\mathcal{D}}(p))$ will generally be empty in dimension above $k(\mathcal{D}) - 2$. (Indeed it can almost always be shown to be empty [**26**].) Theorem 3 can, however, be used to show that the heart converges to the voting equilibrium $E(\sigma_{\mathcal{D}}(p))$, when the latter is non-empty.

Corollary 2 Let $\mathcal{D}$ be a voting rule on a Fan space, W. If $p' \to p$ in the C^1–topology on $P_1^C(W)^N$ and $E(\sigma_{\mathcal{D}}(p)) \neq \phi$, then $\mathcal{H}_{\mathcal{D}}(p') \to E(\sigma_{\mathcal{D}}(p))$ in a set-theoretic sense. Moreover, $\mathcal{H}_{\mathcal{D}}(p) \subset E_N(p)$ for all $p \in P^C(W)^N$.

Proof Since $d : P_1^C(W)^N \to S_1^C(W)^N$ is continuous, it follows that lhc of $\mathcal{H}_{\mathcal{D}}$ on the domain $S_1^C(W)^N$ induces lhc on the domain $P_1^C(W)^N$, using the identity $\mathcal{H}_{\mathcal{D}}(p) \equiv \mathcal{H}_{\mathcal{D}}(d(p))$. If $E(\sigma_{\mathcal{D}}(p)) \neq \phi$, then $E(\hat{h}_{\mathcal{D}}(q)) \neq \phi$. Moreover, using the retraction f_1 between $S_1^C(W)^N$ and $S_1^E(W)$ it is easy to see, in this case, that $E^*(\hat{h}_{\mathcal{D}}(q)) = \phi$. Thus $\mathcal{H}_{\mathcal{D}}(p) = E(\sigma_{\mathcal{D}}(p))$. The set theoretic convergence follows by lhc. The Pareto property follows from Theorem 2, using the fact that the critical Pareto set, $E_N(q)$, and global Pareto set, $E_N(p)$, coincide when $q = d(p)$ and $p \in P^C(W)^N$. □

Observation 5 The set theoretic convergence of the heart, $\mathcal{H}_{\mathcal{D}}(p')$ to $E(\sigma_{\mathcal{D}}(p))$ requires some clarification. Lower hemi-continuity of the heart implies that, for all $z \in \mathcal{H}_{\mathcal{D}}(p)$, there exists a sequence $\{p_t\}$ converging to p (in the C^1-topology) such that a sequence $\{z_t \in \mathcal{H}_{\mathcal{D}}(p_t)\}$ can be found which converges to z in W. In the typical case, $E(\sigma_{\mathcal{D}}(p))$, if it is non-null, will be an isolated point, z. Thus the convergence considered here implies that there are sequences $\{z_t\}$ and $\{p_t\}$, such that

$$inf\ \{\| z - z_t \| : \{z\} = E(\sigma_{\mathcal{D}}(p)),\ z_t \in \mathcal{H}_{\mathcal{D}}(p_t)\}$$

converges to zero as p_t converges to p.

However the correspondence $\mathcal{H}_{\mathcal{D}}$ is not uhc in a neighborhood of z. In particular, failure of uhc means that the Hausdorff distance between $E(\sigma_{\mathcal{D}}(p))$ and $\mathcal{H}_{\mathcal{D}}(p_t)$ does not converge to zero as $p_t \rightarrow p$.

Cox in [**10**], in fact, showed that the uncovered set, regarded as a correspondence, is uhc near $E(\sigma_{\mathcal{D}}(p))$. However, it is not known, at present, whether the uncovered set correspondence admits a continuous selection. (I am grateful to Jeff Banks and John Duggan for helpful observations on this difference between the uncovered set and the heart.) See [**32**] for discussion.

The fact that the heart does admit a continuous selection, and is Paretian on covex preference profiles, allows us to use it to construct a continuous *aggregator.*

4 Existence of a continuous aggregator

We now introduce the idea of an *aggregator* [**6**], namely a continuous function $\Gamma : P_1^*(W)^N \rightarrow P_1^*(W)$. The easiest case to deal with is when Γ has domain based on Euclidean preferences. As before, a profile $p \in P^E(W)^N$ iff each p_i is Euclidean, based on a bliss point $x_i \in W$. Clearly $P_1^E(W)^N$ is homeomorphic to the product space W^N, and when W is convex, this is contractible. We first construct Γ on $S_1^E(W)^N$. As in Theorem 3, $\exists$ a selection $g : S_1^E(W)^N \rightarrow W$ of the heart $\mathcal{H}_{\mathcal{D}}$. By Corollary 2, $g(q) \in E(h_{\mathcal{D}}(p)) = E(\hat{h}_{\mathcal{D}}(q))$ when this set is non-empty. For each q, define the preference field at x as follows: if $x \neq g(q)$, then $\Gamma_g(q)(x) = \lambda(g(q) - x)$ where $\lambda = \| g(q) - x \|^{-1}$, while if $x = g(q)$, then let $\Gamma_g(q)(g(q)) = \phi$. It is obvious that $\Gamma_g(q) : W \rightarrow SW$ is a Euclidean preference field, whose equilibrium $E(\Gamma_g(q)) \in E(\sigma_{\mathcal{D}}(q))$ if this exists. Moreover, this defines a continuous function $\Gamma_q : S_1^E(W)^N \rightarrow S_1^E(W)$ which can clearly be pulled back to define $\Gamma_g : P_1^E(W)^N \rightarrow P_1^E(W)$.

Obviously the equilibrium map $E(\Gamma_g) : P_1^E(W)^N \rightarrow W$, given by $E(\Gamma_g)(p) = g(p)$, is continuous. Moreover, suppose the dual is non-null at x, so $\hat{h}_{\mathcal{D}}(q)^*(x) \neq \phi$. Then $\Gamma_g(dp)(x) \in \hat{h}_{\mathcal{D}}(q)^*(x)$: that is, $\Gamma_g(dp)$ is a "social" gradient vector at x with respect to all vectors of the process. In other words, for any feasible vector $v \in \hat{h}_{\mathcal{D}}(q)(x)$, it is the case that $\Gamma_g(dp)(x) \cdot v > 0$.

For $p \in P^C(W)^N$, we can extend Γ using the deformation retraction of $P_1^C(W)^N$ onto $P_1^E(W)$. As in the proof of Theorem 3, use the retraction to construct a continuous selection $g : P_1^C(W)^N \rightarrow W$ of $\mathcal{H}_{\mathcal{D}} : P_1^C(W)^N \rightarrow 2(W)$. Then define $\Gamma_g(p)$ to be that preference with $E(\Gamma_g(p)) = g(p)$. At any point $x \notin E(\Gamma_g(p))$, choose $\Gamma_g(q)(x)$ to be a gradient vector which belongs to the dual $\hat{h}_{\mathcal{D}}(q)^*(x)$, if this is non-empty. The preference field $\Gamma_g(q)$ can then be pulled back to $P_1^C(W)$ to give a preference, $\Gamma_g(p)$. All steps in this construction are continuous. Thus we have the corollary.

Corollary 3 Let $\mathcal{D}$ be a voting rule on a Fan space, W. Then there is a continuous aggregator $\Gamma_g : P_1^C(W)^N \rightarrow P_1^C(W)$ which is compatible with $\mathcal{D}$ in the following sense:

1. if $E(\sigma_{\mathcal{D}}(p)) \neq \phi$, then $E(\Gamma_g(p)) \in E(\sigma_{\mathcal{D}}(p))$;
2. if $x \notin E(\sigma_{\mathcal{D}}(p))$, then the gradient vector, $d(\Gamma_g(p))(x)$, defined by $\Gamma_g(p)$ at x is a continuous function of the gradient vectors $(dp_1(x), \ldots, dp_n(x))$;
3. moreover, if the dual, $\hat{h}_{\mathcal{D}}(dp)^*(x)$, defined by the preference p at x, is non-empty then the gradient vector $d(\Gamma_g(p))(x)$ at x belongs to this dual.

In general it is not easy to define a continuous aggregator Γ on the full domain $P_1^*(W)^N$. To see this consider, first, the problem of constructing a continous function $\Gamma : S_1^*(W)^W \to S_1^*(W)$ which is compatible with $\mathcal{D}$. Following Theorem 2, the obvious idea would be to choose Γ such that $E(\Gamma(q)) \in \mathcal{H}_{\mathcal{D}}(q)$. For example, suppose $\dim(W) \leq k(\mathcal{D}) - 2$. If we use $E(h_{\mathcal{D}}(q))$ to denote the "critical" voting core of $q \in S_1^*(W)^N$, then it is clear that $E(h_{\mathcal{D}}(q))$ is non-empty, and is a subset of $\mathcal{H}_{\mathcal{D}}(q)$. Consider the equilibrium correspondence

$$E(h_{\mathcal{D}}) : S_1^*(W)^N \to 2(W)$$

given by $E(h_{\mathcal{D}})(q) = E(h_{\mathcal{D}}(q))$. This correspondence will fail lhc at various low dimensional strata in the space $S_1^*(W)^N$. As discussed in [**34**], lhc fails because the equilibria can become degenerate and "disappear". However, these strata are nowhere dense in $S_1^*(W)^N$ so we can infer that $E(h_{\mathcal{D}})$ is "generically" lhc.

Now consider the question of local stability of points in $E(h_{\mathcal{D}}(q))$. Suppose that the dual process $h_{\mathcal{D}}(q)^*$ admits a continuous selection (a gradient field) for points outside $\mathcal{H}_{\mathcal{D}}(q)$. If this preference field can be extended over $W \setminus E(h_{\mathcal{D}}(q))$, then we could use this preference field to define a Hessian, and thus the *index* of the preference field, at critical points in $E(h_{\mathcal{D}})(q)$. As Smale in [**36**] suggested, the induced vector field satisfies the Morse inequalities [**11, 19**] at all these critical points. In particular, if the "gradients" $\{q_i(x) : i \in N\}$ point inwards on the boundary of W, then the Morse inequalities can be used to show existence of a locally stable component $SE(h_{\mathcal{D}}(q))$ of $E(h_{\mathcal{D}}(q))$. Here a component of the equilibrium set is locally stable if it is an attractor of the gradient process. If we use $\Gamma(q)$ to denote the preference field, then the gradient process representing $\Gamma(q)$ has the form $\Gamma(q)(x') = grad(\psi)(x')$. Near an attractor, ψ has a local representation of the form $\psi(x') = x - \sum_{t=1}^{w} t_i^2$. Clearly x is a critical point of the gradient process, with maximal index. Thus x is locally stable. Although we performed the analysis for the equilibrium $E(h_{\mathcal{D}}(q))$, the same argument holds for the heart.

Corollary 4 Let $\mathcal{D}$ be a voting rule on a Fan Manifold, W. Then there is an open set $V \subset S_1^*(W)^N$ such that the following holds, for all $q \in V$.

1. There is a preference field $\Gamma(q) \in S_1^*(W)$ such that $\Gamma(q)(x) \in \hat{h}_{\mathcal{D}}(q)^*(x)$ for each $x \notin \mathcal{H}_{\mathcal{D}}(q)$.
2. $E(\Gamma(q)) = \cup_{i=1}^{r} \{x_i\}$ is a finite union of locally isolated points, such that exactly one point $\{x_i\}$ belongs to each path connected component of $\mathcal{H}_{\mathcal{D}}(q)$.
3. $\Gamma(q)$ can be continuously extended over $W \setminus E(\Gamma(q))$.
4. At least one critical point x_i is a stable attractor (of maximum index) of the preference field $\Gamma(q)$.

Proof It is easy to show that the process $\hat{h}_{\mathcal{D}}(q)^* : W \setminus \mathcal{H}_{\mathcal{D}}(q) \to SW$ is lhc. Moreover, regarded as a cone in T_xW, $\hat{h}_{\mathcal{D}}(q)^*(x)$ is a convex set. Since W is a manifold, it is paracompact (an open cover admits a locally finite subcover). By Michael's Selection Theorem [**18**], $\hat{h}_{\mathcal{D}}(q)^*$ admits a selection, $\Gamma(q)$, which can be continuously extended to give

$$\Gamma(q) : W \setminus \cup\{x_i\} \to SW.$$

Let V be an open set in $S_1^*(W)^N$ such that for each $q \in V$, each $q_i(x)$ points inward at each point x on the boundary W. By the Morse inequalities [**19**], one of the critical points in $E(\Gamma(q))$ is a stable attractor. □

The preference field $\Gamma(q)$ and its locally isolated critical points are offered as a counterpart, for a voting game, of the price equilibria of an exchange economy. Just as price equilibria are generically isolated [**11**] and continuous in the parameters of the economy, we conjecture that the preference field of Corollary 4 is continuous on an open dense subset of $S_1^*(W)^N$.

Conjecture Let $\mathcal{D}$ be a voting rule, on a Fan manifold, W. Then there is an open dense set $V \subset S_1^*(W)$ such that $\Gamma : S_1^*(W)^N \rightarrow S_1^*(W)$ exists, is continuous on V, and is compatible with $\mathcal{D}$ in the sense that $E(\Gamma) : S_1^*(W)^N \rightarrow 2(W)$ is a continuous selection of $\mathcal{H}_\mathcal{D}$ on the domain V.

As we have observed, $\mathcal{H}_\mathcal{D}$ fails lhc on nowhere dense strata in the preference space. Otherwise $\mathcal{H}_\mathcal{D}$ will be lhc. As discussed in [**34**] the components of $\mathcal{H}_\mathcal{D}$ will "locally" admit continuous selections, which can be used as the equilibrium points of $E(\Gamma)$. As in Corollary 4, the preference field $\Gamma(q)$ can be constructed so that it is continuous with respect to the gradient fields $\{q_1, \dots, q_n\}$.

There is a degree of arbitrariness in the selections involved. Nonetheless the construction gives a way of partially solving the Arrow problem in social choice.

In actual applications the selections $E(\Gamma(q))$ may be interpreted as the outcomes of institutional "rules". The last section presents a bargaining process between party leaders resulting in an outcome in a "political heart".

5 Policy choice in representative democracies

Models of representative politics tend to separate into those that concentrate on pre-election strategies designed to maximize electoral votes, and those that analyze post-election bargaining between parties. In this section we will attempt to integrate these two classes of models.

Suppose that there is a set, N, of n parties represented in the Parliament after an election. We shall suppose that each party, i, has a leader, who has a preferred policy position, z_i, in some "policy space", X. Moreover, each party has gained a *share* e_i of seats in the legislature. Without great loss of generality we may suppose that any coalition M of parties controlling a majority of the seats is winning: thus the voting rule, $\mathcal{D}$, is defined by

$$\mathcal{D} = \{M \subset N : \sum_M e_i > \frac{1}{2}\}.$$

Each leader has preferences derived both from policy and from the "prerequisites" of office. Assume that each leader has a utility function of the form $u_i(y, \sigma_i) = -\frac{1}{2} \parallel y - z_i \parallel^2 + \alpha_i \sigma_i$, where σ_i is the perquisite obtained from office, $y \in X$ is a government policy outcome, and α_i is a real-valued constant. This defines a smooth, strictly convex preference correspondence

$$p_i^{\alpha_i} : X \times \Delta_N \rightarrow X \times \Delta_N$$

where Δ_N is the set of all possible distributions of government perquisites.

Let $p^\alpha(z)$ denote the profile of leader preferences on $X \times \Delta_N$, and let $\mathcal{H}_\mathcal{D}(p^\alpha(z))$ denote the "political heart" induced by this profile and the rule, $\mathcal{D}$. If we assume that α is fixed (or common knowledge) and regard z as a variable, then the heart may be viewed as a correspondence

$$\mathcal{H}_\mathcal{D}^\alpha : X^n \rightarrow 2(X \times \Delta_N)$$

using the identity $\mathcal{H}_{\mathcal{D}}^{\alpha}(z) = \mathcal{H}_{\mathcal{D}}(p^{\alpha}(z))$. Thus, when the vector of leader positions $z = (z_1, \ldots, z_n)$ is given, the political heart, $\mathcal{H}_{\mathcal{D}}^{\alpha}(z)$, is a specific subset of $X \times \Delta_N$. Now consider the "policy" heart, in the case that preferences for government perquisites are null (so $\alpha = 0$). Then $\mathcal{H}_{\mathcal{D}}^{0}(z)$ will be a subset of X.

We propose that policy bargaining among the parties results in an outcome within $\mathcal{H}_{\mathcal{D}}^{0}(z)$, and that coalition perquisites will be distributed in a way which is compatible with $\mathcal{H}_{\mathcal{D}}^{\alpha}(z)$. Bargaining within the parties is assumed to result in policy outcomes and distributions of perquisites that lie within the political heart. For example, consider a point (z, σ) lying outside the political heart. Then for some coalition $M \in \mathcal{D}$, the efficient Pareto process for M will lead away from (z, σ) towards the political heart. Institutional rules in the legislature will select either a particular point (z, σ) in the heart, or possibly a probability distribution, ρ, over the heart. Let us write $\tilde{Y}$ for the set of all probability measures over the set Y, when this set is endowed with the weak topology [25]. A probability measure ρ is a "dual" operator $\rho : U(Y) \to \Re$, mapping a utility function $u \in U(Y)$, with domain Y, to a real number $\int u d\rho \equiv \rho(u)$. Here the support, $supp(\rho)$ of ρ is some subset of Y on which ρ is non-zero. A basic neighborhood $Z(\rho, \epsilon)$ of ρ in the weak topology has the form

$$Z(\rho, \epsilon) = \{\rho' \in \tilde{Y} : | \rho'(u) - \rho(u) | < \epsilon \ \forall \ u \in U(Y)\}.$$

In the case under consideration here, let $\tilde{\mathcal{H}}_{\mathcal{D}}^{\alpha}(z)$ be the set of all probability measures, with support $\mathcal{H}_{\mathcal{D}}^{\alpha}(z)$. Let us denote the space of all probability measures over $X \times \Delta_N$ by $\tilde{X} \times \tilde{\Delta}_N$, and use $2(\tilde{X} \times \tilde{\Delta}_N)$ to denote the induced subspaces. Then the heart correspondence induces a correspondence

$$\tilde{\mathcal{H}}_{\mathcal{D}}^{\alpha} : X^n \to 2(\tilde{X} \times \tilde{\Delta}_N).$$

In the same way, we may write $\tilde{\mathcal{H}}_{\mathcal{D}}^{0} : X^n \to 2(\tilde{X})$. It readily follows that these correspondences are lhc.

Corollary 5 The correspondences

$$(\tilde{\mathcal{H}}_{\mathcal{D}}^{\alpha}, \tilde{\mathcal{H}}_{\mathcal{D}}^{0}) : X^n \to (2(\tilde{X} \times \Delta_N), 2(\tilde{X}))$$

are lhc and admit continuous selections

$$(\tilde{g}_{\mathcal{D}}^{\alpha}, \tilde{g}_{\mathcal{D}}^{0}) : X^n \to (\tilde{X} \times \tilde{\Delta}_N, \tilde{X}).$$

Proof Let $\mathcal{H}$ denote either $\mathcal{H}_{\mathcal{D}}^{\alpha}$ or $\mathcal{H}_{\mathcal{D}}^{0}$, and use $\tilde{W}$ to denote $\tilde{X} \times \tilde{\Delta}_n$ or $\tilde{X}$.

To show $\tilde{\mathcal{H}} : X^n \to 2(\tilde{W})$ is lhc, we use the following necessary and sufficient condition: whenever $z_t \to z$ in X^n and $\rho \in \tilde{\mathcal{H}}(z)$, then there exists a sequence $\rho_t \to \rho$, in the weak topology, with $\rho_t \in \tilde{\mathcal{H}}(z_t)$.

To show this property holds, let V be an open set in W, containing $\mathcal{H}(z)$, let $\rho \in \mathcal{H}(z)$ and let $Z(\rho, \frac{1}{t}) = \{\rho' \in \tilde{V} : | \rho'(u) - \rho(u) | < \frac{1}{t}, \forall u \in U(W)\}$. By lhc of $\mathcal{H}$ on X^n, it is clear that whenever $z_t \to z$, then $\exists \, t_0$, such that $\mathcal{H}(z_t) \cap V \neq \phi$, for all $t > t_0$.

For each $t > t_0$, construct a probability measure $\rho_t = \rho_t^0 + \rho_t^1$ with support V such that $supp(\rho_t^0) = V \cap \mathcal{H}(z_t)$ and $supp(\rho_t^1) = V \setminus \mathcal{H}(z_t)$, with $\rho_t \in Z(\rho, \frac{1}{t})$. By choosing $\{\rho_t\}$ such that $\rho_t^1(u) \to 0 \ \forall \ u \in U(W)$, we have constructed a sequence $z_t \to z$ with $\rho_t^0 \in \tilde{\mathcal{H}}(z_t)$ such that $\rho_t^0 \to \rho$. This construction can be performed for any $\rho \in \tilde{\mathcal{H}}(z)$. Thus both $\tilde{\mathcal{H}}_{\mathcal{D}}^{\alpha}$ and $\tilde{\mathcal{H}}_{\mathcal{D}}^{0}$ are lhc correspondences. By Theorem 3, they admit continuous selections. □

We assume that the outcome of policy negotiations between the parties is a particular probability measure $\tilde{g}^0_{\mathcal{D}}(z)$, with support the policy heart $\mathcal{H}^0_{\mathcal{D}}(z)$. Moreover, perquisites are also distributed to party leaders in a way consistent with the selection $\tilde{g}^\alpha_{\mathcal{D}}(z)$, with support $\mathcal{H}^\alpha_{\mathcal{D}}(z) \subset X \times \Delta_N$. Notice in particular that there may be an equilibrium, or policy core point, for some vector z of leader positions. Typically such a core point will coincide with the policy position of one of the parties, say z_i. In this case the probability measure $\rho = \tilde{g}^0_{\mathcal{D}}(z)$ will be a "Dirac" function with the property that $\rho(u) = u(\rho) = u(z_i)$.

Using a model of this type it is possible to incorporate electoral risk (or uncertainty over the election results) to analyze the choice of leader by each party [**35**].

For example when the leaders' positions are declared to the electorate, then the electorate responds in a stochastic fashion, allocating votes or seats to the party leaders. However, before the election takes place the parties may well have beliefs about likely decisive structures. If we let $\{\mathcal{D}_S\}$ denote the various possible coalition structures, and $\{\pi_S(z)\}$ denote the subjective "electoral" probability associated with each structure (conditional on the declarations of the leaders), then the expected outcome of the vector of declarations is a lottery $\{(\pi_S(z), \tilde{g}^\alpha_S(z)\}$. Here $\tilde{g}^\alpha_S(z)$ denotes the expected probability measure associated with the coalition structure $\mathcal{D}_S$. This can be viewed as a *game form*.

If the representative delegate of party i has a utility functon u_i (for both policy and perquisites), then the expected utility, associated with the vector of leader positions, will have the form

$$U_i(z) = \sum \pi_S(z)\, \tilde{g}^\alpha_S(z)(u_i).$$

Here we use $\tilde{g}^\alpha_S(z)(u_i)$ to denote the expectation of u_i under the probability measure $\tilde{g}^\alpha_S(z)$. In the usual way, a Nash equilibrium, z^*, in the choice of party (or leader) positions is a joint optimal vector, such that

$$U_i(z_1^*, \dots, z_i, z_{i+1}^*, \dots) > U_i(z_1^*, \dots, z_i^*, \dots, z_n^*)$$

for no i and no $z_i \in X$.

The stochastic models of voting response that have been developed [**35**] suggest that the electoral probabilities are continuous (even differentiable) in the party positions. By Corollary 5, it is possible to choose continuous selections of the heart to describe the outcome of political bargaining. Thus the utility functions $\{U_i\}$ will be continuous in the strategies $\{z_i\}_{i \in N}$. Standard results imply the existence of mixed strategy Nash equilibria ($z^* \in \tilde{X}^N$). Indeed it is possible to use calculus techniques in situations where the game form is differentiable.

Typically there will only be two or three different coalition structures to consider. In the studies that have been carried out, political leaders tend to bifurcate into those who attempt to control the electoral center, and those on the periphery who attempt to influence policy through bargaining. We conclude this paper with a brief discussion of our understanding of the nature of political bargaining in the Israeli Knesset in recent years.

Example To illustrate the previous analysis, Figures 3 and 4 give estimates of the positions of the parties in the Israeli Knesset after the 1988 and 1992 elections. (Table 1 lists the strengths of the parties in the last three elections.) After the 1988 election the voting core in $\Re^2$ was empty. Although Likud was the largest party (with 40 seats out of 120), there was a potential winning coalition of Labor, together with the smaller parties to the lower left, and Shas. Likud, however, was able to form government coalitions first with Shas and then with Labor. Assuming Euclidean preferences, the policy heart can be seen to be the triangle bounded by the Likud, Labor and Shas positions. After the 1992 election, Labor (with 44 seats) together with the smaller parties on the lower left controlled 61 out of 120 seats. Consequently the Labor position is at the voting equilibrium or "structurally stable" core, and it was eventually able to form a minority government. See [**35**].

In the 1996 election the policy picture becomes more complex with new splinter parties forming, and a separate prime ministerial election won by Netanyahu (of Likud). Likud gained 32 seats (as in 1992) but Labor lost 10 seats (only winning 34 out of 120). The policy heart expanded out to become similar to the 1988 diagram. Because of the prime ministerial election, Likud has led the coalition government, but has had great difficulty retaining Shas and other splinter groups within its coalition. It is clear that there is little pressure for small parties in the Knesset to approach the electoral center. Attempting to model the balance of electoral and policy objectives of the parties in such a situation is clearly very difficult. The analysis that has been done to date suggests that small parties such as Shas may rationally choose leaders who prefer policies that are further from the electoral center than the typical member of the party.

Table 1. Three Elections in Israel

Party (Acronym)	Knesset Seats		
	1988	1992	1996
Labor (LAB)	39	44	34
Meretz (MRZ)		12	9
CRM	5	—	—
Mapan (MPM)	3	—	—
Shinu (S)	2	—	—
PLP	1	—	—
Daroushe (ADL)	1	2	4
Hadash (HS)	4	3	5
SUBTOTAL	**55**	**61**	**52**
Shas (SHAS)	6	6	10
Likud (LIK)	40	32	32
Aguda (AI)	5	4	4
Mafdal (NRP)	5	6	9
Molodet (MO)	2	3	2
Techiya (TY)	3	—	—
Degel Hatora (DH)	2	—	—
Tzomet (TZ)	2	8	—
SUBTOTAL	**65**	**59**	**57**
Israel Ba'aliya (RUS)	—	—	7
Third Way (TW)	—	—	4
TOTAL	**120**	**120**	**120**

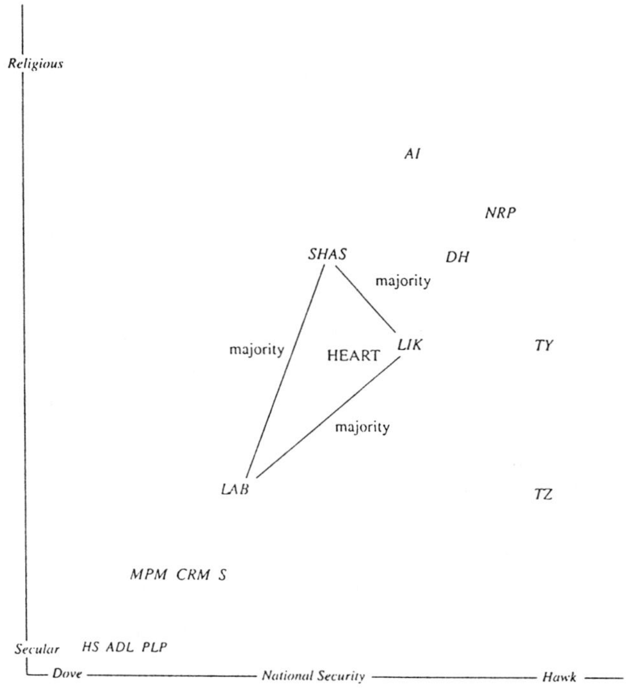

Figure 3 The Knesset in 1988

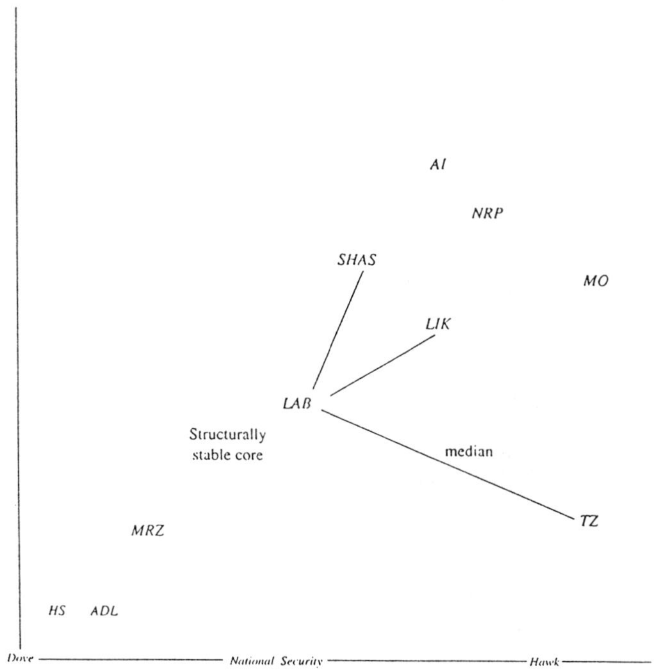

Figure 4 The Knesset in 1992

References

[1] Arrow, K. [1951], *Social Choice and Individual Values*, John Wiley, New York.

[2] Berge, C. [1963], *Topological Spaces*, Oliver and Boyd, Edinburgh.

[3] Bordes, G., LeBreton, M. and Salles, M. [1992], *Gillies' and Miller's sub-relations of a relation over an infinite set of alternatives: General results and applications to voting games*, Math. Op. Res. **17**, 508-518.

[4] Brouwer, L. E. J. [1912], *Uber Abbildung von Mannigfaltikeiten*, Math Annalen **371**, 97-115.

[5] Brown, R. F. [1970] *The Lefschetz Fixed Point Theorem*, Scott-Foresman, Glenview, Illinois.

[6] Chichilnisky, G. [1980], *Social choice and the topology of spaces of preferences*, Ad Math. **37**, 165- 176.

[7] Chichilnisky, G. [1993], *Intersecting families of sets and the topology of cones in economics*, Bull. Am. Math. Soc. **29**, 189-207.

[8] Chichilnisky, G. [1994], *Social Diversity and gains from trade: a unified perspective on resource allocations*, Am. Econ. Rev. Papers and Proceedings **84**, 427-434.

[9] Chichilnisky, G. [1997] *Market Arbitrage, Social Choice and the Core*, Soc Choice and Welfare **14**, 161-198.

[10] Cox, G. [1987], *The uncovered set and the core*, Am. Jour. Pol. Sci. **31**, 408-522.

[11] Dierker, E. [1974], *Topological Methods in Walrasian Economics* (Lecture Notes No. 92), Springer, Berlin.

[12] Fan, K. [1961], *A generalization of Tychonoff's fixed point theorem*, Math. Ann. **42**, 305-310.

[13] Greenberg, J. [1979], *Consistent majority rules over compact sets of alternatives*, Econometrica **47**, 627-636.

[14] Kramer, G. [1973], *On a class of equilibrium conditions for majority rule*, Econometrica **41**, 285-297.

[15] Mas-Colell, A. [1977], *On the continuous representation of preorders*, Int. Rev. of Econ. **18**, 509-513.

[16] Mas-Colell, A. [1979], *A selection theorem for open graph correspondence with star-shaped values*, Jour. Math. Anal. and App. **68**, 273-275.

[17] McKelvey, R. D. [1986], *Covering, dominance and institution-free properties of social choice*, Am. Jour. of Pol. Sci. **30**, 283-314.

[18] Michael, E. [1956], *Continuous selections I*, An. Math. **63**, 361-382.

[19] Milnor, J. [1963], *Morse Theory*, (Annals of Math Studies No. 51), Princeton University Press, Princeton.

[20] Milnor, J. [1978], *Analytic Proofs of the 'hairy ball theorem' and the Brouwer fixed point theorem*, Ann. Math. Monthly **85**, 521-524.

[21] Montiero, P., Page, F. and Wooders, M. H. [1997], *Arbitrage, equilibrium and gains from trade; a counter-example*, J. Math. Econ. **28**, 481-501.

[22] Nakamura, K. [1979], *The vetoers in a simple game with ordinal preference*, Int. Jour. of Game Theory **8**, 55-61.

[23] Page, F. H., Jr. and Wooders, M. H. [1996], *A necessary and sufficient condition for compactness of individually rational and feasible outcomes and existence of an equilibrium*, Economics Letters **52**, 153-162.

[24] Page, F. H., Jr. and Wooders, M. H. [1996], *The partnered core and the partnered competitive equilibrium*, Economics Letters **52**, 143-152.

[25] Parthasarathy, K. R. [1967], *Probability measures on metric spaces*, Academic Press, New York.

[26] Saari, D. [1997], *Generic existence of a core for q-rules*, Ec. Theory **9**, 219-260.

[27] Schofield, N. [1977], *Transitivity of preferences on a smooth manifold*, Jour Econ Th. **14**, 149-172.

[28] Schofield, N. [1978], *Instability of simple dynamic games*, Rev of Econ Studies **45**, 575-594.

[29] Schofield, N. [1984], *Social equilibrium and cycles on compact sets*, Jour Econ Theory **33**, 59-71.

[30] Schofield, N. [1984], *Existence of equilibrium on a manifold*, Math Op Res. **9**, 545-557.

[31] Schofield, N. [1995], *Existence of a smooth social choice functor*, in Social Choice, Welfare and Ethics (W. Barnett, H. Moulin, M. Salles, N. Schofield, eds.), Cambridge University Press, Cambridge.

[32] Schofield, N. [1999a], *The heart and the uncovered set*, Journal of Economics, Suppl. **8**, 79–114.

[33] Schofield, N. [1999b], *The C^1-topology on the space of smooth preference profiles*, Social Choice and Welfare **16**, 445–470.

[34] Schofield, N. [1998], *Aggregation of smooth preferences,* Social Choice and Welfare **15**, 161-185.

[35] Schofield, N., Sened, I. and Nixon, D. [1998], *Nash Equilibrium in Multiparty Competition with 'Stochastic' Voters*, Annals of Operations Research **84**, 3–27.

[36] Smale, S. [1973], *Global Analysis and Economics I: Pareto Optimum and a Generalization of Morse theory*, in Dynamical Systems (M. Peixoto, ed.), Academic Press, New York.

[37] Strnad, J. [1985], *The structure of continuous-valued neutral monotonic social functions*, Social Choice and Welfare **2**, 181-195.

[38] Werner, J. [1987], *Arbitrage and existence of competitive equilibrium*, Econometrica **55**, 1403-1418.

Fields Institute Communications
Volume **23**, 0000

Clubs, Near Markets and Market Games

Martin Shubik
Cowles Foundation for Research in Economics,
Yale University
Box 2125, Yale Station
New Haven, CT O6520 USA
martin.shubik@yale.edu

Myrna H. Wooders
Department of Economics
University of Toronto,
Toronto, Ontario, Canada M5S 3G7
mwooders@chass.utoronto.ca

Abstract. In this paper we develop a new model of replica economies with clubs, with crowding only in production, and with multiple private and public goods. Each agent may belong to many clubs. Our main result is that under apparently mild conditions on the model, economies generate games satisfying per capita boundedness (finiteness of the supremum of per capita payoff). Thus, the results of Wooders (1979,1980b) can be applied to show that (a) for all sufficiently large economies, the game generated by the economy is approximately a market game, and (b) approximate cores are nearly equal treatment in the sense that most agents who are identical are treated approximately identically. When *all* gains to collective activities can be exhausted by groups of agents bounded in size, then stronger results are obtained: (c) for all sufficiently large games, all payoffs in the core have the equal treatment property and (d) approximate cores converge to cores of associated "balanced cover" games. In addition, a discussion of the issues involved in modeling economies as markets is provided. Finally, since much related research has been carried out since the current paper appeared as a 1982 Cowles Foundation Discussion Paper, we provide some relationships to the more recent literature.

1991 *Mathematics Subject Classification.* Primary 90A53; Secondary 90D12.

This paper originally appeared as Cowles Discussion Paper No. 657 (1982c). The nonemptiness result is published in Shubik and Wooders (1986). The results reported in the current version of the paper are identical to those in the 1982 Discussion Paper. We have updated references and revised and shortened some of the exposition. Also, proofs from Wooders (1979,1980b), available elsewhere (cf. Wooders (1994a,b) or the original papers), have been removed.

1 Introduction

1.1 Market games. The market game, both in its sidepayment and nosidepayment versions, has provided a valuable tool for the utilization of game theoretic analysis for the study of exchange economies. It is known that exchange economies map into totally balanced games[1] for both sidepayments and nosidepayments. In addition, totally balanced sidepayment games map into exchange economies and, with special assumptions on consumption sets, totally balanced nosidepayment games also map into exchange economies.[2] Clearly the mappings are not one-to-one in both directions, as the market game contains far less information than does the exchange economy.

The intimate relationship between exchange economies and totally balanced games gives no intimation of the relationship of the competitive equilibria to limit points of the core. In order to study this property we need to be able to define a sequence of exchange economies and a related sequence of totally balanced games. The result that we can define a replication sequence of exchange economies and associate with it a sequence of totally balanced games for which cores converge to competitive payoffs is an economic result arising from having utilized the economic data to construct the sequence of games.

1.2 The purpose of this paper. A natural question to ask is, are there other economic phenomena which give rise to market games or "near-market games" and from which one can construct a sequence of games that exhibits core convergence. In this paper we consider the relationship of market games and near-market games to economies with complexities beyond that of the exchange economy. We show that a broad class of economies generate near-market games, including private goods economies with non-convexities, coalition production economies, and economies with local public goods.[3] We also illustrate that economies with pure public goods do not, without special restrictions, give rise to near-market games.

To make this more precise, some terminology is required. A *sequence of replica games* is a sequence of games with a fixed distribution of a finite number of types. The rth replica game contains r times as many players of each type as the first game in the sequence. The sequence satisfies *minimum efficient scale* (or, equivalently, exhaustion of gains to scale by coalitions bounded in size) if there is some given bound satisfying the condition that all returns to coalition size can be exhausted by coalitions containing fewer members than the bound. A sequence of replica games is *a sequence of near-market games* if the games are superadditive and the sequence satisfies a "near - minimum efficient scale for coalitions" property – for large bounds, almost all returns to coalition size are nearly exhausted by coalitions bounded in size. Equivalently, the sequence satisfies *per capita boundedness*, the property that the supremum of per capita payoffs is finite.

Wooders (1980b) shows that if a sequence of games satisfies the near minimum efficient scale property then the sequence is asymptotically totally balanced – given any epsilon greater than zero and any subgame of any game in the sequence, when we replicate the set of players in that subgame, for all sufficiently large replications the epsilon-core of the replicated subgame is non-empty. Moreover, epsilon-cores of large games have an equal-treatment property, that is, most players who are

[1] We note that every subgame of a totally balanced game has a non-empty core.

[2] See Billera and Bixby (1974) and references therein.

[3] See also Wooders (1979,1980a,b).

identical receive nearly the same payoffs. Under the stronger assumption of minimum efficient scale for coalitions, for all sufficiently large games, all payoffs in the core have the equal treatment property and approximate cores converge to the (equal-treatment) cores of the associated balanced cover games.[4]

In this paper, a model of a sequence of replica economies with coalition production and local public goods, where agents are allowed to be members of possibly more than one club, is developed. There may be multiple public and private goods. No convexity assumptions are required, either in consumption or production. For simplicity, the model allows crowding only in production, that is, the production possibilities open to a club depend on the membership of the club but preferences are unaffected by club memberships.[5] Few restrictions are placed on the model; the major ones are that the asymptotic growth of utility functions is no more than linear and the production correspondences are such that positive outputs do not become virtually free in per-capita terms as the economies become large. We show that the sequence of games derived from the economies is a sequence of near-market games. From the results of Wooders (1980b) it follows that for all sufficiently large economies, approximate cores are nonempty and converge to the equal-treatment core – that is, in large games approximate cores treat almost all identical players nearly equally.

When we further assume that *all* increasing returns to coalition size are realized by some finite economy – there is a minimum efficient scale of coalitions rather than a near minimum efficient scale – we obtain a stronger core convergence result. The ε-cores of the derived sequence of games converge to the cores of the associated balanced cover games. These results depend on Wooders' (1979) results showing that when there is a minimum efficient scale, the limit of ε-cores equals the limit of the cores of the associated balanced cover games.

An investigation of competitive equilibria (or competitive-like equilibria since we are dealing with a class of economies to which the classic definition of the competitive equilibrium does not apply[6]) and convergence of the core to the equilibrium payoffs is beyond the scope of this paper. We note however that since the economies under consideration generate games with side payments, from our results and those of Shapley and Shubik (1969) it follows that large economies can be represented by markets where each player owns one unit of a commodity identified with his

[4]Results in the same spirit, for the special case of economies with a local public good and anonymous crowding – that is, individuals are crowded only by the numbers of individuals in the same jurisdiction and not by the composition of the jurisdiction – were obtained in Wooders (1980a). While the anonymous crowding model is economically interesting, from the perspective of game theory is significantly more restrictive than the class of games defined in Wooders (1980b).

We thank Yakar Kannai for stressing that the difference between these two models should be discussed.

[5]This significantly simplifies the model. If we allowed preferences of an agent to depend on the sizes and compositions of the clubs containing that agent, as in Wooders (1981b), for example, the analysis would be more complicated but, with appropriate modifications of assumptions, the same results would still obtain For some recent related results, see Conley and Wooders (1998b) and Kovalenkov and Wooders (1997a,b).

[6]Some illustrations of such convergence are demonstrated in Boehm (1974) for economies with coalition production and in Wooders (1980a,1981b) for economies with local public goods. See also Bennett and Wooders (1979) where the results of Wooders (1979) are applied to questions of firm formation. For more recent results see, for example, Wooders (1997), Conley and Wooders (1998a) and references in these papers.

"type." Since, under our assumptions, approximate cores treat similar players similarily it follows that approximate cores converge to equilibrium outcomes of the representing markets.[7]

Except under some very special conditions (satiation or "asymptotic" satiation, i.e., marginal utilities go to zero as the amount of the public good increases) replica economies with pure public goods do not generate near-market games. Since games derived from economies with pure public goods may well be market games (i.e., totally balanced),[8] this suggests that, in line with the incentives literature, it is in economies with many players that game-theoretic properties of private-goods or "market-like" economies and those of pure public goods economies differ.

Another question to be addressed is when is an economic situation "adequately" represented by the characteristic function of the derived game? What is "adequate" is, to some extent, a matter of opinion and depends on the ultimate purpose of the model. For our purposes, we say an economic situation has the "c-game" property[9] (i.e., in our view is adequately represented by its derived game), if what a coalition can achieve once it has formed is independent of the actions of the complimentary coalition. Exchange economies obviously have the c-game property; those with pure public goods might not. Those with coalition production and local public goods raise modelling and interpretation problems concerning their "c-gameness"; this will be discussed later.

As indicated above, in this paper we analyze replication sequences of economies – ones with a fixed number of types of agents and increasing numbers of agents of each type. To extend the analysis to economies with a continuum of agents or with simply a "large" number of agents appears to pose different problems than extensions of this nature for private goods exchange economies.[10] The technical results in this paper are based on results concerning non-emptiness and convergence of approximate cores of large replica games (games with a fixed distribution of a finite number of player types). The extension of the analysis to economies with a large number of agents (but not necessarily replica economies) could, we believe, be carried out if results were available for large (not-necessarily replica) games analogous to those for large replica games.[11] Our purpose in this paper is to introduce concepts of near-market economies and near-market games; it is not to obtain the most general results that might be possible.

The models are also restricted to ones with a freely transferable medium of exchange. As has been shown, such a restriction is not necessary.[12]

In Section 2 a mathematical structure and analysis for near-market games is presented. Section 3 contains our model of a sequence of replica economies whose

[7]For the market explicitly constructed by Shapley and Shubik where there is only one player of each type, the equivalence of the core and the equilibrium outcomes holds in finite economies. This is no longer true in the setting of our model. But in the limit, core-equilibrium equivalence obtains. See Wooders (1994a,b) for related results.

[8]See Rosenthal (1976).

[9]See Shubik (1982, pp. 130-131, 354).

[10]Since this paper was written, some such results have been obtained. See especially Kaneko and Wooders (1986), Conley and Wooders (1998b) and further references in these papers.

[11]Some such results have been obtained. See, for example, Wooders and Zame (1984) and Kovalenkov and Wooders (1997a,b).

[12]In this regard, see Wooders (1981a) and Shubik and Wooders (1981a,b). The assumption of a freely transferable medium of exchange simplifies the analysis while allowing quite complex economic sutations. Recent relevant papers include Kovalenkov and Wooders (1997a,b) and Conley and Wooders (1998b).

derived games are near-market games. Section 4 deals with the application and interpretation of near-market games together with a closer examination of the modelling problems involved. Section 5 discusses some of the subsequent literature on market games and on economies with clubs. All proofs are contained in the appendices.

1.3 Near markets. In the remainder of this paper we argue that for large economies, with appropriate conditions essentially limiting the effects of externalities and indivisibilities, near-market games provide a valuable tool for analysis of the first six cases described in Table 1. For nonlocal public goods we suggest that the physical conditions require a direct modelling of the voting or other socio-political mechanism. This is noted in discussion of several special cases in Section 4.

Even though our analysis is carried out in terms of games with sidepayments we conjecture that if there exists a "quasi-transferable" utility (i.e., a good which is divisible, always desirable and a substitute for any other good) our results hold substantially.[13]

2 Near-market games

2.1 Games. We first review some game-theoretic concepts.

A **game** (with sidepayments) is an ordered pair (N, v) where $N = \{1, \ldots, n\}$ is a finite set, called the set of **players** and v is a real-valued function mapping subsets of N into $\mathbf{R}_+$ with $v(\phi) = 0$. Two players i and j are **substitutes** if given any subset S of N where $i \notin S$ and $j \notin S$, we have

$$v(S \cup \{i\}) = v(S \cup \{j\}).$$

A **subgame** (S, v) of (N, v) is an ordered pair consisting of a non-empty subset S of N and the function v restricted to subsets of S. The game (N, v) is **superadditive** if for all disjoint subsets S and S' of N, we have

$$v(S) + v(S') \leq v(S \cup S').$$

A **payoff** for the game is a vector $B = (B^1, \ldots, B^n) \in \mathbf{R}^n_+$, the non-negative orthant of the n-fold Cartesian product of the reals. A **payoff** B is feasible if

$$\sum_{i \in N} B^i \leq v(N).$$

Given $\varepsilon \geq 0$, a payoff is in the (weak) **ε-core**[14] if it is feasible and if, for all non-empty subsets S of N,

$$\sum_{i \in S} B^i \geq v(S) - \varepsilon \mid S \mid$$

where $\mid S \mid$ denotes the cardinal number of the set S. When $\varepsilon = 0$, the ε-core is called simply the **core**.

Given a game (N, v), let $(N, \widetilde{v})$ denote the **totally balanced cover** of (N, v); the function $\widetilde{v}$ is the smallest real-valued function such that, for all non-empty subsets S of N, the subgame $(S, \widetilde{v})$ has a non-empty core and $v(S) \leq \widetilde{v}(S)$.

[13] This conjecture, when made, was based on the fact that analogous development of approximate core theory has been carried out for games without side payments (see Wooders 1981a and Shubik and Wooders 1982b) and some results in the spirit of those herein had been obtained (see Wooders 1980a,1981a). Recent relevant results include Conley and Wooders (1998b).

[14] This concept was introduced by Shapley (1966).

2.2 Sequences of games. A sequence of games $(N_r, v_r)_{r=1}^{\infty}$ is **superadditive** if each game (N_r, v_r) is superadditive. The sequence is **per-capita bounded** if there is a constant K, independent of r, such that

$$v_r(N_r)/ \mid N_r \mid \leq K$$

for all r.

Let $(N_r, v_r)_{r=1}^{\infty}$ be a sequence of games where, for some positive integer T, for each r the set of players N_r contains rT players denoted by

$$N_r = \{(t, q) : t = 1, \ldots, T,\ q = 1, \ldots, r\}.$$

For each r and each t, let $[t]_r = \{(t, q) : q = 1, \ldots, r\}$. The sequence is a **sequence of replica games** if

(1) $N_r \subset N_{r+1}$ for all r;
(2) for each r and all subsets S of N_r, $v_r(S) \leq v_{r'}(S)$ whenever $r' \geq r$;
(3) for each r and each t all players in $[t]_r$ are substitutes for each other.

Note that positive externalities are allowed in the sense that what a coalition can achieve may increase when the total player set increases in size: Negative externalities could also be allowed. For the purposes of this paper, what is crucial is that the sequences $\frac{v_r(N_r)}{|N_r|}$ and $\frac{v_r(s)}{|s|}$ converge to limiting values. This is discussed further in Section 5.

Throughout the following, given a sequence of replica games $(N_r, v_r)_{r=1}^{\infty}$ we define $[t]_r$ as above and call the members of $[t]_r$ **players of type** t. We also assume there are T types of players and denote the set of players N_r as above.

Give a sequence of replica games $(N_r, v_r)_{r=1}^{\infty}$ and a subset S of N_r for some r, define the vector $\rho(S) = (s_1, \ldots, s_T)$ by its coordinates $s_t = \mid S \cap [t]_r \mid$; the vector $\rho(S)$ is called the **profile of** S and is simply a list of the numbers of players of each type contained in S. Let I denote the T-fold Cartesian product of the non-negative integers. Observe that for any r and any subset S of N_r, we have $\rho(S) \in I$. Also, since players of the same type are substitutes, if S and S' are two subsets with the same profiles then for any r such that $S \subseteq N_r$ and $S' \subseteq N_r$, we have $v_r(S) = v_r(S')$. Consequently the function v_r can be completely determined by a mapping from a subset of I to the reals. In the following, given r and a profile s of a subset of N_r, we define $v_r(s)$ as $v_r(S)$ for any $S \subseteq N_r$ with $\rho(S) = s$.[15] Given $s \in I$ we write $\mid s \mid = \sum_{t=1}^{T} s_t$ since when $\rho(S) = s$, we have $\mid S \mid = \sum_{t=1}^{T} s_t$.

We say a sequence of games satisfies the property of "**near minimum efficient scale**" (for coalitions), NMES, if it is per-capita bounded. If the sequence is also superadditive, we say the sequence is a sequence of **near market games**.

We remark that when a sequence of replica games is per-capita bounded and superadditive, both $\widetilde{v}(N_r)/ \mid N_r \mid$ and $v(N_r)/ \mid N_r \mid$ converge and to the same limit. In this case, for r sufficiently large, the per-capita gains to forming a coalition larger than N_r are "small." This motivates our term "near-minimum efficient scale."

A sequence of replica games $(N_r, v_r)_{r=1}^{\infty}$ is **asymptotically totally balanced** if, given any r, any subset S of N_r, and any $\varepsilon > 0$, there is an n^* such that for all $n \geq n^*$ we have

$$\frac{\widetilde{v}_{nr}(S_n)}{\mid S_n \mid} - \frac{v_{nr}(S_n)}{\mid S_n \mid} < \varepsilon,$$

[15]This abuse of notation should create no confusion. We note that we typically denote subsets by upper case letters and profiles by lower case ones.

where v_{nr} denotes the function $v_{r'}$ with $r' = nr$ and S_n is any subset of N_r, with $\rho(S_n) = n\rho(S)$. It can easily be verified (and follows from well-known results, cf. Shapley (1967)), that given any game (N, v), we have

$$\frac{\widetilde{v}\,(N)}{\mid N \mid} - \frac{v(N)}{\mid N \mid} < \varepsilon$$

if and only if the ε-core of the game is non-empty. Consequently, given any subset S of N_r for some r and any sequence of subsets (S_n) satisfying the properties required above, for all n sufficiently large the subgames (S_n, v_{nr}) have non-empty ε-cores.

The following theorem provides sufficient conditions for asymptotic total balancedness of sequences of replica games.

Theorem 2.1 *(Wooders 1980b).*[16] *Let $(N_r, v_r)_{r=1}^{\infty}$ be a sequence of near market games. Then the sequence is asymptotically totally balanced.*

For a proof, we refer the reader to Wooders (1980b, Theorem 3) or Wooders (1994b, Theorem 3.2).

2.3 Convergence of cores. In the remainder of this section, we consider convergence properties of ε-cores of sequences of replica games. These results are from Wooders (1979,1980b).[17]

The next theorem shows that as r goes to infinity and ε goes to zero, in "the limit", only equal-treatment payoffs are in the ε-cores where an **equal treatment payoff** x for N_r has the property that, for each t, $x^{tq} = x^{tq'}$ for all q and q'.

Theorem 2.2 *Wooders (1980b). Let $(N_r, v_r)_{r=1}^{\infty}$ be a sequence of near-market games. Then, given any $\delta > 0$ and any $\lambda > 0$, there is an ε^* and an r^* such that for all $\varepsilon \in [0, \varepsilon^*]$ and for all $r \geq r^*$, if B is in the ε-core of (N_r, v_r) then*

$$\mid \{(t, q) \in N_r : \ \parallel B^{tq} - \overline{B}_t \parallel \ > \delta\} \mid /r < \lambda$$

where $\overline{B}_t = \sum_{q=1}^{r} B^{tq}/r$ and $\parallel \cdot \parallel$ denote the absolute value.

Less formally, Theorem 2.2 states that for small ε and large r, any payoff in the ε-core has the property that the percentage of players whose payoffs differ significantly (by more than δ) from the average payoff for their type can be made arbitrarily small (less than λ). We remark that since the theorem holds for $\varepsilon \geq 0$, it follows that for all sufficiently large r any payoff in the core of the rth balanced cover game has "nearly" the equal-treatment property. For a proof of the result, we refer the reader to Wooders (1980b, Theorem 5) or to Wooders (1994b, Proposition A.1.2).

Given a sequence of replica games $(N_r, v_r)_{r=1}^{\infty}$, let B be an equal-treatment payoff in the ε-core of (N_r, v_r) for some r and some $\varepsilon \geq 0$. Since, for each t, $B^{tq} = B^{tq'}$ for all q and q', B can be completely described by a vector $\overline{B} \in \mathbf{R}^T$ where, for each t, $\overline{B}_t = B^{tq}$ for (any) q. We say that $\overline{B}$ **represents** an equal-treatment payoff in the ε-core of (N_r, v_r). Given $\varepsilon > 0$, for each r let

$$\mathcal{B}_r(\varepsilon) \quad = \quad \{\overline{B} \in^T : \overline{B} \text{ represents an equal treatment payoff in the } \varepsilon - \text{core of } (N_r, v_r)\}.$$

[16] This result also follows from Wooders (1981a) since games with side payments are a special case of games without side payments.

[17] In Wooders (1992) the results of this section are shown to hold for private goods exchange economies with quasi-linear utilities,

Given $\varepsilon > 0$, let $L(\mathcal{B}(\varepsilon))$ denote the closed limit[18] of the sequence of sets $(\mathcal{B}_r(\varepsilon))_{r=1}^{\infty}$. In the appendix it is shown that this limit exists. Define

$$A^* = \cap_{\varepsilon>0} L(\mathcal{B}(\varepsilon)); \tag{2.1}$$

then A^* is the **asymptotic core** of (N_r, v_r). Informally, A^* is the limit of the ε-cores and $B^* \in \mathbf{R}^T$ is in the asymptotic core if, given any $\varepsilon > 0$, there is a sequence $(\overline{B}_r^{\varepsilon})_{r=1}^{\infty}$ where $\overline{B}_r^{\varepsilon} \in \mathcal{B}_r(\varepsilon)$ for each r and $\overline{B}_r^{\varepsilon} \to B^*$ as r goes to infinity.

Given a sequence of replica games, $(N_r, v_r)_{r=1}^{\infty}$, let $\mathcal{C}(r)$ denote the subset of $\mathbf{R}^T$ where each member C of $\mathcal{C}(r)$ represents an equal-treatment payoff in the core of the balanced cover game $(N_r, \widetilde{v}_r)$. Let $L(\mathcal{C})$ denote the closed limit of the sequence of sets $(\mathcal{C}(r))_{r=1}^{\infty}$.

An example in Wooders (1980b) illustrates that approximate cores do not necessarily converge to cores of the corresponding balanced cover games (or the cores of the games themselves, if nonempty). The example is simple: Let $T = 2$. For $s \in I$ define

$$v(s) = \left\{ \begin{array}{ll} \min\{s_1 - 1, s_2\} & \text{when } s_1 \geq 1 \\ 0 & \text{otherwise.} \end{array} \right\}$$

Let the profile of N_1 be $m = (1,1)$. The only equal-treatment payoff in the core of the rth replica game is represented by $(\frac{r-1}{r}, 0)$. It follows that $L(\mathcal{C}) = (1,0)$. It can be verified, however, that A^* is the set $\{(p_1, p_2) : p_1 \geq 0,\ p_2 \geq 0$ and $p_1 + p_2 = 1\}$. Informally, for the ε-core, the fact that there is an "excess" player of type 1 is virtually irrelevant, but the cores of the games depend crutially on the exact proportions of players of each type. As shown in Wooders (1980b), under the assumption of per capita boundedness, the asymptotic core is equivalent to the core of a certain continuum limit game, where a set of measure zero can be ignored. For the purposes of the current paper, since we do not consider continuum economies, we will only appeal to the following convergence result.

We now state conditions under which $L(\mathcal{C})$ exist and equal A^*. We say a sequence of superadditive replica games $(N_r, v_r)_{r=1}^{\infty}$ has a **minimum efficient scale** for coalitions, MES, if there is an r^* such that for all $r \geq r^*$ and $r' \geq r^*$ we have $\widetilde{v}_r\ (N_r)/r = \widetilde{v}_{r'}\ (N_{r'})/r$.[19] We call the replication number r^* an **MES bound** (a minimum efficient scale bound). We remark that when r^* is an MES bound for the sequence, given $r > r^*$ if B is a payoff in the core of (N_r, v_r) then B has the equal-treatment property; this is easily verified (and shown in Wooders (1979, Theorem 6) and Wooders (1983, Theorem 3) for games without side payments).

Theorem 2.3 *(Wooders 1979b). Let $(N_r, v_r)_{r=1}^{\infty}$ be a sequence of near-market games. If the sequence has the MES property, then $L(\mathcal{C})$ exists and equals A^*.*

See Wooders (1979) or (1980b, Theorem 7) for a proof.

3 Near-market economies

3.1 Introduction to Near-market economies. In this section we develop a model of a sequence of replica economies with private goods, local public goods, and coalition production. Minimal restrictions are imposed on the model yet we are able to show that the sequence of derived games is superadditive and per-capita

---

[18]See Hildenbrand (1974) , pp. 15-17 for a definition and some properties of the closed limit of a sequence of sets.

[19]Alternative formulations of MES are contained in Wooders (1979,1981a) and Shubik and Wooders (1982a).

bounded and thus the sequence is asymptotically totally balanced. Therefore, the class of replication economies we consider are near-market economies — for large replications the derived games are near-market games.

3.2 The model. Let us begin with an informal discussion of the model. There are a finite number T of player types and, in the rth game, there are r players of type t.[20] There are ℓ private commodities and m public goods. The utility functions of agents are defined over $\mathbf{R}_+^{\ell+m}$. (This implies that preferences of agents are unaffected by the memberships of clubs.) A club structure of a set S of agents is a collection of subsets of S, called *clubs*. For each subset of agents, there is a given admissible set of club structures. It may be, for example, that the only admissible club structure of S is a partition of S. Alternatively, a club structure may allow an agent to belong to many clubs. Admissible club structures are required to satisfy certain natural properties. These properties imply that the game derived from the economy is superadditive. Given an admissible club structure, each club in the structure may produce public goods for the benefit of its members. The production possibilities available to a club depend on the club structure. Any two clubs with the same number of players of each type have the same production technology available to them. Similarily, private goods are produced by firms. A firm structure of a set S of agents is a collection of subsets of S. Admissible firm structures are required to satisfy the same properties as admissible club structures.

A **sequence of replica economies** $(\mathsf{E}_r)_{r=1}^{\infty}$ is defined as a sequence of septuples

$$\mathsf{E}_r = (N_r, \mathbf{R}^{\ell}, \mathbf{R}_+^m, U_r, W_r, (\mathcal{J}_r, Z_r), (\mathcal{F}_r, Y_r))$$

where:

$N_r = \{(t,q) \ : \ t = 1, \dots, T, \ q = 1, \dots, r\}$ is the set of agents;

$\mathbf{R}_+^{\ell}$ is the private commodity space;

$\mathbf{R}_+^{m}$ is the public commodity space;

$U_r = \{u^{tq} \ : \ (t,q) \in N_r\}$ is an indexed collection of utility functions mapping $\mathbf{R}_+^{m+\ell}$ into $\mathbf{R}_+^1$ with the property that for some linear function L and some real number c we have $u^{tq}(x,y) \leq L(x,y) + c$ for all (x,y) in $\mathbf{R}_+^{m+\ell}$ and for all (t,q) in N_r and $u^{tq} = u^{tq'}$ for all $q, q' \in 1, \dots, r$.

$W_r = \{w^{tq} \in \mathbf{R}^{\ell} \ : \ (t,q) \in N_r\}$ is an indexed collection of endowment vectors, each in $\mathbf{R}_+^{\ell}$ (no public goods are initially endowed) and $w^{tq} = w^{tq'}$ for all $q, q' \in 1, \dots, r$;

$(\mathcal{J}_r, Z_r)$ is a pair of correspondences, where

$\mathcal{J}_r$, called the **allowable club structure correspondence**, maps non-empty subsets S of N_r into non-empty collections of subsets of S and

Z_r, called the **public goods production correspondence**, maps subsets S of N_r into subsets of $\mathbf{R}_+^m \times \mathbf{R}^{\ell}$; and

$(\mathcal{F}_r, Y_r)$ is another pair of correspondences where

$\mathcal{F}_r$, called the **allowable firm structures correspondence,** maps non-empty subsets S of N_r into collections of subsets of S, and

Y_r, the **private goods production correspondence**, maps non-empty subsets S of N_r into subsets of $\mathbf{R}^{\ell}$.

---

[20]This is not as restrictive as it might seem, since players of the different types may be identical, and is not made in several paper, for examples, Wooders (1979).

A number of further specifications are made on the components of a sequence of replica economies:

(1) $N_r \subset N_{r+1}$ for each r;

(2) For each t, each q and q' in $\{q'' : q'' = 1, \dots r\}$, and all r, $u^{tq} = u^{tq'}$; i.e., all agents of the same type have the same utility functions and the same initial endowments. Also, $u^{tq}(w^{tq}, 0) > 0$ for all $(t, q) \in N_r$ and for all r (this assumption is for technical convenience);

(3) Given r and $S \subseteq N_r$, and $J(S) \in \mathcal{J}_r(S)$, $J(S)$ is **an allowable club structure of** S. Allowable club structures $J(S)$ are required to satisfy the properties that

(a) $S \subseteq \cup_{S' \in J(S)} S'$ (allowable club structures of S cover S)

(b) if $J(S) \in \mathcal{J}_r(S)$ and $J(S') \in \mathcal{J}_r(S')$, where S and S' are non-empty, disjoint subsets of agents, then $\{S'' \subseteq N_r : S'' \in J(S) \cup J(S')\} \in \mathcal{J}_r(S \cup S')$;

(c) given $S \subseteq N_r$ and $r' \geq r$, if $J(S)$ is in $\mathcal{J}_r(S)$; then $J(S)$ is in $\mathcal{J}_{r'}(S)$;

(d) if S and S' are non-empty subsets of N_r with the same profiles, then there is a one-to-one mapping, say ψ, of $\mathcal{J}_r(S)$ onto $\mathcal{J}_r(S')$ such that if $\psi(J(S)) = J(S')$, then the collection of profiles of members of $J(S)$ (not all necessarily distinct), equals those of $J(S')$.

Condition (a) simply ensures that allowable club structures of a subset S of agents cover S. Condition (b) ensures that the game derived from the economy is superadditive. Condition (c) ensures that the possible club structures of a subset of agents does not decrease when the size of the economy increases, and finally, (d) implies that admissible club structures depend only on profiles and not names of individual agents.

We remark that in Wooders (1980a) the allowable club structures of a set of agents is the set of all partitions of agents. The present formulation of allowable club structures is sufficiently general to include the possibility that $\mathcal{J}_r(S)$ is the set of all partitions of S and also to include the possibility of "over-lapping" clubs (an agent could possibly belong to several clubs simultaneously).

(4) The public goods production correspondence Z_r is required to satisfy the properties that

(a) given $S \subseteq N_r$ and $r' > r$, $Z_r(S) \subseteq Z_{r'}(S)$;

(b) if, for any r, S and S' are non-empty subsets of N_r with the same profiles then $Z_r(S) = Z_r(S')$, i.e., the production possibility set for a subset of N_r depends only on the profile of that coalition;

(c) $0 \in Z_r(S)$ for all non-empty subsets S of N_r and for all r.

(5) Given a subset S of N_r, **allowable firm structures** $F(S) \in \mathcal{F}_r(S)$ are required to satisfy the same properties as allowable club structures, i.e., (a)-(d) of (3) above. Also, the private goods production correspondence is assumed to satisfy the same properties as the public goods production correspondence, (a)- (c) of (4) above.

For private goods, we also define the **aggregate production correspondence**. Given r and $S \subseteq N_r$, define

$$\overline{Y}_r(S) = \bigcup_{F(S) \in \mathcal{F}_r(S)} \sum_{S' \in F(S)} Y_r(S') \ :$$

then $\overline{Y}_r(\cdot)$ is the rth **aggregate production correspondence**. Note that $\overline{Y}_r(\cdot)$ is **superadditive**; given any two disjoint, non-empty subsets S and S' of N_r, we have $\overline{Y}_r(S) + \overline{Y}_r(S') \subseteq \overline{Y}_r(S \cup S')$.

The above consists of a description of the components of a sequence of replica economies and members of the sequence. In the following, we introduce additional definitions which enable us to relate production decisions to consumption decisions and to define feasible states of an economy.

An N_r-**allocation** is a vector

$$(x,y) = (x^{11}, \dots, x^{Tr}, y^{11}, \dots, y^{Tr}) \in \mathbf{R}_+^{Tr(m+\ell)}$$

where (x^{tq}, y^{tq}) is a commodity bundle for the $(t,q)^{th}$ agent. Given any r and any non-empty subset S of N_r, an S-**allocation** is an N_r-allocation where $x^{tq} = 0$ and $y^{tq} = 0$ if $(t,q) \notin S$.

Given r and a non-empty subset S of N_r, an S-**private goods production plan** is a vector $y \in \overline{Y}_r(S)$.

It is possible, and obvious, that we could define an aggregate public goods production correspondence analogously to the definition of the aggregate private goods production correspondence. However, because of the "non-transferability" of public goods produced in one club to the members of another club, we proceed differently in defining public goods production plans. In particular, we keep track of the club structure associated with a public goods production plan. Given a non-empty subset S of N_r, an S-**public goods production plan** is an ordered pair,

$$\varphi(S) = (J(S), \{(x_{S'}, z_{S'}) \in Z_r(S') \ : \ S' \in J(S)\})$$

where $J(S) \in \mathcal{J}_r(S)$.

Given r and a non-empty subset S of N_r, and S-**state of the economy**, $e(S)$, is an ordered triple,

$$e(S) = (\overline{y}, \varphi(S), (x,y))$$

where $\overline{y}$ is a private goods production plan for S, $\varphi(S)$ is an S-public goods production plan, and (x,y) is an S-allocation where

$$x^{tq} = \sum_{\substack{tq \in S' \\ S' \in J(S)}} x'_S, \text{ for all } (t,q) \in S$$

(the $(t,q)^{th}$ agent consumes the total outputs of public goods produced in all clubs of which he is a member). The state is S-**feasible** if

$$\sum_{(t,q) \in S} (y^{tq} - w^{tq}) + \sum_{s' \in J(S)} z_{S'} = \overline{y}.$$

An N_r-state of the economy is called simply **a state of the rth economy**, or, when no confusion is likely to arise, simply a **state of the economy**.

Given $S \subseteq N_r$, let

$$A_r(S) = \{(x,y) \ : \text{there is an } S\text{-feasible state of the economy with associated } S\text{-allocation } (x,y)\}.$$

The set $A_r(S)$ is called the set of S-**attainable allocations**.

$$\mathrm{Proj}_S A_r(S) \subseteq \mathrm{Proj}_S A_{r'}(S)$$

where $\text{Proj}_S A_{r''}(S)$ denotes the projection of the set $A_{r''}(S)$ onto the subset of $\mathbf{R}^{|S|(m+\ell)}$ associated with the members of S for any replication number r''.

3.3 The derived games. We now define the sequence of games derived from the sequence of economies.

Given r and $S \subseteq N_r$, define

$$v_r(S) = \sup_{(x,y)\in A_r(S)} \sum_{(t,q)\in S} u^{tq}(x^{tq}, y^{tq}) \text{ when } S \neq \phi \text{ and}$$

$$v_r(S) = 0 \text{ otherwise.}$$

Observe that the pair (N_r, v_r) is a game with sidepayments. The generation of a game with sidepayments presumes, as usual, the existence of a freely transferable medium of exchange. There is no need to introduce this explicitly.[21] It is straightforward to verify that the sequence of derived games is a sequence of replica games.

3.4 Near-market economies. Without further restrictions on the economies, in particular on production, there is no assurance that the derived sequence of games is a sequence of near-market games. The restrictions required are, informally, that positive production does not become virtually free as the economies become large. Formally, we assume

A1. There is a closed convex cone $Y^* \subset \mathbf{R}^\ell$, with $-\mathbf{R}^\ell_+ \subseteq Y^*$ and $Y^* \cap \mathbf{R}^\ell_+ = \{0\}$, such that $Y_r(S) \subseteq Y^*$ for all subsets S of N_r and for all r.

A2. There is a closed convex cone $Z^* \subseteq \mathbf{R}^m_+ \times \mathbf{R}^\ell$, with $\{0\} \times -\mathbf{R}^\ell \subseteq Z^*$ and $Z^* \cap \mathbf{R}^m_+ \times \mathbf{R}^\ell_+ = \{0\}$, such that for any r, any non-empty subset $S \subseteq N_r$, and any allowable club structure $J(S) \in \mathcal{J}_r(S)$, we have

$$\sum_{S'\in J(S)} (x_{S'}, z_{S'}) \in \{(x,z) \in R^{m+\ell} \ : \ (\mid S \mid x, z) \in Z^*\}.$$

The first assumption is clear. The second is that there is some set Z^* satisfying the properties of a standard private goods production set and public goods are never cheaper per capita than private goods would be if they were produced with the public set Z^*.

An example of production correspondences which satisfy A2 in the one-private-good, one-public-good case is given by the production functions $x + z/ \mid S \mid = 0$ for each subset S of N_r and for all r with $\mathcal{J}_r(S)$ equal to the set of partitions of S. Here the per capita costs of the public good, in terms of the inputs, is constant and independent of the size and composition of the club structure.

To see what A2 rules out, suppose all coalitions have the production set Z determined by the production function $x + z = 0$, there is only one private good and one public good, and again $\mathcal{J}_r(S)$ is the set of all partitions of S. To show that A2 is not satisfied, let $x_r = 2/r$ and $z_r = -(2/r)$ for each positive integer r. Observe that $(x_r, z_r) \in Z(S)$ for all $S \subseteq N_r$ and for all r. Choose a sequence of subsets $S_r \subseteq N_r$ for each r such that $\mid S_r \mid = r$. We then have $\lim_{r\to\infty} \mid S_r \mid x_r = 2$ and $\lim_{r\to\infty} z_r = 0$. This contradicts A2 since Z^* is closed and $Z^* \cap \mathbf{R}^2_+ = \{0\}$.

[21] To explicitly introduce the medium of exchange, add another private good to the consumption set, but allow this to be achieved in negative as well as positive amounts, so that the consumption set is $\mathbf{R}^\ell_+ \times \mathbf{R}^m_+ \times \mathbf{R}$ and take the utility functions as $\widehat{u}^{tq}(x, y, \xi) = u^{tq}(x, y) + \xi$.

Together, the assumptions on Z_r and Y_r and the boundedness above of the utility functions (by a linear function plus a constant) imply that the sequence of derived games is per-capita bounded. Since the sequence of derived games is also superadditive, it is a sequence of near-market games and asymptotically totally balanced.

Theorem 3.1 *Let $(\mathsf{E}_r)_{r=1}^{\infty}$ be a sequence of replica economies satisfying A1 and A2. Then the derived sequence of replica games is a sequence of near-market games and is asymptotically totally balanced.*

In Appendix 2, we show that the derived sequence of games is per capita bounded and superadditive. Theorem 3.1 then follows as a Corollary to Theorem 2.1. (All theorems in this section are proven in Appendix 1.)

3.5 Convergence of ε-cores to cores. We now impose a further restriction on production which will have the result that the derived sequence of games has the MES property. Informally, the assumption ensures that any state of the economy that can be improved upon can be improved upon by a coalition bounded in size bounded in size by $\rho(N_{r^*})$ for some r^*.

We require the following definition: given r and $(x, y) \in A_r(N_r)$ a **permutation by types** of (x, y) is an N_r-allocation, say (x', y') where for each type t there is a one-to-one mapping of $\{(x^{tq}, y^{tq}) \ : \ q = 1 \dots, r\}$ onto itself. Note that since (x, y) is an N_r-attainable allocation, (x', y') is an N_r-attainable allocation.

A3. There is an r^* such that for all positive integers n, given $r' = nr^*$ for some n and $(x, y) \in A_{r'}(N_{r'})$, there is a permutation by types of (x, y), say (x', y'), and an $(x^*, y^*) \in A_{r^*}(N_{r^*})$ where, for each $(t, q) \in N_{r^*}$, we have $u^{tq}(x^{*tq}, y^{*tq}) \geq u^{t'q'}(x'^{t'q'})$ for all $(t', q') \in N_r$, with $t' = t$ and $q' = q, 2q \dots, nq$.

Informally, A3 is the assumption that agents can do as well for themselves in coalitions with profiles less than or equal to $\rho(N_{r^*})$ as they can in coalitions with profiles less than or equal to $n\rho(N_{r^*})$–increasing returns to coalition size are exhausted by economies with profiles less than or equal to that of N_{r^*}.

Theorem 3.2 *Let $(\mathsf{E}_r)_{r=1}^{\infty}$ be a sequence of replica economies satisfying assumptions A1 to A3. Then the derived sequence of replica games $(N_r, v_r)_{r=1}^{\infty}$ has the MES property.*

It is an immediate consequence of Theorems 2.2, 2.3, and 3.2, that the ε-cores of the derived games $(N_r, v_r)_{r=1}^{\infty}$ converge to the cores of the balanced cover games as r becomes large, i.e., for the derived sequence of games, $A^* = L(\mathcal{C})$. (Since we have made an MES assumption, only equal-treatment payoffs are in the cores of the balanced cover games for sufficiently large replications. Theorem 2.2 ensures the convergence of the ε-cores to equal-treatment payoffs.)

4 The framework of the models

In essence, when the core of a market game associated with an exchange economy is studied the technical and institutional assumptions which are made are that individual ownership of all economic goods is recognized and that all goods can be transferred, without costs or technical difficulty, among all individuals.[22] In particular, implicit in these assumptions is that the economic reality can be well

[22]When we discuss the core here, we refer to the core of the market game as constrasted with the set of undominated distributions of resources in the distrbution space of the economy which

represented by the characteristic function. Shapley and Shubik in Shapley (1974) have suggested the term "c-game" for a model that is adequately represented as a game by its characteristic function and we use that term here. An exchange economy has the c-game property since what a coalition S can achieve via exchange of goods among the members of S is independent of exchanges carried out among the members of the complementary coalition $\overline{S}$.

The introduction of indivisibilities or other nonconvexities does nothing to influence the c-game property. It remains reasonable to assume that a set S can exchange among its members in any way they desire, even if the items are indivisible.

If we assume that production processes manifest constant returns, then no distinction need be made between individual or joint ownership of the technology. The c-game property is preserved. A group S can produce and exchange as it pleases regardless of the actions of $\overline{S}$.

Production and exchange without constant returns poses new problems. If the technology is commonly owned who gets to use it first now matters. This can be avoided by making the reasonable assumption that it is owned by groups, cf. Boehm (1974), Hildenbrand (1974).

In the models used previously and in this paper, the c-game property is preserved between clubs but not necessarily within given clubs. The feasible consumptions, productions, and exchanges of the members of a collection of clubs, acting as a coalition, is independent of the actions of the complementary collection of clubs; this is because there are no "spill-overs" of the public goods between clubs. Within a club, what a subset of agents can consume can be affected by the actions of the other members of the club; the public goods are "pure" public goods within the club. However, from the viewpoint of the study of the core, it is the c-gameness between clubs that is relevant.

As we discussed in the introduction, the modelling of private goods exchange economies as games in characteristic function form is "natural" – the economies have the *c*-game property. For club economies, because production technologies are associated with groups and because exclusion is possible (and also may well be desirable from the point of view of optimality) in consumption of local public goods, we have the set-up of a *c*-game; from the point of view of the study of the core, the appropriate construction of the characteristic function is obvious.

For pure public goods, as we have argued, additional specifications must be made to construct the characteristic function. It needs to be specified who has access to what technology (one could use a coalition production model, for example) and what a coalition can consume of the public goods produced by the complementary coalition.

4.1 The firm and replication. Modelling difficulties that arise in treating production (and also public goods) can be illustrated by examining the alternative ways of replicating an extremely simple example.

We begin by utilizing a simple example to illustrate modelling difficulties finessed by the assumption of group production and how to construct a sequence of replicated economies given group production.

Suppose that an individual i has an initial endowment of $(1, 0, s_i)$ where the first item is an individually owned good which can be used as an input to produce the

---

map into the core of the market game. This distinction may not appear to be of much importance when considering exchange economies, but it is helpful when we consider more general economies.

second good. There is one firm which produces the second good with a production function given by $y = \sqrt{x}$ where y is the output and $x = \sum_{i=1}^{n} x_i$ is the total input. The utility function of an individual i is given by $U_i(x_i, y_i) = y_i$ where $y = \sum_{i=1}^{n} y_i$. The ownership share of the firm by individual i is s_i and $\sum_{i=1}^{n} s_i = 1$. Let p be the price of the output and fix the price of the input as one.

The manager of the firm will attempt to maximize

$$\Pi = p\sqrt{x} - x \quad \text{where} \quad 0 \leq x_i \leq 1$$

and each individual i will attempt to maximize

$$y_i \quad \text{subject to} \quad x_i + s_i\Pi = py_i.$$

Thus $x_i = 1$ and $p = 2\sqrt{n}$; hence $\Pi = n$ and $y = \sqrt{n}$.

Now $0 \leq s_i \leq 1$; thus, depending upon the distribution of shares an individual i could obtain y_i where

$$\frac{1}{2\sqrt{n}} \leq y_i \leq \frac{n+1}{2\sqrt{n}}.$$

Suppose that the original economy was of size $n = 9$ with only one type of agent. If a type has both utility functions and endowments the same for all members we require that $s_i = 1/9$. At this point we have to decide how to describe the characteristic function for this game and furthermore we need to describe how to calculate the characteristic functions for the games arising from the replicated economy.

Alternative 1. Shares are ignored, any subgroup may use the technology. Here the game is immediately inessential, the characteristic function[23] is given by $f(s) = s, s = 0, 1, \ldots, 9$. This is equivalent to creating n independent technologies.

Alternative 2. A simple majority at least is required to operate the firm. The game becomes

$$\begin{aligned} f(s) &= 0 \quad \text{for} \quad s \leq n/2 \\ &= \sqrt{s} \quad \text{for} \quad n/2 < s \leq n. \end{aligned}$$

This game has no core as can be seen from Figure 1.

Alternative 3. Unanimity is required to operate the firm. The game becomes

$$\begin{aligned} f(s) &= 0 \quad \text{for} \quad 0 \leq s \leq n-1 \\ &= \sqrt{n} \quad \text{for} \quad s = n. \end{aligned}$$

Here all imputations are in the core.

Alternative 4. Any simple majority can use the production function, but it can only take the product in proportion to its shares

$$\begin{aligned} f(s) &= 0 \quad \text{for} \quad s \leq n/2 \\ &= \frac{s}{n}\sqrt{s} \quad \text{for} \quad \frac{n}{2} < s \leq n. \end{aligned}$$

It is easy to check that this has a core.

---

[23]For a type symmetric game instead of using the notation $v(S)$ to stand for the amount obained by the sets of agents we can use $f(s)$ where $s = |S|$ as all coalitions of the same size obtain the same amount.

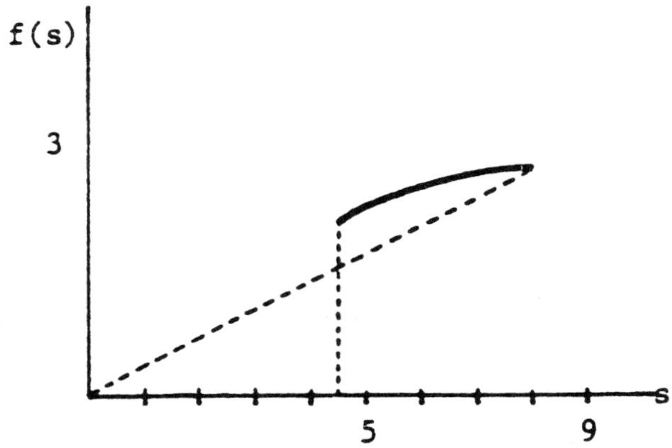

Figure 1

We now turn to replication. What do we mean by a replication of the 9 agents, 1 firm economy? We may consider at least two types of replication. In the first there is still one firm but we replicate the number of agents and shares. Thus the endowments in the rth replication become $(1, 0, 1/9r)$ where although all individuals may have the same number of shares as previously, the percentage ownership of the firm has gone down by l/r. In the second form of replication the firms are increased as well. At this point we may wish to consider the distinction between giving all agents a portfolio in each firm, or enlarging the number of types. For example, for one replication do we consider one type with $(1, 0, 1/2n, 1/2n)$ where the last two entries indicate shares in the two firms or do we consider two types with $(1, 0, 1/n, 0)$ and $(1, 0, 0, 1/n)$?

Replication 1. There is only one firm, regardless of the replication number. The characteristic function becomes

$$\begin{aligned} f(s) &= 0 \quad \text{for} \quad s \leq nr/2 \\ &= \sqrt{s} \quad \text{for} \quad nr/2 < s \leq nr \end{aligned}$$

for the simple majority case. There is no core. Moreover, given any $\varepsilon > 0$ there is an n_0 such that for all n greater than n_0, the game has an empty ε-core. The sequence of games with $n = 1, 2, \ldots$ is not a sequence of replica games since the payoff to a given coalition eventually decreases as the set of players containing that coalition becomes large.

Replication 2a. ("balanced portfolio")

$$\begin{aligned} f(s) &= 0 \quad \text{for} \quad s \leq nr/2 \\ &= r\sqrt{s/r} \quad \text{for} \quad nr/2 < s \leq nr \end{aligned}$$

This sequence of games has cores or ε-cores and is a sequence of replica games.

Replication 2b. (different types)

$$f(s_1, s_2, \dots, s_r) = p\sqrt{\frac{\sum_{j=1}^{r} \delta_j s_j}{p}} \qquad \text{where} \quad \delta_j \quad = 1 \quad \text{if} \quad s_j < n/2$$
$$= 0 \quad \text{if} \quad s_j \leq n/2$$
$$\text{and} \quad p = \sum_{j=1}^{r} \delta_j .$$

This sequence of games has a core or ε-core.

4.2 The firm as a local public good. We could interpret the example in 4.1 in terms of a local public goods and endogenous clubs. Consider an economy of size n with a minimal size club of 9 where each individual has an initial endowment of $(1, 0)$, the first item is a private good which can be converted to a public good according to $y = \sqrt{\sum x_i}$. The utility function for each individual i is $U_i(x_i, y, s) = y/s$ where s is the number of individuals in a club.

It is easy to see that this formulation leads to the formation of as many 9 person clubs as possible. If $n = 0 \mod 9$ there is a core, otherwise an ε-core. Furthermore implicit in this formulation is the idea that all individuals of the same type have an equal share of the club to which they belong. Thus the analogue with the corporate economy can be completed by considering shares in clubs.

4.3 The pure public good. The example noted in 5.2 can be extended to an example illustrating a pure public good by replacing $U_i(x_i, y_i) = y_i$, by $U(x_i, y) = y$ where $y = \sqrt{\sum_i x_i}$.

Suppose that no subgroup smaller than 9 can produce. Then

$$f(s) = 0 \quad \text{for} \quad s < 9$$
$$= s\sqrt{s} \quad \text{for} \quad s \geq 9.$$

This game has a large nonconverging core. Note also that the sequence of derived games does not satisfy the near-minimum efficient scale property.

4.4 The framework revisited. The above discussion may indicate that when we consider sequences of economies, the c-game property within a fixed economy does not necessarily imply *self sufficiency*, that what a coalition can achieve is independent of the economy in which it is embedded. In fact, the model of this paper does not necessarily satisfy self-sufficiency – what a coalition can achieve may increase as the size of the economy in which it is embedded increased. What is crucial for our results is that, for each coalition S, the limit $\lim v_r(N_r)/r$ exists and, for each r and each $S \subseteq N_r$ the limit $\lim_{r'} v_{r'}(S)$ exists. Thus, widespread negative externalities could be allowed. For example, consider a game where all players are identical but for any given coalition S and for r sufficiently large so that $S \subseteq N_r$, $\frac{v_r(S)}{|S|}$ decreases to $|S|$ as r grows large. Such situations could be accommodated at the cost of more complexity of the model.

5 Recent literature

While the convergence results in this paper for economies with clubs are still novel, since this paper was written a number of related results have appeared in the literature and the reader may wish some some updated discussion and references.

The nonemptiness of ε-cores of large games with side payments and has been partially[24] extended to hold uniformly for all games derived from pregames with a compact metric space of player types (Wooders and Zame 1984). A *pregame* is a pair consisting of a compact metric space of player types and a worth function, ascribing a payoff to any coalition of players, described by the number of players of each type in the coalition. The worth of a coalition depends continuously on the types of the coalition members. The framework of Wooders and Zame (1984), as that of Wooders (1979), however, imposes *self-sufficiency* – what a coalition can achieve must be independent of the total player set in which it is embedded. This is essential for the uniformity[25] of the result of Wooders and Zame (1984) and Wooders (1992a).

The current paper and the prior game-theoretic results used suggest that there is a relationship between minimum efficient scale – that *all* gains to collective activities can be realized by groups of players bounded in size – and per capita boundedness.[26] In fact, Wooders (1992a,1994a) formulates an "approximate" version of MES called small group effectiveness – that all or *almost all* gains to collective activites can be realized by groups of players bounded in size – and demonstrates an equivalence between small group effectiveness and per capita boundedness. Wooders (1994a,b) shows that under the condition of small group effectiveness, the per capita payoff function $\frac{v(N)}{|N|}$ converges uniformly to the concave "utility function" of a market game, and for small ε and large player sets N, the core of the market game approximates the ε-core of the game.

There have been further advances in the club-theoretic literature, especially in the area of equilibrium and the equivalence of the core and the set of competitive outcomes. We refer the reader to Scotchmer and Wooders (1986) for an illustrative and expository discussion of economies with clubs and to Conley and Wooders (1998a) and references therein for a more up-to-date treatment. As the work in the current paper suggests, when small groups of participants can realize all or almost all gains to collective activities, then large economies are competitive.

The convergence of approximate cores reported in this paper is extended in Wooders (1992b) to demonstrate that in economies with side-payments (quasi-linear utilities, in economics terminology) core payoffs converge to competitive prices. Another sort of core convergence result was previously demonstrated in Engl and Scotchmer (1991). Roughly, the authors consider a situation where players are characterized by their attributes, represented by points in a finite-dimensional Euclidean space. Attributes may be, for example, endowments of commodities. They show that under the assumption of per capita boundedness as in this paper and other prior research, and with an additional differentiability assumption in per capita terms approximate core payoffs to large groups of players are approximated by a linear function on attribute space. To illustrate the differences in approaches between Engl and Scotchmer (1991) and the convergence results of Wooders and the current paper, consider an n-person game where all people are identical and as

[24]Wooders and Zame (1984) require a stonger assumption of boundedness of marginal contributions to coalitions.

[25]That is, the result is for all sufficently large games derived from a pregame rather than for sequences of games with a fixed distribution of player types.

[26]This relationship is very clear in the proof technique of Wooders (1980b). Essentially, sequences of games satisfying per capita boundedness are approximated by sequences satisfying minimum efficient scale.

n becomes large, $\frac{v(n)}{n}$ converges to $\overline{x}$.[27] Engl and Scotchmer's result implies that for all sufficiently large coalitions S if x is in the ε-core, then

$$\text{Engl-Scotchmer: } \left|\frac{x(S)}{|S|} - \overline{x}\right| \text{ is "small".}$$

In contrast, the results of the current paper and Wooders' results all show that for most players i,

$$\text{Wooders' } |x_i - \overline{x}| \text{ is "small".}$$

Examples illustrating further distinctions between the convergence results of Engl and Scotchmer and those reported in this paper are provided in Wooders (1992b). There is a large literature on convergence of cores to equilibrium outcomes in economies with private goods only which is outside the scope of the current paper; we refer the reader to Anderson (1992) for a recent survey.

The result of Engl and Schotchmer (1991) may suggest that it is possible to hedonically price attributes of players, that is, to infer from core payoffs values of attributes. However, hedonic pricing requires *hedonic independence* – that the worth of a bundle of attributes equals the sum of the worths of the attributes in the bundle. Conley and Wooders (1995) provide examples showing that in Tiebout economies (or club eonomies) in general hedonic independence is not satified.

An aspect of economies satisfying minimum efficient scale that is not addressed in the current paper is "monotonicity," that is, when the abundance of players of a given type is increased then the core payoff to players of that type does not increase and may decrease. This is shown in Scotchmer and Wooders (1988). A similar result, allowing the percentage of a player type to increase, is presented in Engl and Scotchmer (1991). Wooders (1992b,1994a) also provides related results, allowing the abundances of all player types to change simultaneously, and demonstrates that a stronger condition of "cyclic monotonicity" is satisfied.

A difficulty with the pregame framework, used in the above papers, is that self-sufficiency rules out economy-wide externalities. Moreover, the pregame structure has hidden consequences. For example, the equivalence of small group effectiveness and per capita boundedness does not necessarily hold outside the pregame framework. To overcome these restrictions, recently, Kovalenkov and Wooders (1997a) introduced the framework of parameterized collections of games with and without side payments. This framework treats classes of games that can be described by certain parameters: (a) the number of approximate types of players and a bound on the distance between players of the same approximate type, and (b) the size of near-effective groups of players and a bound on how close such groups must be to being completely effective. In this framework, the payoff to a group of players need not depend on the size of the economy in which the group is embedded. Thus, widespread positive or negative externalities are allowed. Moreover, uniform results (that is, results for all large games, rather than for sequences) are obtained. Kovalenkov and Wooders (1997b) applies their game-theoretic results to a model with clubs. Unlike the current paper, Kovalenkov and Wooders (1997b) allows agents to be affected by the characteristics of the memberships of the clubs to which they belong.

[27] In spirit, the Engl-Scotchmer result shows convergence in mean while our results show convergence in probability.

Another issue discussed in the current paper is convergence of the core to competitive outcomes in economies with public goods. Conley (1994) demonstrates a sufficient condition.

6 Appendix

In this appendix, Theorems 3.1 and 3.2 are proven.

Proof of Theorem 3.1

To prove the theorem, we need only prove that the sequence of games derived from the economies is superadditive and per-capita bounded. It is straightforward to verify that the sequence is superadditive; thus we omit the proof.

To prove per-capita boundedness of the sequence of games, we construct another sequence of economies, say the *-economies, so that the sequence of games derived from the *-economies, denoted by $(N_r, v_r^*)_{r=1}^{\infty}$, is per-capita bounded and has the property that for all r and for all non-empty subsets S of N_r, we have $v_r(S) \leq v_r^*(S)$.

For the *-economies, we let Y^* be the private goods production possibility set available to all coalitions $S \subseteq N_r$ for all r. Observe that for any firm structures, say $F_r(S)$ and $F_r^{'}(S)$ we have

$$\sum_{S' \in F_r(S)} Y^* = \sum_{S' \in F_r'(S)} Y^*$$

so the firm structure will be irrelevant. Of course, $Y_r(S) \subseteq Y^*$ and for any firm structure $F(S)$, $\sum_{S' \in F(S)} Y^* \subseteq Y^*$.

Given r and a subset S of N_r, define $Z^*(S) = \{(x, z) \in \mathbf{R}^{m+\ell} : (\mid S \mid x, z) \in Z^*\}$. Note that $Z^*(S)$ is a closed convex cone with vertex $\{0\}$ and $\{0\} \times -\mathbf{R}_+^{\ell} \subseteq Z^*(S)$; this follows from the assumptions on Z^*. Let $\varphi(S) = (J(S), \{(x'_S, z'_S) \in Z_r(S') : S' \in J(S)\})$ be an S-public goods production plan. From assumption A.2, there is an $(x, z) \in Z^*(S)$ such that

(a) for each $(t, q) \in S$, we have $\displaystyle\sum_{\{S' \in J(S) : (t,q) \in S'\}} x_{S'} \leq x$,

and

(b) $\displaystyle\sum_{S' \in J(S)} z'_S \leq z.$

Informally, there is an (x, z) in $Z^*(S)$ "at least as good as" any S-public goods production plan in the sense that with the production possibility set $Z^*(S)$, the agents can consume as much of the local public goods while using no more of the inputs. (Note, however, because agents do not necessarily have monotonic increasing preferences for the local public goods, they might not prefer to have more of them.)

For the purposes of this theorem, we can restrict our attention to states of the *-economies with associated club structures $\{\{(t, q)\} : (t, q) \in N_r\}$ since with this club structure all agents can be made "at least as well-off" as with any other club structure. To see this, given any r and any non-empty subset S of N_r, let $(x, z) \in Z^*(S)$ so $(\mid S \mid x, z) \in (Z)^*$ and $(x, z/ \mid S \mid) \in Z^*(\{(t, q)\})$ for each $(t, q) \in S$; thus with the club structure $\{\{(t, q)\} : (t, q) \in N_r\}$ each agent in S can consume x and total inputs are unchanged.

In the *-economies, the agents will all have the same utility functions. For each r and for all $(t, q) \in N_r$, define $u^{*tq}(x, y) = L(x, y)$ where L is a linear function such that $u^{tq}(x,y) \leq L(x,y) + c$ for some constant c for all $(x,y) \in \mathbf{R}^m_+ + \mathbf{R}^\ell_+$ and for all $(t,q) \in N_1$.

For each r and each non-empty subset S of N_r, let $A^*_r(S)$ denote the set of S-attainable allocations for the *-economies.

For each r and all non-empty subsets S of N_r, define

$$v^*_r(S) = \sup_{(x,y)\in A^*_r(S)} \sum_{(t,q)\in S} L^*(x^{tq}, y^{tq}).$$

The finiteness of this "sup" is ensured by the linearity of the function L^* and boundedness of $A^*_r(S)$. Also,

$$v^*_r(S) = \sup_{(x,y)\in A^*_r(S)} L\left(\sum_{tq\in S} x^{tq}, \sum_{tq\in S} y^{tq}\right).$$

Let K be a real number such that $K > v^*_1(N_1)$. We will show that K is a per-capita bound for the sequence of games (N_r, v^*_r). Suppose not. Then there is an r' and a feasible state of the r'^{th} $*$-economy, say $e^*(N_{r'}) = (\overline{y}, J(N_{r'}), \{x_{S'}, z_{S'}) \in Z^*(S') : S \in J(N_{r'})\})), (x,y))$, such that $L(\sum_{tq\in N_{r'}} x^{tq}, \sum_{tq\in N_{r'}} y^{tq}) > Kr'$. We can assume without any loss that $F(N_{r'}) = \{N_{r'}\}$ and $J(N_{r'}) = \{\{(t,q) : (t,q) \in N_{r'}\}$. Also we can assume that $x^{tq} = x^{t'q'}$ and $y^{tq} = y^{t'q'}$ for all (t,q) and (t',q') in $N_{r'}$. We claim that there is an $(x',y') \in A^*_1(N_1)$. with $x'^{tq} = x^{tq}$ and $y'^{tq} = y^{tq}$ for all $(t,q) \in N_1$, which will yield a contradiction. Since $e^*(N_{r'})$ is feasible (for the r'^{th} economy), we have $\overline{y} \in Y^*$. Observe that $\overline{y}/r' \in Y^*$. Also, for some z^{tq} for each $(t,q) \in N_{r'}$, we have $(x^{tq}, y^{tq}) \in Z^*(\{(t,q)\})$, and

$$\sum_{tq\in N_{r'}} (y^{tq} - w^{tq}) + \sum_{tq\in N_{r'}} z^{tq} = \overline{y} \tag{6.1}$$

so

$$\sum_{tq\in N_1} (y^{tq} - w^{tq}) + \sum_{tq\in N_1} z^{tq} = \overline{y}/r. \tag{6.2}$$

This proves our assertion that $(x',y') \in A^*_1(N_1)$. But then

$$\begin{aligned} &L(\textstyle\sum_{tq\in N_1} x'^{tq}, \sum_{tq\in N_1} y'^{tq}) = \\ &\tfrac{1}{r'} L(\textstyle\sum_{tq\in N_{r'}} x^{tq}, \sum_{tq\in N_{r'}} y^{tq}) > K \end{aligned} \tag{6.3}$$

which is a contradiction. Therefore the sequence $(N_r, v_r)_{r=1}^\infty$ is per capita bounded.

Since $u^{tq}(x,y) \leq L(x,Y)+c$ for all $(x,y) \in \mathbf{R}^m_+ \times \mathbf{R}^\ell_+$, and from the construction of the production possibilities set for the *-economies, we have

$$v_r(N_r) \leq v^*_r(N_r) + c|N_r|$$

for all r. Therefore the sequence $(N_r, v_r)_{r=1}^\infty$ is per-capita bounded. □

Proof of Theorem 3.2

Given $\delta > 0$, r^* satisfying the conditions of A3, and $r' = nr^*$ for some positive integer n, there is an $(x,y) \in A_{r'}(N_{r'})$ such that

$$v_{r'}(N_{r'}) \leq \sum_{tq\in N_{r'}} u^{tq}(x^{tq}, y^{tq}) + \delta.$$

From A3, there is an $(x^*, y^*) \in A_{r^*}(N_{r^*})$ such that

$$\sum_{tq \in N_{r'}} u^{tq}(x^{tq}, y^{tq}) \leq n \sum_{tq \in N_{r^*}} u^{tq}(x^{tq}, y^{tq}).$$

Therefore, it follows that

$$\frac{v_{r'}(N_{r'})}{r'} \leq \frac{v_{r^*}(N_{r^*})}{r^*}$$

and, from superadditivity, that

$$\frac{v_{r'}(N_{r'})}{r'} \geq \frac{v_{r^*}(N_{r^*})}{r^*}$$

so

$$\frac{v_{r'}(N_{r'})}{r'} = \frac{v_{r^*}(N_{r^*})}{r^*}$$

Since this holds for all positive integers n, we have

$$\lim_{r \to \infty} \frac{v_r(N_r)}{r} = \frac{v_{r^*}(N_{r^*})}{r^*}.$$

Since

$$\lim_{r \to \infty} \frac{\widetilde{v}_r\ (N_r)}{r} = \lim_{r \to \infty} \frac{v_r(N_r)}{r}$$

we have

$$\lim_{r \to \infty} \frac{\widetilde{v}_r\ (N_r)}{r} = \frac{v_{r^*}(N_{r^*})}{r^*}.$$

Since $\frac{v_r(N_r)}{r} \leq \frac{\widetilde{v}_r(N_r)}{r}$ for all r and since $\frac{\widetilde{v}_r(N_r)}{r}$ is non-decreasing, we have

$$\frac{\widetilde{v}_r\ (N_r)}{r} = \frac{\widetilde{v}_{r^*}\ (N_{r^*})}{r} \quad \text{for} \quad r \geq r^*.$$

□

References

Anderson, R. J. (1992) *The core in perfectly competitive economies*, Handbook of Game Theory, Volume 1 eds. R.J. Aumann and S. Hart, North Holland Amsterdam/London/New York/Tokyo, 413-457.

Bennett, E. and Wooders, M. H. (1979) *Income distribution and firm formation*, Journal of Comparative Economics **3**, 304-317.

Billera, L. J and Bixby, R. E. (1974), *Market representation of n-person games*, Bulletin of the American Mathematical Society **80**, 522-526.

Boehm, V. (1974), *A limit theorem on the core of an economy with production*, International Economic Review **15**, 143-148.

Conley J. (1994) *Convergence theorems on the core of a public goods economy: Sufficient conditions*, Journal of Economic Theory **62**, 161-85

Conley, J. and Wooders, M. H. (1995), *Hedonic independence and the taste homogeneity of optimal jurisdictions in a Tiebout economy with crowding types*, University of Illinois, Champaign, Office of Research Working Paper 95-118, 1995.

Conley, J. and Wooders, M. H. (1998a), *Anonymous pricing in Tiebout economies and economies with clubs*, Topics in Public Finance, D. Pines, E. Sadka and I. Zilcha, eds. Cambridge University Press, Cambridge, 89-120.

Conley, J. and Wooders, M. H. (1998b) *Tiebout economies with differential genetic types and endogenously chosen crowding chracteristics*, Journal of Economic Theory (to appear).

Engl, G. and Scotchmer, S. (1991) *The core and hedonic core: Equivalence and comparative statics*, mimeo.

Hildenbrand, W. (1974), *Core and Equilibria of a Large Economy*, Princeton University Press, Princeton, NJ.

Ischiishi, T. (1977), *Coalition structure in a labour-managed market economy*, Econometrica **45**, 341-360.

Kaneko, M. and Wooders, M. H. (1986) *The core of a game with a continuum of players and finite coalitions: The model and some results* Mathematical Social Sciences **12**, 105-137.

Kovalenkov, A. and Wooders, M. H. (1997a) *Three theorems on non-emptiness of approximate cores; Part 1. Game-theoretic results* Autonoma University of Barcelona Working Paper No. 390.97.

Kovalenkov, A. and Wooders, M. H. (1997b) *Three theorems on non-emptiness of approximate cores; Part 2. Economies with clubs* Autonoma University of Barcelona Working Paper No. 391.97.

Rosenthal, R. W. (1971), *External economies and cores*, Journal of Economic Theory **3**, 182-188.

Rosenthal, R. W. (1976), *Market games with production and public commodities*, Discussion paper No. 156 (revised), The Center for Mathematical Studies in Economics and Management Science.

Scotchmer, S and Wooders, M. H. (1986) *Optimal and equilibrium groups* Harvard Discussion Paper 1251, on-line at http://www.chass.utoronto.ca/~wooders2.

Scotchmer, S. and Wooders, M. H. (1988) *Monotonicity in games that exhaust gains to scale*, IMSSS Technical Report No. 525, Stanford University, to appear on-line at http://www.chass.utoronto.ca/~wooders2.

Shapley, L. S. (1967), *On balanced sets and cores*, Naval Research Logistics Quarterly **14**, 453-460.

Shapley, L. S. (1974), *Characteristic function, core, and stable sets*, Rand Memorandum R904/6.

Shapley, L. S. and Shubik, M. (1966), *Quasi-cores in a monetary economy with nonconvex preferences*, Econometrica **34**, 805-827.

Shubik, M. (1959) *Edgeworth market games*, Contributions to the Theory of Games (A. W. Tucker and R. D. Luce, eds.), Vol. IV, Princeton University Press, Princeton, NJ, 267-278.

Shubik, M. (1982), *Game Theory in the Social Sciences, Concepts and Solutions*, The MIT Press, Cambridge, MA and London, England.

Shubik, M. and Wooders, M. H. (1982a) *Approximate cores of a general class of economies: Part 1, Replica games, externalities, and approximate cores*, Cowles Foundation Discussion Paper No. 619, *Mathematical Social Sciences* **6** (1983), 27-48.

Shubik, M. and Wooders, M. H. (1982b), *Approximate cores of a general class of economies: Part II. Set-up costs and firm formation in coalition production economies*, Cowles Foundation Discussion Paper No. 620. *Mathematical Social Sciences* **6** (1983), 285-306.

Shubik, M. and Wooders, M. H. (1982c) *Near-Markets and Market-Games*, Cowles Foundation Discussion Paper No. 657.

hubik, M. and Wooders, M. H. (1986). *Near-markets and market-games*, Economic Studies Quarterly **37**, 289-299.

Wooders, M. H. (1979) *A Characterization of approximate equilibria and cores in a class of coalition economies*, (A revision of Stony Brook Department of Economics Working Paper No. 184), on line at http://www.chass.utoronto.ca/~wooders2.

Wooders, M. H. (1980a), *The Tiebout hypothesis; near optimality in local public good economies*, Econometrica **48**, 1467-1485.

Wooders, M. H. (1980b) *Asymptotic cores and asymptotic balancedness of large replica games* (Stony Brook Working Paper No. 215, Revised and extended) on line at http://www.chass.utoronto.ca/~wooders2.

Wooders, M. H. (1981a), *The epsilon core of a large game*, Cowles Foundation Discussion Paper No. 612 (Revised), Journal of Mathematical Economics **11** (1983) 277-300.

Wooders, M. H. (1981b), *A limit theorem on the ε-core of an economy with local public goods*, National Tax Institute of Japan Paper No. 20, published in part in Mathematical Social Sciences **18** (1989) 33-55.

Wooders, M. H. (1983), *Stability of club structures in economies with local public goods*, University of Toronto Discussion Paper. Published in *Mathematical Social Sciences 15* (1988), 29-49.

Wooders, M. H. (1992a) *Inessentiality of large groups and the approximate core property; An equivalence theorem'*, *Economic Theory* **2**, 129-147.

Wooders, M. H. (1992b) *The attribute core, core convergence, and small group effectiveness; the effects of property rights assignments on attribute games*, to appear in *Essays in Honor of Martin Shubik*, Dubey, P. and Geanakoplos, J. eds.

Wooders, M. H. (1994a) *Large games and economies with effective small groups*, in Game-Theoretic Methods in General Equilibrum Analysis, eds. J-F. Mertens and S. Sorin, Kluwer Academic Publishers Dordrecht/Boston/London 145-206.

Wooders, M. H. (1994b) *Equivalence of games and markets* Econometrica **62**, 1141-1160.

Wooders, M. H. and Zame, W. R. (1984) *Approximate cores of large games*, Econometrica **52**, 1327-1350.

Fields Institute Communications
Volume **23**, 1999

On Economic Applications of the Kuhn-Fourier Theorem

Arja H. Turunen-Red
Department of Economics and Finance
University of New Orleans
New Orleans, LA 70148 USA
ahte?@mindspring.com

Alan D. Woodland
Department of Econometrics
University of Sydney
N.S.W. 2006 Australia
a.woodland@econ.usyd.edu.au

Abstract. We discuss economic applications of a result on linear systems due to [**30**]. Kuhn's theorem, the proof of which can be based on an algorithm by [**22**], yields a convenient means of obtaining the dual of any finite linear system. We show that the Kuhn-Fourier theorem can be regarded as a unifying principle that underlies many seemingly disparate formulations of economic rationality: nonexistence of arbitrage in financial markets, aggregation of preferences in social choice, linear and concave rationalizability of preferences for consumer choice and the dual characterizations of the core for transferable utility games.

1 Introduction

The Kuhn-Fourier theorem [**30**] and ([**42**], Chapter 1) gives abstract conditions which, if satisfied, guarantee that an arbitrary finite-dimensional system of linear equations and inequalities can be solved. The solvability conditions are expressed in terms of dual relations which are derived from the primal equality and inequality system by forming linear dependencies. The Kuhn-Fourier theorem can be regarded as a "meta - theorem of the alternative" which states: *either the primal linear system is solvable or there is a solution to the dual linear relations but never both.*[1] All established theorems of the alternative, such as those by [**18**] and [**33, 34**], are special cases of the Kuhn-Fourier duality statement.[2]

The purpose of this paper is twofold. We first describe the basic mathematical facts pertaining to the Kuhn-Fourier theorem and apply the theorem to establish

1991 *Mathematics Subject Classification.* Primary 90A14; Secondary 90A08, 90A09, 90D12.

[1]The first economic applications of such dual relations appeared in the context of linear programming and zero sum games (e.g., [**23**]). Motzkin's theorem of the alternative has become an invaluable tool in the policy reform literature ([**12, 13**], [**50**], [**44, 45, 46**]).

[2]Many more theorems of the alternative have been stated by [**31**] and [**23**].

new linear duality statements. The extremely simple analytics of these derivations demonstrate the power of the Kuhn-Fourier theorem as a tool: it can be used to automatically generate the required duality statement whenever a theorem of the alternative is needed.[3] We then consider a number of applications of theorems of the alternative in various, ostensibly disparate, areas of economic theory. We show that using the Kuhn-Fourier theorem, several fundamental concepts of finance, social choice, game theory and consumer theory can be given dual interpretations at a stroke. Even if individual instances of such dualities have been observed before, the connections of the various concepts have remained much less understood. The Kuhn-Fourier duality principle serves to unify the economic applications that we discuss and establishing this unity is our main contribution.

We begin our discussion of the economic applications of the Kuhn-Fourier theorem in Section 3 by stating two results pertaining to asset portfolios and their returns. The first proposition – a version of the Fundamental Theorem of Finance [**39, 40**] – provides a characterization of asset markets in which arbitrage portfolios (i.e., sure gains and losses) do not exist. Our second, apparently new, theorem augments the Fundamental Theorem by characterizing all portfolio returns that are attainable in arbitrage-free asset markets. Our interest in the two results springs from their very special interpretations: in each theorem, a linear aggregation equation is established as the dual of a notion of economic rationality, i.e., infeasibility of unexplored arbitrage opportunities. In the remainder of our paper, we apply the duality theorems of finance to analyze analogous dual relationships of aggregation and arbitrage in other, quite different, areas of economic theory.

In Section 3, we apply the Kuhn-Fourier theorem and derive a version of Harsanyi's social aggregation theorem for expected utility preferences [**25, 26**]. We show that Harsanyi's proposition, which establishes an affine aggregation equation for individual and social preferences, is a linear duality statement exactly analogous to the Fundamental Theorem of Finance. Accordingly, we reinterpret the Pareto conditions that appear in the statement of Harsanyi's theorem as characterizations of economic rationality: the Pareto conditions are satisfied if and only if preferences of individuals and society do not expose these individuals and society to arbitrage losses in exchanges with arbitrary outside opponents. Turning to individual choice in Section 4, we first discuss a theorem of Buehler [**8**] on linear rationalizability of choice under uncertainty and then state a version of Afriat's theorem [**1, 2**] on concave rationalizability of consumer choice when the consequences of actions are not uncertain. In both instances, definitions of economic rationalizability can be given interpretations in terms of aggregation and infeasibility of arbitrage.

Finally, we consider the Bondareva [**6**] - Shapley [**41**] theorem on cooperative transferable utility (TU) games. As is well-known, the Bondareva - Shapley theorem characterizes the nonemptiness of the core of a finite TU-game in terms of balancedness of the game. We establish the Bondareva - Shapley theorem as a corollary of the Kuhn-Fourier theorem and apply the duality theorems of finance to reinterpret balancedness of a coalitional game. We establish that the core of an arbitrary TU-game is nonempty if and only if no "arbitrage allocations" are necessary for the coalitional payoffs of the game to become attainable.

Section 2 introduces the Kuhn-Fourier theorem and serves as the basis of the ensuing discussion.

---

[3]Thus, one does not have to search the literature for the appropriate result or to expend energy in proving such a result (e.g., via geometric arguments) if it does not already exist.

2 Linear systems and the Kuhn-Fourier theorem

In this section, we review the basic theory relating to the Kuhn-Fourier theorem. We establish two new nonhomogeneous theorems of the alternative which, along with a theorem of the alternative by [**23**], are applied in the subsequent sections.

2.1 The Kuhn-Fourier theorem. An arbitrary (finite) system of linear equations and inequalities is of the form[4]

$$Ax >> A_0, Bx \geq B_0, Cx = C_0, \tag{2.1}$$

where A, B and C are matrices and x is a vector of appropriate dimension. The Kuhn-Fourier theorem relates the solvability of system (2.1) to linear combinations of the equations and inequalities in (2.1), characterized by multipliers for which[5]

$$\alpha^T A + \beta^T B + \gamma^T C = 0^T, \ \alpha \geq 0, \beta \geq 0, (\alpha^T, \beta^T, \gamma^T) \neq 0^T. \tag{2.2}$$

The equations in (2.2) are formed by taking linear combinations of the left-hand sides of the relations in (2.1). If (2.2) has a solution, a linear dependence exists among the left-hand sides of the linear relations in (2.1).[6]

Each choice of multipliers (α, β, γ) that is a solution to (2.2) implies a statement

$$0\rho(\alpha^T A_0 + \beta^T B_0 + \gamma^T C_0), \tag{2.3}$$

where ρ is an equality/inequality symbol determined according to the rule

$$\rho = \begin{cases} > & \text{if } \alpha > 0, \\ \geq & \text{if } \alpha = 0 \text{ but } \beta > 0, \\ = & \text{if } \alpha = 0 \text{ and } \beta = 0. \end{cases} \tag{2.4}$$

Statements such as (2.3) can only be true or false. The Kuhn-Fourier theorem establishes that the consistency of an arbitrary system such as (2.1) can be completely characterized by statements (2.3) in which the multipliers (α, β, γ) satisfy (2.2). Theorem 2.1 is the Kuhn-Fourier theorem.

Theorem 2.1 ([**30**], [**42**]) *The system of linear relations defined in (2.1) is solvable if and only if each set of multipliers that solves (2.2) implies a statement (2.3) which is true. If there are no solutions to (2.2), then the system (2.1) is solvable.*

According to Theorem 2.1, the solvability of a system of linear equations and inequalities depends on whether there are any linear dependencies within the system and, if so, on the form that they take. If no such dependencies exist, then there is a solution to the system.[7] If linear dependencies do exist among the left-hand sides of (2.1), for the system to be solvable the dependencies must lead to logically truthful statements defined by (2.3) and (2.4).

An interpretation of the Kuhn-Fourier theorem more familiar to economists is obtained if the result is regarded as a "meta-theorem of the alternative". This form of the theorem states:

---

[4]Notation: $x >> 0$ means that every component of the vector is positive; $x \geq 0$ means that every component of the vector is nonnegative; $x > 0$ means that every component of the vector is nonegative and at least one component is positive.

[5]Notation: the superscript "T" refers to the transpose of a vector.

[6]The sign restrictions on α and β are imposed in order to preserve the direction of the inequalities in (2.1).

[7]This case arises, for example, if the A and B relations are vacuous and the matrix C is of full rank. Then, the unique solution for the system (2.1) equals $x = C^{-1}C_0$.

Corollary 2.2 *Either the primal system (2.1) has a solution or there is a solution to the dual system (2.2) such that the corresponding statement (2.3) is false, but not both.*

Corollary 2.2 identifies the linear dual of any finite-dimensional linear system.

2.2 Individual Theorems of the Alternative. We apply Theorem 2.1 and its corollary and derive two nonhomogenous theorems of the alternative. The elementary (nongeometric) proofs of the lemmas exhibit the convenience of the Kuhn-Fourier theorem as an analytic tool.[8]

Lemma 2.3 *(Nonhomogenous Gordan) Either* $\{Ax << 0, c^T x \geq 0$ *or* $Ax \leq 0, c^T x > 0\}$ *or* $A^T y = c, y > 0$.

Proof Consider the system $-Ax >> 0$, $c^T x \geq 0$. By Theorem 2.1, this system is not solvable if and only if there exist $(\alpha^T, \beta) > 0^T$ such that $-\alpha^T A + \beta c^T = 0^T$ and the implied statement $0\rho 0$ is false. Hence, by (2.4), for the system $-Ax >> 0, c^T x \geq 0$ not to have a solution, there must exist $\alpha > 0$ and $\beta \geq 0$ such that $-\alpha^T A + \beta c^T = 0^T$. Analogously, the system $-Ax \geq 0, c^T x > 0$ has no solution if and only if there exist $\alpha \geq 0$ and $\beta > 0$ such that $-\alpha^T A + \beta c^T = 0^T$. Therefore, neither of the primal inequality systems has a solution if and only if $y \equiv \alpha/\beta > 0$ does not solve $A^T y = c$. □

Lemma 2.4 *(Nonhomogenous Motzkin) Either* $Ax \geq b, c^T x = a$ *(with* $c \neq 0$*) or* $A^T y_1 + c y_2 = 0, b^T y_1 + a y_2 > 0, y_1 > 0$.

Proof By Theorem 2.1, the system $Ax \geq b, c^T x = a$ is solvable if and only if there are no multipliers $\beta \geq 0$ and γ such that the statement $0\rho(\beta^T b + \gamma a)$ is false while $(\beta^T, \gamma) \neq 0^T$ and $\beta^T A + \gamma c^T = 0^T$. Because $c \neq 0$, we must have $\beta > 0$. Thus, by (2.4), the primal system has a solution if and only if there are no $\beta(\equiv y_1) > 0$ and $\gamma(\equiv y_2)$ such that $\beta^T A + \gamma c^T = 0^T, \beta^T b + \gamma a > 0$. □

Finally, we state without proof a theorem of the alternative by [**23**].

Lemma 2.5 ([**23**], Theorem 2.10) *Either* $Ax << 0, x \geq 0$ *or* $A^T y \geq 0, y > 0$.

Economic applications of the Kuhn-Fourier theorem and Lemmas 2.3-2.5 are discussed below.

3 Linear aggregation in finance and social choice

Ross' Fundamental Theorem of Asset Pricing [**39, 40**] has inspired a large literature in finance (e.g., [**17**]); Harsanyi's social aggregation theorem [**25, 26**] has similarly inspired a sizable body of work in social choice (e.g., [**51**]). In this section, we show that the two acclaimed propositions are formally identical and state a similar linear aggregation equation. Whether concerned with the aggregation of asset returns (Ross) or the aggregation of preferences (Harsanyi), in each instance, the characterization of linear aggregation is intimately related to an expression of economic rationality, i.e., infeasibility of arbitrage outcomes. Applying the Kuhn-Fourier approach, we derive a dual characterization of attainable returns in financial markets and obtain a remarkably simple proof for a version of Harsanyi's theorem. The set of feasible portfolio returns is later applied to establish an intuitive interpretation for balancedness of finite TU-games.

[8] Lemma 2.3 and 2.4 are versions of [**24**] and [**33, 34**] theorems of the alternative.
Gordan's Lemma: Either $Ax >> 0$ or $A^T y = 0, y > 0$.
Motzkin's Lemma: Either $Ax >> 0, Bx \geq 0, Cx = 0$ or $A^T y_1 + B^T y_2 + C^T y_3 = 0, y_1 > 0, y_2 \geq 0$.

3.1 Attainable Returns in Asset Markets. We assume that there exists a finite number of possible states of nature, $s = 1, \ldots, S$, a finite number of marketed assets, $a = 1, \ldots, A$, and one future time period. We denote the $S \times A$ matrix of asset returns by R and let x be a portfolio of investments in these assets.[9] We denote the prices of the marketed assets by p, and obtain $p^T x$ as the cost of a portfolio with the state-specific returns equal to Rx.

3.1.1 *Arbitrage portfolios.* An arbitrage portfolio, if it can be found, is a particularly attractive object.

Definition 3.1 *An arbitrage portfolio, x, is a solution to the linear system*

$$Rx \geq 0, p^T x < 0 \quad or \quad Rx >> 0, p^T x \leq 0. \tag{3.1}$$

An arbitrage portfolio thus yields at least a nonnegative return in every state of nature (and possibly a strictly positive return in all states) while imposing a negative (or, at least, nonpositive) cost on the lucky investor. All arbitrage opportunities can be characterized by exchanges of portfolios that contain no short sales.

Lemma 3.2 *A portfolio x solves (3.1) if and only if there exist portfolios $x_1 \geq 0$ and $x_2 \geq 0$ such that*

$$Rx_1 \geq Rx_2, p^T x_1 < p^T x_2 \quad or \quad Rx_1 >> Rx_2, p^T x_1 \leq p^T x_2. \tag{3.2}$$

Proof Let the portfolio x satisfy (3.1). Then, as the portfolio x is unrestricted, there exist $x_1 \geq 0$ and $x_2 \geq 0$ such that $x \equiv x_1 - x_2$ and (3.2) holds. Conversely, if $x_1 \geq 0$ and $x_2 \geq 0$ solve (3.2), then $x \equiv x_1 - x_2$ satisfies (3.1). □

By Lemma 3.2 an arbitrage opportunity corresponds to an exchange of a portfolio x_2 for another portfolio x_1 which yields uniformly higher returns at a lower cost.

Are there arbitrage portfolios? The first characterization of arbitrage-free asset markets is the following.

Lemma 3.3 *There are no arbitrage portfolios, i.e., the linear system (3.1) has no solution, if and only if for all portfolios $x_1 \geq 0$ and $x_2 \geq 0$*

$$Rx_1 \geq Rx_2 \rightarrow p^T x_1 \geq p^T x_2 \tag{3.3}$$

and

$$Rx_1 >> Rx_2 \rightarrow p^T x_1 > p^T x_2. \tag{3.4}$$

Proof Conditions (3.3) and (3.4) are the negation of (3.2). □

Accordingly, infeasibility of arbitrage opportunities is equivalent to a natural condition on portfolios that contain no short sale elements: every such portfolio that dominates another in its returns across all contingencies must be more costly when evaluated at the prevailing market prices. Formulae (3.3) and (3.4) state the no arbitrage conditions that correspond to the definition of an arbitrage portfolio in (3.1).

The Fundamental Theorem of Asset Pricing characterizes arbitrage-free asset markets using the dual of Definition 3.1. We state the Fundamental Theorem as Theorem 3.4.

[9] The vector x may contain both nonnegative and nonpositive (short sale) components.

Theorem 3.4 *There are no arbitrage portfolios, i.e., the linear system (3.1) has no solution, if and only if there exists a vector* π *such that*

$$R^T\pi = p, \quad \pi > 0. \tag{3.5}$$

Proof Apply Lemma 2.3 to the linear system (3.1) to obtain (3.5). □

According to Theorem 3.4, nonexistence of arbitrage portfolios is dual to a linear aggregation equation (3.5) that connects the returns of financial assets to their prices. The aggregation weights, π, are commonly interpreted as the values of a monetary unit across all states of nature ("state prices"); by this interpretation, Theorem 3.4 shows that the market prices of assets (whether marketed or not marketed) must equal their prospective returns evaluated at the state prices, given that the asset markets are arbitrage-free.[10]

Alternative versions of the Fundamental Theorem have been stated by [**39, 40**], [**17**], [**48**], [**29**], [**15**] and [**16**] among others. Varian and Ingersoll have formulated the definition of an arbitrage portfolio dual to (3.5) with nonnegative state prices ($\pi \geq 0$), whereas Ross, Dybvig and Ross, Dothan and Duffie have determined the equivalent of (3.1) dual to (3.5) with strictly positive state prices ($\pi >> 0$). Our version of the Fundamental Theorem establishes the nonarbitrage condition dual to the aggregation equation (3.5) with semipositive weights.[11]

Combining Lemma 3.3 and Theorem 3.4 we obtain a reformulation of Theorem 3.4.

Theorem 3.4′ *There is a solution to the linear system (3.5) if and only if conditions (3.3) and (3.4) hold for all nonnegative portfolios* x_1 *and* x_2.

This version of the Fundamental Theorem provides an alternative characterization for linear pricing of financial assets and establishes a close connection to Harsanyi's social aggregation theorem which we will explore shortly.

3.1.2 *Attainable Returns for Fixed Investments.* Arbitrage portfolios and conditions for their (non)existence have occupied a central stage in finance. A somewhat more general question concerns the attainability of *a priori* specified returns for investment, given a fixed initial monetary outlay.

Definition 3.5 *Given the asset returns* R *and asset prices* $p(\neq 0)$, *a portfolio* x *attains the state-specific returns* r *for the initial monetary investment* t *if and only if*

$$Rx \geq r \ \text{ and } \ p^T x = t. \tag{3.6}$$

By (3.6), the arbitrage portfolios defined in (3.1) correspond to choosing either $r = 0$ and $t < 0$ or $r >> 0$ and $t \leq 0$.

A characterization of feasible returns is obtained using the dual of the linear system (3.6). For simplicity, we assume that the returns of marketed assets for all states of nature form a set of S linearly independent vectors.

---

[10] Another interpretation of (3.5) is obtained if the state prices π are normalized to sum to unity. In this case, the aggregation weights can be regarded as a vector of probabilities for the S states of nature. Theorem 3.4 then observes that market prices of all assets must be equal to the expected return of the assets, evaluated at the probabilities π.

[11] Dybvig and Ross, Dothan and Duffie have employed separation theorems for convex sets to estabish the Fundamental Theorem while Varian and Ingersoll have applied the Duality Theorem of Linear Programming for this purpose. Ross [**39**] transformed the nonhomogenous definition of state prices, (3.5), into a homogenous linear system and applied Farkas' Lemma to obtain the theorem.

Theorem 3.6 *Let the rows of the matrix R be linearly independent. Then, there is a portfolio that attains the returns r given the initial investment t, i.e., the system (3.6) has a solution, if and only if for all $\pi > 0$*

$$R^T \pi = p \rightarrow \pi^T r \leq t. \tag{3.7}$$

Proof Lemma 2.4 implies that (3.6) has a solution if and only if there are no $y_1 > 0$ and y_2 such that

$$R^T y_1 + py_2 = 0, \; r^T y_1 + ty_2 > 0. \tag{3.8}$$

Because $y_1 > 0$ and the rows of R are linearly independent, (3.8) cannot have a solution for which $y_2 = 0$.[12] If $y_2 > 0$, (3.8) implies that there must not exist $\overline{y}_1 \equiv y_1/(-y_2) < 0$ such that $R^T\overline{y}_1 = p$ and $r^T\overline{y}_1 < t$. If $y_2 < 0$, we obtain from (3.8) that there must not exist $\overline{y}_1 \equiv y_1/(-y_2) > 0$ such that $R^T\overline{y}_1 = p$ and $r^T\overline{y}_1 > t$. Combining the conclusions for the two cases yields (3.7). □

Theorem 3.6 establishes a simple condition which must be satisfied for a vector of portfolio returns to be attainable, given the returns and prices of marketed assets. By the weak inequality in (3.7), the desired portfolio returns across all states of nature must aggregate to at most the value of the initial investment and by the equations in (3.7), the semipositive aggregation weights, π, must satisfy the characterization of state prices established in the Fundamental Theorem of Asset Pricing.

By the Fundamental Theorem, there are semipositive π that solve the equations in (3.7) if and only if asset markets admit no arbitrage. In conjunction with Theorem 3.4, Theorem 3.6 thus yields two conclusions:

(i) If asset markets admit no arbitrage, i.e., there are solutions to (3.5), portfolio returns that are attainable must be compatible with the absence of arbitrage gains and, by (3.7), the total value of such portfolio returns, evaluated at any state prices, must not exceed the initial investment. Portfolio returns for which the value of the returns across all states is larger than the initial investment can only be realized if arbitrage portfolios are to be found.

(ii) If there are arbitrage opportunities, i.e., (3.5) admits no solutions, then (3.7) holds trivially for *any* r; in such a market, arbitrarily large returns can be attained.[13]

3.2 Harsanyi's Social Aggregation Theorem. Harsanyi's social aggregation theorem establishes a very specific connection between individual and social (expected utility) preferences when there is uncertainty: given only a seemingly

---

[12]The assumption that the rows of R are linearly independent could be replaced by the following condition: *There is a state s such that the returns to all assets in state s are nonnegative and some are positive.* Such a state s could be called a "Diamond-Mirrless-Weymark state" in reference to **[10]** and **[50]** who applied an analogous condition on consumer preferences to establish the existence of Pareto improving commodity tax reforms.

[13]Geometrically, by (3.7), attainable portfolio returns r are found in the intersection of the half-spaces defined by the state prices that solve (3.5) and the initial investment t.

weak Pareto assumption, the social utility function must be an affine combination, i.e., a linear aggregate, of individual utility functions.[14,15] In order to state Harsanyi's theorem, we adopt the following notation.

We assume that the set of feasible social alternatives, denoted by P, consists of probability distributions (lotteries) on a fixed set of basic outcomes $X \equiv \{x_0, x_1, \dots, x_m\}$ where $m \geq 2$. The set of individuals is denoted by $N \equiv \{1, \dots, n\}$ with $n \geq 2$. Individual and social preferences on P are represented by expected utility functions, $V_i, i \in N$, and V for which

$$V_i(p) = \sum_{j=0}^{m} p_j V_i(e_j),\ \forall i \in N,\ \forall p \in P;\ V(p) = \sum_{j=0}^{m} p_j V(e_j), \forall p \in P. \tag{3.9}$$

In (3.9), e_j denotes the probability distribution in which the probability of the outcome $x_j \in X$ is one, $j = 0, 1, \dots, m$.

We apply the following Pareto condition.

Pareto Assumption For all social alternatives $p, q \in P$,

(i) if $V_i(p) \geq V_i(q)$ for all individuals $i \in N$, then $V(p) \geq V(q)$;
(ii) if $V_i(p) > V_i(q)$ for all individuals $i \in N$, then $V(p) > V(q)$.

This Pareto Assumption requires society to weakly prefer one lottery, p, to another, q, if everyone weakly prefers p over q (*Semistrong Pareto Assumption*); strict social preference is imposed if there is unanimous strict preference for p over q (*Weak Pareto Assumption*).

Theorem 3.7 is the Weak Pareto version of Harsanyi's social aggregation theorem.

Theorem 3.7 *Given the expected utility functions, $V_i, i \in N$, and V, there is a solution to the linear system*

$$V(p) = \sum_{i=1}^{n} \lambda_i V_i(p) + \mu,\ \lambda > 0, \forall p \in P, \tag{3.10}$$

if and only if the Pareto Assumption is satisfied.

Proof We transform the individual and social expected utility functions according to

$$\hat{V}_i(p) \equiv V_i(p) - V_i(e_0), \forall i \in N, \forall p \in P; \hat{V}(p) \equiv V(p) - V(e_0), \forall p \in P. \tag{3.11}$$

Given (3.11), we have $\hat{V}_i(e_0) = 0$ for all $i \in N$ and $\hat{V}(e_0) = 0$ for society. Applying the definitions (3.11), a solution to (3.10) exists if and only if the linear system

$$\hat{V}(p) = \sum_{i=1}^{n} \lambda_i \hat{V}(p),\ \lambda > 0,\ \forall p \in P, \tag{3.12}$$

[14] Harsanyi's social aggregation theorem is a "single profile" social choice result in that it is being assumed that there is a single preference ordering for each individual and a single preference ordering for society. Harsanyi's theorem concerns the relationship of the particular individual and social utility functions representing these preferences. The Impossibility Theorem of Arrow in [**3**] on the other hand deals with the infeasibility of aggregation of individual preferences when there are no restrictions on the particular preferences that each member of society may have. Arrow's theorem is therefore a "multiprofile" social choice result.

[15] Alternative versions of Harsanyi's theorem have been derived under various Pareto conditions in [**14**], [**20**], [**9**] and [**52, 53**] among others.

has a solution as these λ will solve (3.10) with $\mu \equiv V(e_0) - \sum_i \lambda_i V_i(e_0)$.

Using (3.9), we have

$$\hat{V}_i(p) = \sum_{j=1}^{m} p_j \hat{V}_i(e_j), \forall i \in N; \;\; \hat{V}(p) = \sum_{j=1}^{m} p_j \hat{V}(e_j). \tag{3.13}$$

By further defining

$$v(p) \equiv \left[\hat{V}_1(p), \dots, \hat{V}_n(p)\right]^T,$$

$$\hat{v}_i \equiv \left[\hat{V}_i(e_1), \dots, \hat{V}_i(e_m)\right]^T, i = 1, \dots, n; \; \hat{v} \equiv \left[\hat{V}(e_1), \dots, \hat{V}(e_m)\right]^T, \tag{3.14}$$

$$\hat{p} \equiv [p_1, \dots, p_m]^T, \; H^T \equiv [\hat{v}_1, \dots, \hat{v}_n],$$

and noting that (3.12) has a solution for all p if and only if it has a solution for $p = e_1, \dots, e_m$, we may express (3.12) in the matrix form

$$\hat{v} = H^T \lambda, \; \lambda > 0. \tag{3.15}$$

By definitions (3.14), the Pareto Assumption is equivalent to the nonexistence of a solution to the linear system

$$Hx \geq 0, \hat{v}^T x < 0 \quad \text{or} \quad Hx >> 0, \hat{v}^T \leq 0. \tag{3.16}$$

The proof is as follows. According to the Pareto Assumption, $v(p) \geq v(q)$ implies $\hat{V}(p) \geq \hat{V}(q)$ and $v(p) >> v(q)$ implies $\hat{V}(p) > \hat{V}(q)$, where p and q are arbitrary lotteries in P. By (3.13), we have $v(p) = H\hat{p}$, while $\hat{V}(p) = \hat{v}^T\hat{p}$. Thus, the Pareto Assumption may be expressed as

$$\begin{gathered} \hat{p} \geq 0, \; \hat{q} \geq 0, \; H\hat{p} \geq H\hat{q} \to \hat{v}^T\hat{p} \geq \hat{v}^T\hat{q}, \\ \hat{p} \geq 0, \; \hat{q} \geq 0, H\hat{p} >> H\hat{q} \to \hat{v}^T\hat{p} \geq \hat{v}^T\hat{q}, \end{gathered} \tag{3.17}$$

for all $p, q \in P$, or, defining $x \equiv p - q$, as

$$\begin{gathered} Hx \geq 0 \to \hat{v}^T x \geq 0, \\ Hx >> 0 \to \hat{v}^T x > 0 \end{gathered} \tag{3.18}$$

for all x. The satisfaction of (3.17) for all $p, q \in P$ is equivalent to nonexistence of solutions to (3.16).

By Lemma 2.3, the linear system (3.15) is solvable if and only if the system (3.16) is not. Thus, (3.10) is solvable if and only if the Pareto Assumption holds, i.e., we have (3.17) for all $p, q \in P$. □

According to Theorem 3.7, the Pareto Assumption is equivalent to a linear aggregation equation between individual and social utility functions in which the aggregation (welfare) weights of all individuals form a semipositive vector. This aggregation equation was first formulated by Weymark [**52, 53**]. He showed that, given the Weak Pareto Assumption and assuming that there exists a lottery which is strictly preferred to another lottery by all individuals, the aggregation equation (3.10) has a solution. Weymark further established that the Semistrong Pareto Assumption is an implication of the Weak Pareto Assumption and unanimous strict preference among some pair of lotteries. In contrast to these results, Theorem 3.7 proves directly and without additional assumptions that the Semistrong and Weak Pareto Assumptions *in combination* are equivalent to the Harsanyi equation (3.10).

Individual theorems of the alternative have been applied by [**7**] and Weymark to obtain versions of Harsanyi's social aggregation theorem. The above proof of Theorem 3.7, based on the Kuhn-Fourier theorem, is considerably shorter and simpler than these earlier applications of duality principles in social choice. The essential difference is that, armed with the Kuhn-Fourier theorem, we are able to derive the particular nonhomogenous duality principle (Lemma 2.3) that yields Theorem 3.7.[16] This provides two advantages. First, we do not have to conform to the particulars of the existing theorems of the alternative - thus, we are spared difficulties arising from the discrepancy between the nonhomogenous aggregation equation (3.10) and the almost universally homogenous theorems of the alternative. Secondly, our derivation clearly demonstrates the linear dual character of Harsanyi's theorem: *for each alternative Pareto Assumption (which defines a linear system) there is a dual, and for each version of Harsanyi's aggregation equation there is a corresponding primal condition.* However, not every Pareto condition necessarily yields an aggregation equation such as (3.10) and not every aggregation equation yields a known Pareto condition as its dual.[17] The Kuhn-Fourier theorem is the most convenient tool for exploring the dual relations.

3.3 Linear Aggregation in Finance and Social Choice. The Fundamental Theorem of Asset Pricing and Harsanyi's social aggregation theorem both establish a linear aggregation equation. The former shows that there are state prices which can be applied to evaluate the returns of any asset and the latter proves the existence of welfare weights that connect individual and social utility functions. In each theorem, the existence of the aggregation weights (state prices and welfare weights, respectively) is characterized by a dual condition: infeasibility of arbitrage opportunities in one and the Pareto Assumption in the other. Formally, the dualities contained in the two theorems are identical - both Theorems 3.4 and 3.7 coincide with a particular nonhomogenous theorem of the alternative which we have stated in Lemma 2.3.[18]

The similarity of the two results is particularly apparent in Theorems 3.4′ and 3.7, and this yields a reinterpretation of Harsanyi's social aggregation theorem.[19] Just as, by Lemmas 3.2 and 3.3, any and all violations of conditions (3.3) and (3.4) indicate the existence of arbitrage opportunities in exchanges of asset portfolios,

---

[16]Earlier versions of this manuscript contain the Kuhn-Fourier proof of the Strong Pareto version of Harsanyi's theorem. Previous proofs for this aggregation theorem were given by [**52**] and [**14**]).

[17]Weymark in [**53**] has shown that, if only the Weak Pareto Assumption is applied, an aggregation relationship exists between individual and social preferences but this aggregation equation contains an additional parameter multiplying the social utility function. Thus, the Weak Pareto Assumption on its own is not dual to (3.10). On the other hand, Turunen-Red and Weymark in [**43**] have derived new versions of Pareto principles which are dual to Harsanyi aggregation equations which apply when individual and social preferences do not satisfy expected utility axioms but conform to Fishburn's axioms [**19, 21**] for choice under uncertainty.

[18]Other versions of the Fundamental Theorem and Harsanyi's aggregation theorem can be derived by altering the Pareto Assumption and the definition of an arbitrage portfolio. Given congruent changes in the statements of the theorems, the resulting versions of the Fundamental Theory and Harsanyi's aggregation theorem remain formally identical.

[19]According to Theorem 3.4′, there are solutions to (3.5) if and only if all pairs of nonnegative portfolios obey conditions (3.3) and (3.4); by Theorem 3.7, there are solutions to (3.15) if and only if conditions (3.17) are satisfied. Because the individual and social utility functions may assume different origins of utility scales, the appearance of equation (3.10) differs from (3.5). However, once the utility scales are normalized to admit a common origin (as defined in (3.11)), the Harsanyi equation, (3.15), is exactly analogous to (3.5).

violations of conditions (3.17) (or, equivalently, violations of the Pareto Assumption) indicate opportunities for arbitrage in exchanges of lotteries between the set of individuals and any unspecified outside opponent. If, for instance, all individuals should weakly prefer lottery p to lottery q, i.e., $V_i(p) \geq V_i(q)$ for all $i \in N$, but social preference for q were strict in violation of the Pareto Assumption, i.e., $V(p) < V(q)$, then each individual on his own would willingly exchange the lottery q for the lottery p with an outside opponent, while society (all individuals, N) would exchange p for q and would even be willing to make the payment somewhat larger than the lottery p. Such trades of lotteries would yield an arbitrage loss to all individuals and an arbitrage gain to the outside opponent. Harsanyi's social aggregation theorem, which demands the Pareto Assumption to be satisfied, thus shows that *minimal economic rationality from the part of society, i.e., refusing to be a losing party to arbitrage exchanges, is equivalent to a strong aggregation relation between individual and social utility functions.*

4 Individual rationality

The Fundamental Theorem of Asset Pricing and Harsanyi's social aggregation theorem are duality results that characterize linear aggregation of asset returns and individual and social preferences by the absence of arbitrage gains. Analogous instances of dualities appear in the theory of individual choice. In this section, we consider dual characterizations of rationalizable behavior in two contexts: when uncertainty is present and when consequences of actions are certain. In the first instance, we discuss a theorem of Buehler [**8**] on linear rationalizability of choice under uncertainty and in the second, we state a version of Afriat's theorem [**1, 2**] on rationalizable consumer choice.

4.1 Choice under Uncertainty. We apply the notation of Buehler [**8**] and assume that there are S states of nature, indexed by $s \in \{1, \dots, S\}$. A *prospect*, $g(s)$, is a real-valued (bounded) function on the set of states, and its values represent the possible utility (or monetary) outcomes associated with an uncertain action. We assume that $\succeq$ denotes an individual's weak preference ordering on the set of prospects. We further assume that the individual has expressed n preference rankings $g'_t \succeq g_t$ among some feasible prospects g_t and g'_t, indexed by $t = 1, \dots, N$.

We call the n preference rankings *linearly rationalizable* (*p-coherent* by Buehler) if there exists a vector of prior probabilities which could have given rise to the decision maker's preferences according to the following definition.

Definition 4.1 *The preferences $g'_t \succeq g_t$, are linearly rationalizable if there exists a probability vector $[p(1), \dots, p(S)] > 0$ such that*

$$\sum_{s=1}^{S} p(s)(g'_t(s) - g_t(s)) \equiv v_t \geq 0, \quad t = 1, \dots, N. \tag{4.1}$$

According to Definition 4.1, the decision maker's preferences must be derivable from an expected utility calculation with $v_t, t = 1, \dots, N$, representing the preference comparison between any two prospects, and any preference ranking of prospects arising from such a calculation must coincide with the expressed preferences, i.e.,

$v \equiv [v_1, \dots, v_n] \geq 0$. Each of these two aspects of Definition 4.1 has consequences as to the dual characterization of linearly rationalizable behavior.[20]

The first requirement directs us to consider the linear system

$$B^T p = v, p > 0; \; B \equiv [g_1'(s) - g_1(s), \dots, g_N'(s) - g_N(s)], \tag{4.2}$$

which is equivalent to (4.1). If (4.2) has a solution for given $v_t, t = 1, \dots, N$, then the n preference rankings are based on probability weighted aggregates of state-specific consequences of prospects g_t and g_t'.[21] A direct application of Lemma 2.3 yields

Lemma 4.2 *The linear system (4.2) has a solution p if and only if for all $\gamma_1 \geq 0$ and $\gamma_2 \geq 0$,*

$$B\gamma_1 \geq B\gamma_2 \rightarrow v^T\gamma_1 \geq v^T\gamma_2, \tag{4.3}$$

$$B\gamma_1 >> B\gamma_2 \rightarrow v^T\gamma_1 > v^T\gamma_2. \tag{4.4}$$

Lemma 4.2 can be given a simple interpretation. We may think of the decision maker as consisting of a number of separate facets, one for each state of nature; the $S \times n$ matrix B expresses the preference rankings of prospects g_t and g_t' by each of these potentially quite independent "individuals". The definition of linear rationalizability imposes a very specific connection between the evaluations of the state-specific individuals and the actual decision maker: the linear system (4.2) must hold, i.e., relative evaluations of g_t and g_t' by the decision maker must be semipositive linear aggregates of his state-specific evaluations of the prospects.

By Harsanyi's theorem, an aggregation equation such as (4.2) has a solution if and only if a Pareto condition is satisfied. Lemma 4.2 identifies the Pareto condition necessary for linear rationalizability under uncertainty. These conditions are analogous to our earlier Pareto Assumption and involve comparisons of "individual" and "social" preferences among lotteries over basic prospects.[22] According to (4.3), if the decision maker's state-specific evaluations of the prospects g_t and g_t' are such that, in each state of nature, the decision maker prefers a lottery γ_1 to another lottery γ_2 over the same prospects, then the decision maker's overall preference should be (weakly) in favor of the lottery γ_1 (*Semistrong Pareto*) and by (4.4), in the case of a unanimous strict preference, the overall preference for γ_1 should be strict (*Weak Pareto*).

We also obtain an arbitrage interpretation of conditions (4.3) and (4.4). Analogously to the Pareto Assumption of Theorem 3.7, conditions (4.3) and (4.4) can be regarded as characterizations of a minimal degree of economic rationality: *an individual's behavior under uncertainty can be linearly rationalizable only if he is not a willing party to arbitrage exchanges of lotteries which are sure to benefit others.*

---

[20] Buehler's definition in [**8**] of p-coherent preferences did not include the definition of v_t and did not differentiate between the two aspects of linear rationalizability.

[21] The vector p in (4.2) can be normalized into a probability vector as the sum of its elements is positive.

[22] Here, the individuals of Harsanyi's theorem are the state-specific facets of the decision maker while the decision maker corresponds to Harsanyi's society. The Harsanyi matrix H in (3.15) expresses all individual evaluations of basic outcomes with respect to a common origin, while the vector $\hat{v}$ contains the corresponding social evaluations. In (4.2), the matrix B contains all state-specific evaluations of basic outcomes, now defined by $(g_t' - g_t), t = 1, \dots N$, with respect to an origin of zero utility (or, monetary value) and the vector v consists of the evaluations of the aggregate individual.

In order to obtain the full characterization of linear rationalizability, we combine Lemma 4.2 and the requirement that the aggregate preference comparisons must correspond to the decision maker's expressed preference rankings ($v \geq 0$).

Theorem 4.3 ([**8**]) *The preferences $g'_t \succeq g_t, t = 1, \ldots, N$, are linearly rationalizable, i.e., there is a probability vector $p > 0$ such that (4.1) holds, if and only if there are no $\gamma \geq 0$ such that*

$$B\gamma << 0. \tag{4.5}$$

Proof Apply Lemma 2.5 to (4.2) when $v \geq 0$. □

Theorem 4.3 states the dual characterization of linearly rationalizable preferences under uncertainty: *for linear rationalizability, it is necessary and sufficient that the decision maker's evaluations of prospects g_t and g'_t in all states of nature are such that no lottery such as γ makes him a willing party to an arbitrage loss.*[23]

Buehler in [**8**] has presented an alternative interpretation for Theorem 4.3 based on preference reversals and Nau and McCardle [**36**] have interpreted a result analogous to Theorem 4.3 in terms of "acceptable gambles".[24] The interpretations of both Buehler and Nau and McCardle indicate a connection between linear rationalizability and arbitrage: according to Buehler, linear rationalizability means absence of sure gains (suitably defined) and according to Nau and McCardle absence of sure losses (suitably defined) is of the essence. We have interpreted the dual characterization of linear rationalizability using Harsanyi's aggregation theorem; in Lemma 4.2 we have shown that the nonarbitrage dual of linear rationalizability arises precisely because of the aggregation aspect of Definition 4.1.

Lemma 4.2 and Theorem 4.3 can be applied also in more general contexts. It can be shown that the definitions of a correlated equilibrium [**4, 5**] and an incentive compatible probabilistic mechanism (e.g., [**35**]) are analogous to the linear system (4.1). The dual characterizations of linear rationalizability in (4.3), (4.4) and (4.5) can thus be applied to obtain nonarbitrage interpretation of these concepts as well.[25]

4.2 Concave rationalizability of choice. Afriat in [**1, 2**], as well as references [**11**], [**47**] and [**32**], have determined conditions under which a set of consumer choices can be rationalized by a concave (as opposed to linear) utility function. Aside from the curvature of the objective function, this problem is analogous to the

[23] The inequalities (4.5) indicate that the lottery γ (on the basic prospects $(g'_t - g_t), t = 1, \ldots, N$), if it were to exist, would yield a loss to the individual in each state and would therefore provide an arbitrage gain to any outside opponent of the decision maker. Yet, because $v \geq 0$ and $\gamma \geq 0$, we have $v^T\gamma \geq 0$, i.e., the decision maker would accept the lottery. Theorem 4.3 shows that linear rationalizability, i.e., the existence of the semipositive aggregation weights p in (4.1), is not compatible with such arbitrage opportunities when the decison maker's preferences are such that $v \geq 0$. For a γ, for which (4.5) holds, the Pareto conditions (4.3) and (4.4) (which are dual to (4.2)) require $v^T\gamma < 0$, i.e., no solution of (4.5) is compatible with $g'_t \succeq g_t, t = 1, \ldots, N$.

[24] According to Buehler, any nonnegative solution to (4.5) indicates the existence of a uniform gain to the individual: the individual only needs to invest the "stakes" γ for the "bets" $(g'_t - g_t)$ and reverse his state-specific preferences on those alternatives t for which $\gamma_t > 0$. Absence of such uniform gains through preference reversals characterizes linear rationalizability according to Buehler. The Nau and McCardle acceptable gambles correspond to vectors $B\gamma, \gamma \geq 0$, in our notation. According to this viewpoint, linear rationalizability means the absence of acceptable gambles that yield uniform losses.

[25] See [**27**], [**37**] and earlier versions of this manuscript.

question of linear rationalizability of choice under uncertainty. We explore the similarity and present a version of Afriat's Theorem on concave rationalizability (e.g., [**49**], Chapter 8.8)). This result is the counterpart of Theorem 4.3 by Buehler.

We consider a set of price and consumption observations, $(p_t, x_t), t = 1, \dots, N$. According to [**47**], there is a locally nonsatiated, continuous, concave and monotonic utility function that rationalizes the data (p_t, x_t) if and only if the data satisfy the Generalized Axiom of Revealed Preference (GARP).[26] Afriat, on the other hand, has shown that concave rationalizability of consumer data can be characterized by linear inequalities. Theorem 4.4 is a version of Afriat's Theorem.

Theorem 4.4 *The following conditions are equivalent:*

(i) *there exists a continuous, concave and monotonic utility function that rationalizes the data, $(p_t, x_t), t = 1, ..., N$;*
(ii) *there exist vectors $u \equiv [u_t; t = 1, \dots, N]$ and $\lambda \equiv [\lambda_t; t = 1, \dots, N]$ such that*

$$u_i \leq u_j + \lambda_j p_j^T (x_i - x_j) \ \forall i, j = 1, \dots, N; \ u >> 0, \lambda > 0. \tag{4.6}$$

Proof The result can be established using the arguments of [**47, 49**]. □

By Theorem 4.4, concave rationalizability of consumer data can be checked by showing that inequalities (4.6) have a solution $(u^T; \lambda^T)$. The multipliers u and λ, respectively, can be interpreted as the utility levels and marginal utilities of income associated with the N observations; these two sets of multipliers completely characterize the concave objective function of the consumer.[27]

We have assumed that the multipliers λ form a semipositive vector as opposed to being strictly positive ($\lambda >> 0$) as in other versions of Afriat's Theorem. This implies that the utility function established in Theorem 4.4 is not necessarily locally nonsatiated at every data point; however, Theorem 4.4 does still provide a possible rationalization of consumer data, this rationalization being more easily attainable than if local nonsatisation were insisted upon in the entire data set.[28]

A dual of the Afriat inequalities can be found by transforming the system (4.6) into an equivalent form in which the utilities (u_i, u_j) have been eliminated. This yields the linear system[29]

$$\lambda_i D_{ij} + \lambda_j D_{ji} \geq 0, \ \forall i, \ j = 1, \dots, N, \tag{4.7}$$

in which[30]

$$D_{ij} \equiv p_i^T x_j - p_i^T x_i = p_i^T x_j - 1, \quad \forall i, j = 1, \dots, N. \tag{4.8}$$

---

[26] Varian in [**47**] has determined an algorithm which may be applied to check whether or not the data satisfy the GARP.

[27] [**47, 49**].

[28] Matzkin and Richter in ([**32**], Theorem 2) have shown that a stronger version of Afriat's Theorem can be obtained if, in addition to requiring $v >> 0$, $\lambda >> 0$, the weak inequalities in (4.6) are replaced by strict inequalities; then, the utility function that rationalizes the data must be strictly concave, strictly monotonic and differentiable.

[29] Inequalities (4.6) imply (4.7) by the elimination of v. To obtain the converse, we note that, given (4.7), we may construct an equivalent inequality system $v_{ji} \leq \lambda_j D_{ji}, v_{ij} \leq \lambda_i D_{ij}, v_{ij} + v_{ji} = 0$. Because the signs of v_{ij} are not restricted, we may define $v_{ij} \equiv u_j - u_i$, where the u_i form a strictly positive vector.

[30] We assume that the data is normalized so that $p_t^T x_t = 1$ for all $t = 1, \dots, N$.

Inequalities (4.7) contain all the information of (4.6): *the data (p_t, x_t) can be rationalized by a continuous, concave and monotonic utility function if and only if there is a solution to (4.7).*

The system (4.7) can be rewritten in the matrix form

$$A^T\lambda \geq 0, \lambda > 0; \quad A^T \equiv [D_{ij}Q_i^T + D_{ji}Q_j^T;\ i, j = 1, \dots, N], \tag{4.9}$$

where the $N^2 \times N$ matrix A^T expresses all pairwise comparisons of the consumption bundles as defined in (4.7) and (4.8). In (4.9), the vector Q_i picks the element λ_i from the vector λ and the ij^{th} row of A^T is zero, except for the ith element which is D_{ij} and the jth element which equals D_{ji}. The dual of the linear system (4.9) is defined by the following theorem.

Theorem 4.5 *There exists a continuous, concave and monotonic utility function that rationalizes the data, $(p_t, x_t), t = 1, \dots, N$, if and only if there is no $\gamma \geq 0$ such that*

$$Av << 0. \tag{4.10}$$

Proof Apply Lemma 2.5 to obtain the dual of (4.9) (which is equivalent to (4.6)). □

Theorem 4.5 is of interest on two accounts. First, it identifies an additional practical test which can be applied to determine whether or not a set of price and consumption data conforms to the model of utility maximization: if the linear system (4.10) has no nonnegative solution, then there is a continuous, concave and monotonic utility function which rationalizes the data.[31] Secondly, Theorem 4.5 shows that there are fundamentals that characterize individually rationalizable choice, whether uncertainty is present or not. According to Theorems 4.3 and 4.5, *individual economic rationality, irrespective of context, is equivalent to not making oneself a willing victim of arbitrage seekers.*[32]

5 Balanced TU-games

In this section, we discuss our final application of the Kuhn-Fourier theorem, the Bondareva [**6**] - Shapley [**41**] theorem on cooperative transferable utility (TU) games. As is well-known, the Bondareva - Shapley theorem characterizes nonemptiness of the core of a finite TU-game in terms of balancedness of the game. We will show that the concept of arbitrage yields a simple and intuitive interpretation of such balancedness.

We make the standard assumptions and let $N \equiv [1, \dots, n]$ denote the set of players in a finite TU-game, G. A payoff allocation among the players is any vector $x \in R^n$ with the ith component of x being interpreted as the payoff of the player $i \in N$. The characteristic function of the game G, defined on all subsets (coalitions) of N, is $v(s), s \subset N$. The function $v(s)$ indicates the total amount of transferable utility that the members of the coalition $s \subset N$ are able to secure themselves without the cooperation of players that are not included in the coalition; the payoff available to the grand coalition (the set of all players) is denoted by $v(N)$.

[31] Alternative versions of Theorem 4.5 can be derived to identify data sets that can be rationalized by utility functions that are locally nonsatiated and (possibly) differentiable.

[32] Analogously to Lemma 4.2 and Theorem 4.3, Theorem 4.5 can be used to define Pareto conditions similar to (4.3) and (4.4). Any violation of these (unstated) Pareto conditions defines an arbitrage exchange of lotteries (on basic prospects identified by the columns of A) which is acceptable to the decision maker and brings him a certain loss.

We apply the vectors $1_s \in R^n, s \subset N$, to indicate the membership of various coalitions. The vectors 1_s consist of elements equal to zero (corresponding to those players not in the coalition s) and one (for the players that are members of s). Core allocations are defined as follows.

Definition 5.1 *An allocation x is in the core of the game G if*

$$Mx \geq v, 1_N^T x = v(N); \; M \equiv [1_s^T; s \subset N, s \neq N], v \equiv [v(s); s \subset N, s \neq N]. \quad (5.1)$$

In (5.1), the matrix M defines the membership of all coalitions, excluding the grand coalition. By the weak inequalities in (5.1), a core allocation must yield each of these coalitions at least the collective payoff that the coalitions could have secured on their own. According to the equality in (5.1), a core allocation must also be feasible, i.e., the total payoff of all players must equal the payoff of the grand coalition. The core allocations define outcomes of the coalitional game that no coalition can improve by unilaterally deviating from the specified allocation of payoffs. As such the core allocations represent the stable outcomes of the TU-game.

When do core allocations exist? Bondareva [**6**] and Shapley [**41**] have derived the following characterization.

Theorem 5.2 *The coalitional TU-game G has a nonempty core if and only if the game is balanced, i.e., for all $\alpha > 0$*

$$M^T\alpha = 1_n \rightarrow \alpha^T v \leq v(N). \quad (5.2)$$

Proof Apply Lemma 2.4 to the system (5.1) and replicate the proof of Theorem 3.6. □

According to Theorem 5.2, there are core allocations if and only if, for any collection of weights, α, that satisfies the equations in (5.2), the weighted sum of the coalitional payoffs, v, is less or equal to the payoff of the grand coalition, $v(N)$.

There have been few efforts to interpret the meaning of balancedness of coalitional games.[33] We have established Theorem 5.2 as a corollary of a nonhomogenous theorem of the alternative.[34] This proof strategy displays clearly the linear duality statement implicit in the definition of core allocations and the definition of balancedness. It also suggests an interpretation: *the duality of the linear systems (5.1) and (5.2) may be regarded as an aggregation - arbitrage result akin to the Fundamental Theorem of Asset Pricing and Harsanyi's social aggregation theorem.*

To explore this interpretation, we turn to Definition 3.5 and Theorem 3.6 of Section 3. We interpret the game G in terms of an equivalent portfolio investment problem. We regard the payoff allocation, x, in (5.1) as a portfolio of investments selected by the set of players as they allocate the total payoff, $v(N)$, amongst themselves; in this reallocation of transferable payoff, each player corresponds to an "asset" in which the grand coalition may invest at the cost of one unit of the payoff. Each player may join a number of coalitions; in each of these "states of nature", a player yields the return of one unit of payoff for each payoff unit invested in him.

[33] Hildenbrand and Kirman in ([**28**], Chapter 3.2) have called balancedness a concept which has no intuitive economic interpretation: indeed, the Hildenbrand and Kirman interpretation of (5.2) is expressed in terms of centers of gravity of various simplexes. Myerson in ([**35**], Chapter 9) has interpreted (5.2) using a dynamic matching process of players, while Osborne and Rubinsten in ([**38**], Chapter 13) have regarded (5.2) as a feasibility condition on the players' allocation of time.

[34] Proofs of Theorem 5.2 have relied on the Duality Theorem of Linear Programming (e.g. [**6**], [**41**], ([**35**], Chapter 9)) or on separation theorems for convex sets (e.g., [**38**], Chapter 13)).

By (5.1), a "core portfolio" is analogous to a portfolio of investment which must yield at least the *a priori* specified return, $v(s)$, in every state of nature while the total investment of all players equals $v(N)$.[35]

According to Theorem 3.6, a vector of portfolio returns is attainable if and only if the desired state-specific returns aggregate to at most the amount of the initial investment, when evaluated at the appropriate state prices, π. In the context of the investment problem equivalent to the game G, we therefore obtain Theorem 5.2: for core allocations to exist, the coalitional payoffs must aggregate to at most the value of the game, when evaluated at the weights, α, that satisfy the equations of (5.2). These equations define the appropriate state prices in the current portfolio investment problem and, according to them, the value of a unit of payoff invested in any player (which is unity) must equal the aggregate value of the unit of payoff across all coalitions that the player may join.

By the Fundamental Theorem of Asset Pricing, there are state prices that solve equations such as those in (5.2) if and only if there are no arbitrage opportunities in exchanges of the relevant asset portfolios. Accordingly, dual to the definition of the aggregation weights in (5.2), we may define "arbitrage allocations" in the game G. These arbitrage allocations yield positive coalitional payoffs even when the value of the game is negative.[36] By its definition, the game G does not admit arbitrage allocations[37] and it is the infeasibility of such arbitrage gains that leads to our interpretation of Theorem 5.2: *core allocations exist if and only if the coalitional payoffs, v, are compatible with the value of the game, $v(N)$, and the nonexistence of free lunches via coalition formation. Conversely, coalitional payoffs that do not solve (5.2) are unattainable precisely because in order to attain them, arbitrage coalitions would have to exist.*

6 Conclusions

We have explored dual interpretations and interconnections of a number of notions of economic rationality. The meta-theorem of the alternative by [**30**] has been applied as the tool of analysis and, by the power and generality of the Kuhn-Fourier theorem, we have been able to expose the similarities among the various concepts. We have seen that the important linear aggregation theorems of Ross in [**39, 40**] and Harsanyi [**25, 26**] coincide with the same linear duality statement (i.e., Lemma 2.3); we have applied this connection to reinterpret the Pareto assumptions of social choice in terms of arbitrage in exchanges of lotteries. We have further demonstrated that similar dualities of linear aggregation and arbitrage form the foundation of individually rational behavior - the theorems of Buehler [**8**] and Afriat [**1, 2**] on linear and concave rationalizability of individual choice bear this out. As our final application of the Kuhn-Fourier theorem, we have established the precise theorem of the alternative (Lemma 2.4) that underlies the dual concepts of the core and the balanced game. This has allowed us to present a simple interpretation of balancedness of a coalitional TU-game in terms of absence of arbitrage allocations and coalitions.

---

[35]Using the notation of Section 3, we have $R = M, p = 1, t = v(N)$, and $r = v$.

[36]An arbitrage allocation solves $\{Mx \geq 0, 1_n^T x < 0 \text{ or } Mx >> 0, 1_n^T x \leq 0 \}$.

[37]Consider, e.g., the inequalities $Mx \geq 0$. By the definition of the matrix M (which includes all coalitions consisting of single players), we obtain $x \geq 0$; therefore, the inequality $1_N^T x < 0$ cannot hold. The other case is analogous.

Acknowledgements

We thank James Brander for suggesting the applications of theorems of the alternative as a topic of study. Seppo Honkapohja, Preston McAfee, Dale Stahl, John Weymark and Myrna Wooders have offered guidance in finance, game theory and social choice. Participants of the Canadian Theory Meetings in Montreal, especially Camille Bronsard, Henry van Egteren and Barton Lipman, had many useful comments. We thank John Weymark, Murray Kemp, Robert Nau, Peter Hammond and David Sibley for comments on an earlier draft. Participants of the Canadian Economic Association Meetings in Calgary, the Econometric Society North-American Summer Meetings in Quebec City and the Second International Conference of the Society for Social Choice and Welfare in Rochester were helpful. Our thanks to the participants of the seminars at Tulane University (especially Irving LaValle) and the University of New Orleans. The authors gratefully acknowledge the hospitality and financial support of their respective universities during their travels.

References

[1] Afriat, S. [1967], *The Construction of a Utility Function from Expenditure Data*, International Economic Review **8**, 67-77.

[2] Afriat, S. [1973], *On a System of Inequalities in Demand Analysis: An Extension of the Classical Method*, International Economic Review **14**, 460-472.

[3] Arrow, K. J. [1951], *Social Choice and Individual Values*, Wiley, New York.

[4] Aumann, R. [1974], *Subjectivity and Correlation in Random Strategies*, Journal of Mathematical Economics **1**, 67-96.

[5] Aumann, R. [1987], *Correlated Equilibrium as an Expression of Bayesian Rationality*, Econometrica **55**, 1-18.

[6] Bondareva, O. [1963], *Some Applications of Linear Programming Methods to the Theory of Cooperative Games*, (in Russian), Problemy Kibernetiki **10**, 119-139.

[7] Border, K. [1981], *Notes on v. Neumann-Morgenstern Social Welfare Function*, unpublished manuscript, Division of the Humanities and Social Sciences, California Institute of Technology.

[8] Buehler, R. [1976], *Coherent Preferences*, Annals of Statistics **4**, 1051-1064.

[9] Coulhon, T. and Mongin, P. [1989], *Social Choice Theory in the Case of v. Neumann - Morgenstern Utilities*, Social Choice and Welfare **6**, 175-187.

[10] Diamond, P. and Mirrlees, J. [1971], *Optimal Taxation and Public Production, I*, American Economic Review **61**, 8-27.

[11] Diewert, W. E. [1973], *Afriat and Revealed Preference Theory*, Review of Economic Studies **40**, 419-426.

[12] Diewert, W. E. [1978], *Optimal Tax Perturbations*, Journal of Public Economics **10**, 139-177.

[13] Diewert, W. E. [1983], *Cost-Benefit Analysis and Project Evaluation*, Journal of Public Economics **22**, 265-302.

[14] Domotor, Z. [1979], *Ordered Sum and Tensor Product of Linear Utility Structures*, Theory and Decision **11**, 374-399.

[15] Dothan, M. [1990], *Prices in Financial Markets*, Oxford University Press.

[16] Duffie, D. [1992], *Dynamic Asset Pricing Theory*, Princeton University Press, Princeton.

[17] Dybvig, P. and Ross, S. [1987], *Arbitrage*, in The New Palgrave: A Dictionary of Economics (J. Eatwell, M. Milgate and P. Newman, eds.), Vol. 1, McMillan, London.

[18] Farkas, J. [1902], *}Uber die Theorie der Einfachen Ungleichungen*, J. Reine Angew. Math. **124**, 1-24.

[19] Fishburn, P. [1982], *Nontransitive Measurable Utility*, Journal of Mathematical Psychology **26**, 31-67.

[20] Fishburn, P. [1984], *On Harsanyi's Utilitarian Cardinal Welfare Theorem*, Theory and Decision **17**, 21-28.

[21] Fishburn, P. [1988], *Nonlinear Preference and Utility Theory*, Johns Hopkins University Press, Baltimore.

[22] Fourier, J.-B. J. [1826], *Solution d'une Question Particuliere du Calcul des Inegalites*, Oeuvres II, 317-328.

[23] Gale, D. [1960], *The Theory of Linear Economic Models*, McGraw-Hill, New York.

[24] Gordan, P. [1873], *Über die Auflösungen Linearer Gleichungen mit Reelen Coefficienten*, Math. Ann. **6**, 23-28.

[25] Harsanyi, J. [1955], *Cardinal Welfare, Individualistic Ethics and Interpersonal Comparisons of Utility*, Journal of Political Economy **63**, 309-321.

[26] Harsanyi, J. [1977], *Rational Behavior and Bargaining Equilibrium in Games and Social Situations*, Cambridge University Press, Cambridge.

[27] Hart, S. and Schmeidler, D. [1989], *Existence of Correlated Equilibria*, Mathematics of Operations Research **14**, 18-25.

[28] Hildenbrand, W. and Kirman, A. [1976], *Introduction to Equilibrium Analysis*, North-Holland.

[29] Ingersoll, J. [1987], *Theory of Financial Decision Making*, Rowman & Littlefield, Totowa, New Jersey.

[30] Kuhn, H. [1956], *Solvability and Consistency for Linear Equations and Inequalities*, American Mathematical Monthly **63**, 217-232.

[31] Mangasarian, O. [1969], *Nonlinear Programming*, McGraw-Hill, New York.

[32] Matzkin, R. and Richter, M. [1991], *Testing Strictly Concave Rationality*, Journal of Economic Theory *53*, 287-303.

[33] Motzkin, T. [1936], *Beitrage zur Theorie der Linearen Ungleichungen*, Inaugural Dissertation, University of Basel, Azriel Printers, Jerusalem, 1936; an English translation published in Theodore S. Motzkin: Selected Papers (D. Cantor, B. Gordon and B. Rothschild, eds.), Birkhauser, Boston, 1983.

[34] Motzkin, T. [1951], *Two Consequences of the Transposition Theorem of Linear Inequalities*, Econometrica **19**, 185-186.

[35] Myerson, R. [1991], *Game Theory*, Harvard University Press, Cambridge, Mass.

[36] Nau, R. and McCardle, K. [1990], *Coherent Behavior in Noncooperative Games*, Journal of Economic Theory **50**, 424-444.

[37] Nau, R. and McCardle, K. [1991], *Arbitrage, Rationality, and Equilibrium*, Theory and Decision **31**, 199-240.

[38] Osborne, M. and Rubinstein, A. [1994], *A Course in Game Theory*, MIT Press, Cambridge, Mass.

[39] Ross, S. [1976], *Return, Risk and Arbitrage*, in Risk and Return in Finance (I. Friend and J. Bicksler, eds.), Ballinger, Cambridge, Mass.

[40] Ross, S. [1978], *A Simple Approach to the Valuation of Risky Streams*, Journal of Business **51**, 453-475.

[41] Shapley, L. [1967], *On Balanced Sets and Cores*, Naval Research Logistics Quarterly **14**, 453-460.

[42] Stoer, J. and Witzgall, C. [1970], *Convexity and Optimization in Finite Dimensions, I*, Springer Verlag, Berlin.

[43] Turunen-Red, A. H. and Weymark, J. [1998], *Paretian Aggregation of Asymmetric Utility Functionals*, Theory and Decision, forthcoming.

[44] Turunen-Red, A. H. and Woodland, A. D. [1988], *On the Multilateral Transfer Problem*, Journal of International Economics **25**, 249-269.

[45] Turunen-Red, A. H. and Woodland, A. D. [1991], *Strict Pareto Improving Multilateral Reforms of Tariffs*, Econometrica **59**, 1127-1152.

[46] Turunen-Red, A. H. and Woodland, A D. [1995], *Multilateral Reform of Quantity Restrictions on Trade*, Discussion Paper, Department of Economics and Finance, University of New Orleans.

[47] Varian, H. [1982], *The Nonparametric Approach to Demand Analysis*, Econometrica **50**, 945-973.

[48] Varian, H. [1987], *The Arbitrage Principle in Financial Economics*, Journal of Economic Perspectives **1**, 55-72.

[49] Varian, H. [1992], *Microeconomic Analysis*, 3rd. edition, Norton, New York.

[50] Weymark, J. [1979], *A Reconciliation of Recent Results in Optimal Taxation Theory*, Journal of Public Economics **12**, 171-189.

[51] Weymark, J. [1991], *A Reconsideration of the Harsanyi - Sen Debate on Utilitarianism*, in Interpersonal Comparisons of Well-Being (J. Elster and J. Roemer, eds.), Cambridge University Press, Cambridge.

[52] Weymark, J. [1994], *Harsanyi's Social Aggregation Theorem with Alternative Pareto Principles*, in Models and Measurement of Welfare and Inequality (W. Eichhorn, ed.), Springer, Berlin.

[53] Weymark, J. [1993], *Harsanyi's Social Aggregation Theorem and the Weak Pareto Principle*, Social Choice and Welfare **10**, 209 - 221.

Fields Institute Communications
Volume **23**, 1999

Coalition-Proofness of the Competitive Allocations in an Indivisible Goods Market

Jun Wako
Department of Economics
Gakushuin University
1-5-1 Mejiro, Toshima-ku, Tokyo 171, Japan
jun.wako@gakushuin.ac.jp

Abstract. The competitive allocations in the Shapley-Scarf market model with indivisible goods are characterized as the Coalition-Proof Nash outcomes in a simple normal form game. In this game the set of Coalition-Proof Nash outcomes is nonempty and coincides with the set of Strong Nash outcomes. The competitive allocations are also regarded as the core defined by a modified weak domination.

1 Introduction

In this paper we reconsider the competitive equilibrium for the exchange model of Shapley and Scarf in [**7**].

Shapley and Scarf present an exchange model with a finite number of commodities and the same number of agents as commodities. Each agent is endowed with one indivisible commodity. All the commodities are differentiated. Furthermore, each agent is assumed to have no use for more than one indivisible commodity. Thus a feasible allocation consists of a permutation of the endowments.

Because of the simplified structure describing coalition formation for transactions among agents, the Shapley-Scarf model has been investigated intensively (see [**6**]). Especially, the strict core, i.e., the core defined by weak domination has been noted as a solution that illustrates specific features of the Shapley-Scarf model. Roth and Postlewiate in [**5**] and Wako in [**9**] and [**10**] show that the nonempty strict core is a unique von Neumann-Morgenstern solution and gives a unique point with respect to utility. Ma in [**4**] proves that the strict core social choice correspondence is the unique mechanism satisfying Pareto optimality, individual rationality, and strategy-proofness. However, the strict core may be empty. It is difficult to apply the strict core to a prescriptive analysis of the model.

Competitive equilibrium is a robust solution concept for an analysis of the Shapley-Scarf model. It always exists, and a simple algorithm will find all the competitive allocations [**7**]. The set of competitive allocations is a superset of the strict core and subset of the core [**9**].

1991 *Mathematics Subject Classification.* Primary 90D12; Secondary 90A60.

The equilibrium has some unexpected features, however. Because of the discreteness of the model, competitive prices are purely ordinal. If the price orderings are the same, a given competitive allocation is supported by multiple price systems. In addition, a competitive allocation is not always Pareto efficient.

In this paper we give a game-theoretic interpretation to the competitive equilibrium of the Shapley-Scarf model. We represent the Shapley-Scarf model as a simple normal form game and prove that the set of competitive allocations is equivalent to the set of Coalition-Proof Nash outcomes in the game. Furthermore, it is proved for this game that the set of Coalition-Proof Nash outcomes coincides with the set of Strong Nash outcomes. Finally, we discuss that the competitive allocations can be regarded as the core defined by a modified weak domination.

2 The market and its normal form game

There are n agents and n differentiated indivisible commodities in the market. Let $N = \{1, ..., n\}$ be the set of agents. The respective commodities are given the index numbers 1 through n. Each agent $i \in N$ is endowed with *commodity* i, which we formally describe by the ith unit vector e^i in $\Re^N$. Furthermore, it is assumed that each agent wants to consume exactly one of the indivisible commodities and has preferences over all those commodities. We then define the consumption set of each agent i to be $X := \{e^1, \ldots, e^n\}$ and his preference ordering R_i to be a complete, reflexive and transitive binary relation over X. Let P_i and I_i denote agent i's strict preference and indifference relations associated with R_i, respectively.

To make notations simple, we use index number i to denote element $e^i \in X$. With this convention, we define a **feasible allocation,** or simply, **allocation** x to be a vector $x = (x_i)_{i \in N} \in \prod_{i \in N} N$ such that $\{x_i | \ i \in N\} = N$, where x_i denotes the index number of the commodity assigned to agent i. An allocation is a permutation of the endowments. Let A denote the set of allocations. We refer to the market above as market $\mathcal{M}(N, (R_i)_{i \in N})$, or shortly, $\mathcal{M}$.

Shapley and Scarf in [**7**] define a competitive equilibrium for market $\mathcal{M}$ and prove its existence.

Definition 2.1 *Let p_i denote a price of commodity i. A* ***competitive equilibrium*** *of market $\mathcal{M}$ is a pair (x, p) consisting of an allocation x and a positive price vector $p = (p_1, ..., p_n)$ with the properties that for each $i \in N$,*

(1) $p_{x_i} \leq p_i$, and

(2) $x_i \ R_i \ k$ for any $k \in N$ with $p_k \leq p_i$.

A ***competitive allocation*** *is an allocation attained at some competitive equilibrium. Let $CA(\mathcal{M})$ denote the set of competitive allocations of market $\mathcal{M}$.*

The equilibrium conditions above specify only an order of prices of the respective commodities. Thus a given competitive allocation is supported by multiple price vectors with the same price ordering. The ordinality of prices is a feature of a competitive equilibrium of market $\mathcal{M}$. We give an example of an inefficient competitive allocation later in Example 1.

Theorem 2.2 *([**7**]) A competitive equilibrium exists for any market $\mathcal{M}$.*

This paper tries to interpret a competitive equilibrium of market $\mathcal{M}$ as an equilibrium of a normal form market game. A normal form game for market $\mathcal{M}(N, (R_i)_{i \in N})$ is defined to be a four-tuple $G(\mathcal{M}) = (N, S^N, f, (R_i)_{i \in N})$. The first element is a player set, which is the agent set N of market $\mathcal{M}$. The second element $S^N = \prod_{i \in N} S^i$ is a strategy space. Each agent i's strategy set S^i is defined by $S^i := N$. Thus each agent chooses an index number of a commodity. The third element f is an outcome function $f : S^N \to A$. For each $s = (s_i)_{i \in N} \in S^N$, function f determines an allocation $f(s) = (f_i(s))_{i \in N}$ as follows: For each $i \in N$, if there exists a set of agents $\{i = i_1, ..., i_k\}$ such that $s_{i_k} = i_1$ and $s_{i_h} = i_{h+1}$ for $h = 1, ..., k-1$, then $f_i(s) := s_i$; otherwise, $f_i(s) := i$. Under this rule, each agent gets the commodity he chooses if he is in a cycle of strategies. Otherwise, he keeps his endowment.[1] The last element $(R_i)_{i \in N}$ is a preference profile, which is the same profile as in market $\mathcal{M}$. Agent i evaluates his assigned commodity $f_i(s)$ according to the preference ordering R_i.

For each nonempty subset $T \subseteq N$ and $s \in S^N$, let $S^T := \prod_{i \in T} S_i$, $s_T := (s_i)_{i \in T} \in S^T$ and $s_{-T} := (s_i)_{i \in N \setminus T} \in S^{N \setminus T}$. Here, s_{-T} denotes a null vector[2] if $T = N$.

A strategy vector $s^* \in S^N$ is said to be a **Nash equilibrium** of game $G(\mathcal{M})$ if $f_i(s^*)\ R_i\ f_i(s_i, s^*_{-\{i\}})$ for each $i \in N$ and $s_i \in S_i$. Although a Nash equilibrium exists for any $G(\mathcal{M})$, some Nash equilibria may be trivial. The strategy vector s with $s_i = i$ for each $i \in N$, for example, is a Nash equilibrium that gives the initial allocation regardless of preference profiles. Therefore, we consider a Strong Nash equilibrium and a Coalition-Proof Nash equilibrium of game $G(\mathcal{M})$. A Strong Nash equilibrium is defined as follows:

Definition 2.3 *A strategy vector $s^* \in S^N$ is a **Strong Nash equilibrium** of game $G(\mathcal{M})$ if no nonempty subset $D \subseteq N$ has a strategy vector $s_D \in S^D$ with $f_j(s_D, s^*_{-D})\ P_j\ f_j(s^*)$ for each $j \in D$.*

For any nonempty subset $D \subseteq N$ and strategy vector $s^o \in S^N$, we define a **reduced game** $G(D, s^o)$ for D at s^o of game $G(\mathcal{M})$ to be a normal form game $(D, S^D, \bar{f}, (R_i)_{i \in D})$ such that outcome function $\bar{f} : S^D \to A$ is defined by $\bar{f}(s_D) := f(s_D, s^o_{-D})$ for each $s_D \in S^D$. Although a notation $G(N, s^o)$ denotes game $G(\mathcal{M})$, it is useful to define a Coalition-Proof Nash equilibrium.

Definition 2.4 *A strategy vector $s^* \in S^N$ is a **Coalition-Proof Nash equilibrium (CPNE)** of $G(\mathcal{M})$ if s^* is a CPNE of $G(N, s^*)$, which is defined recursively as follows:*

(1) *For any reduced game $G(\{i\}, s)$ with $i \in N$ and $s \in S^N$, a **CPNE** of $G(\{i\}, s)$ is a strategy $s^*_i \in S_i$ with $\bar{f}_i(s^*_i)\ R_i\ \bar{f}_i(s'_i)$ for all $s'_i \in S^i$.*[3]

(2) *Let $G(D, s)$ be any reduced game with $\#D \geq 2$ and $s \in S^N$. Assume that a CPNE is defined for any reduced game $G(T, s)$ with $\#T < \#D$ and $s \in S^N$. Then,*

   (a) *$s^*_D \in S^D$ is **self-enforcing** in $G(D, s)$ if, for any nonempty proper subset $T \subset D$, strategy vector $s^*_T \in S^T$ is a CPNE of $G(T, (s^*_D, s_{-D}))$.*

   (b) *s^*_D is a **CPNE** of $G(D, s)$ if s^*_D is self-enforcing in $G(D, s)$ and there is no self-enforcing strategy vector $s'_D \in S^D$ in $G(D, s)$ with $\bar{f}_i(s'_D)\ P_i\ \bar{f}_i(s^*_D)$ for all $i \in D$.*

---

[1] Any agent always can get his endowed commodity by choosing it. We can find an origin of this outcome function in ([**8**], Chapter 5).

[2] Thus vector (s_N, s'_{-N}) means vector s_N. We write s for s_N when there is no confusion.

[3] This defines a CPNE of a single player game.

Let $SN(\mathcal{M})$ be the set of Strong Nash equilibria and $CPN(\mathcal{M})$ the set of CPNE's of game $G(\mathcal{M})$. Since only self-enforcing deviations are considered in Coalition-Proof Nash equilibrium, it holds for any game $G(\mathcal{M})$ that $SN(\mathcal{M}) \subseteq CPN(\mathcal{M})$.

3 A characterization of the competitive allocations

The following propositions show that a competitive allocation of market $\mathcal{M}$ is attained at a Strong and also Coalition-Proof Nash equilibrium of market game $G(\mathcal{M})$.

Proposition 3.1 *For any market $\mathcal{M}$, if allocation x is competitive, then there exists a Strong Nash equilibrium s^* of game $G(\mathcal{M})$ with $f(s^*) = x$.*

Proposition 3.2 *For any market $\mathcal{M}$, if strategy vector $s^* \in S^N$ is a Coalition-Proof Nash equilibrium of game $G(\mathcal{M})$, then allocation $f(s^*)$ is competitive.*

Proposition 3.1 means that the set of competitive allocations is a subset of the set of Strong Nash allocations: $CA(\mathcal{M}) \subseteq f(SN(\mathcal{M}))$. Proposition 3.2 means that the set of Coalition-Proof Nash allocations is a subset of the set of competitive allocations: $f(CPN(\mathcal{M})) \subseteq CA(\mathcal{M})$. Since $f(SN(\mathcal{M})) \subseteq f(CPN(\mathcal{M}))$, the three sets are equivalent and nonempty by Theorem 2.2, that is, $f(SN(\mathcal{M})) = f(CPN(\mathcal{M})) = CA(\mathcal{M}) \neq \emptyset$.

We can see from the proof of Proposition 3.2 that the sets of Strong and Coalition-Proof Nash equilibria coincide, i.e., $SN(\mathcal{M}) = CPN(\mathcal{M})$. In game $G(\mathcal{M})$, if a subcoalition deviates after a deviation of its super-coalition, the subcoalition can deviate from the beginning on their own, or, together with some other agents in the super-coalition. This makes the two equilibria equivalent.

Konishi, Le Breton and Weber in [**3**] recently showed that "Partial Rivalry" condition is sufficient for the equivalence of Coalition-Proof and Strong Nash equilibria in a class of normal form games. Partial Rivalry means that if the number of agents taking the same strategy declines, then the payoff of each agent weakly increases. Although our game $G(\mathcal{M})$ does not belong to the class studied in [**3**], if $k \ R_i \ i$ for all $k, i \in N$, then game $G(\mathcal{M})$ satisfies Partial Rivalry.

The Strong Nash equilibrium is related to the core. For game $G(\mathcal{M})$, however, the set of Strong Nash allocations, i.e., competitive allocations, does not necessarily coincide with the core or the strict core of market $\mathcal{M}$. If we apply a modified concept of weak domination, then this discrepancy disappears.

Definition 3.3 *(i) Allocation x is said to* ***antisymmetrically weakly dominate*** *allocation y if the following three conditions hold:*

(a) there is a nonempty subset $S \subseteq N$ with $\{x_i | i \in S\} = S$;

(b) $x_i \ R_i \ y_i$ for each $i \in S$, and $x_j \ P_j \ y_j$ for some $j \in S$; and

(c) for each $i \in S$, if $x_i \ I_i \ y_i$, then $x_i = y_i$.

(ii) the core defined by antisymmetric weak domination, denoted by $C_{aw}(\mathcal{M})$, is the set of allocations that are not antisymmetrically weakly dominated by any allocations.

Observation. *The set of competitive allocations coincides with the core defined by antisymmetric weak domination, $C_{aw}(\mathcal{M})$.*

In antisymmetric weak domination, condition (c) is added to conditions (a) and (b) for weak domination. Condition (c) reflects a natural idea that an agent will not change the commodity he currently plans to obtain unless he becomes better off by doing so. Since the set of Strong Nash allocations coincides with $C_{aw}(\mathcal{M})$, the observation above is clear.

Finally we give an example to review our discussion.

Example 1 *Let* $N = \{1, 2, 3\}$ *and assume the following preference profile:*

(1) $2\ p_1\ 3\ p_1\ 1$

(2) $1\ I_2\ 3\ p_2\ 2$

(3) $2\ p_3\ 1\ I_3\ 3.$

This market $\mathcal{M}$ has two competitive allocations $x = (2, 1, 3)$ and $y = (1, 3, 2)$, which are supported by prices $p_1 = p_2 > p_3$ and $p_2 = p_3 > p_1$, respectively. However, y is not Pareto efficient.

Consider market game $G(\mathcal{M})$. The initial allocation $w = (1, 2, 3)$ is attained at two Nash equilibria: $s^o = (1, 2, 3)$ and $s' = (3, 2, 3)$. Let's consider coalitional deviations from equilibrium s^o, for example. We see that coalition N can be better off with deviation $\bar{s} = (3, 1, 2)$. Thus s^o is not a Strong Nash equilibrium. This fact does not necessarily imply that s^o is not a CPNE either. In game $G(\mathcal{M})$, however, if any deviation can occur, then a self-enforcing deviation also exists. In fact, strategy vector $(\tilde{s}_1, \tilde{s}_2) = (2, 1)$ of coalition $\{1, 2\}$ is self-enforcing in reduced game $G(\{1, 2\}, (\tilde{s}_1, \tilde{s}_2, s_3^o))$ and makes agents 1 and 2 better off, i.e., $f_i(\tilde{s}_1, \tilde{s}_2, s_3^o)\ P_i\ f_i(s_1^o, s_2^o, s_3^o)$ for $i = 1, 2$. Therefore, s^o is not a CPNE. This feature of game $G(\mathcal{M})$ makes the two equilibrium concepts equivalent. In this example, $f(SN(\mathcal{M})) = f(CPN(\mathcal{M})) = \{x, y\}$.

Allocations x and y weakly dominate each other through coalitions $\{1, 2\}$ and $\{2, 3\}$. These two allocations, however, are equivalent of $C_{aw}(\mathcal{M})$.

4 Proofs

We first give some definitions. For each nonempty subset $S \subseteq N$ and agent $i \in S$, define $B(i, S) := \{j \in S \mid j\ R_i\ k \text{ for any } k \in S\}$. Each $B(i, S)$ gives the commodities that agent i likes best in the endowments of S, so it is nonempty. Given an allocation x, a nonempty subset $T \subseteq N$ is called a **trading cycle in** x if $\{x_i | i \in T\} = T$ and $\{x_i | i \in T'\} \neq T'$ for any nonempty proper subset $T' \subset T$. A trading cycle T in x is a minimal set such that $(x_i)_{i \in T}$ is a redistribution of the endowments of T. The set of trading cycles in x constitutes a partition of N.

Proof of Proposition 3.1 Let (x, p) be any competitive equilibrium and s^* be the strategy vector with $s_i^* = x_i$ for each $i \in N$. Since $f(s^*) = x$, it suffices to prove that s^* is a Strong Nash equilibrium. To show this, suppose that s^* is not a Strong Nash equilibrium. Then there must be a nonempty subset $D \subseteq N$ and a strategy vector $\tilde{s}_D \in S^D$ with

$$f_j(\tilde{s}_D, s_{-D})\ P_j\ x_j \quad \text{for all } j \in D. \tag{4.1}$$

Let $\Pi = \{T_1, \ldots, T_K\}$ be the set of trading cycles in x. At competitive equilibrium (x, p), the commodities in each trading cycle T_k have the same price r_k, i.e., $p_i = r_k$ for all $i \in T_k$. We may assume that $r_1 \geq \ldots \geq r_K$. It is then implied that

$$x_i \in B(i, N \backslash \bigcup_{h=1}^{k-1} T_h) \quad \text{for each} \quad i \in T_k \quad \text{and} \quad T_k \in \Pi,$$

where $N \backslash \bigcup_{h=1}^{k-1} T_h = N$ for $k = 1$. Let $T_m \in \Pi$ be the trading cycle with $T_m \bigcap D \neq \emptyset$ and $T_k \bigcap D = \emptyset$ for $k = 1, \ldots, m-1$. Since Π is a partition of N, such a trading cycle must exist. The definition of f implies that $f_i(\tilde{s}_D, s^*_{-D}) = f_i(s^*)$ for $i \in \bigcup_{h=1}^{m-1} T_h$, and that $\{f_i(s^*)|i \in \bigcup_{h=1}^{m-1} T_h\} = \bigcup_{h=1}^{m-1} T_h$. Therefore, $f_j(\tilde{s}_D, s^*_{-D}) \in N\backslash \bigcup_{h=1}^{m-1} T_h$ for $j \in T_m \bigcap D$. Since $x_j \in B(i, N\backslash \bigcup_{h=1}^{m-1} T_h)$ for $j \in T_m$, it must hold that $x_j \; R_j \; f_j(\tilde{s}_D, s^*_{-D})$ for each $j \in T_m \bigcap D$. This contradicts (4.1). Thus, strategy vector s^* is a Strong Nash equilibrium. □

Proof of Proposition 3.2 Let s^* be any Coalition-Proof Nash equilibrium. In the following, we show that the set Π of trading cycles in $f(s^*)$ can be ordered in such a way that $\Pi = \{T_1, \ldots, T_K\}$ and

$$f_i(s^*) \in B(i, N\backslash \bigcup_{h=1}^{k-1} T_h) \quad \text{for each } i \in T_k \text{ and } T_k \in \Pi. \tag{4.2}$$

This completes the proof because, under the price vector p^* such that, for each $k = 1, \ldots, K$, $p^*_i = r_k$ for all $i \in T_k$ and $r_1 > \ldots > r_K > 0$, allocation $f(s^*)$ becomes competitive.

Step 1. Construct a strategy vector $s \in S^N$ that satisfies (i) $s_i \in B(i, N)$ for $i \in N$ with $f_i(s^*) \notin B(i, N)$ and (ii) $s_i = s^*_i$ for $i \in N$ with $f_i(s^*) \in B(i, N)$. Then pick any trading cycle T with the property[4] that

$$f_i(s) = s_i \quad \text{for all } i \in T, \tag{4.3}$$

and name it T_1. We show that $s_i = s^*_i$ for all $i \in T_1$. Suppose that the set $D := \{j \in T_1 | \, s_j \neq s^*_j\}$ is nonempty. By the definition of s, if $s_j \neq s^*_j$, then $f_j(s^*) \notin B(j, N)$. Thus, $s_j \in B(j, N)$ and $s_j \; P_j \; f_j(s^*)$ for $j \in D$. From (4.3), each $j \in D$ satisfies that $s_j = f_j(s) = f_j(s_{T_1}, s^*_{-T_1}) = f_j(s_D, s^*_{-D})$. Therefore,

$$f_j(s_D, s^*_{-D}) \in B(j, N) \quad \text{and} \quad f_j(s_D, s^*_{-D}) \; P_j \; f_j(s^*) \quad \text{for each } \; j \in D.$$

Since $f_j(s_D, s^*_{-D})$ is agent j's most preferred commodity in N, strategy vector s_D is clearly a self-enforcing one in game $G(D, s^*)$ that makes each agent $j \in D$ better off than strategy vector s^*_D. This contradicts Coalition-Proofness of s^*. This contradiction comes from the supposition that $D \neq \emptyset$. Thus, $s_i = s^*_i$ for all $i \in T_1$.

From (4.3), the fact that $s_i = s^*_i$ for all $i \in T_1$ implies that T_1 is also a trading cycle in $f(s^*)$ with $f_i(s^*) = s^*_i = s_i$ for all $i \in T_1$. Thus, from the definition of s, we have $f_i(s^*) \in B(i, N)$ for all $i \in T_1$. It also holds that $\{s^*_i | i \in T_1\} \subseteq T_1$.

Step $t \, (\geq 2)$**.** Suppose that $f(s^*)$ has trading cycles $T_1, \ldots, T_{t-1}$ with the properties that, for each $k = 1, \ldots, t-1$,
(A1) $f_i(s^*) \in B(i, N\backslash \bigcup_{h=1}^{k-1} T_h)$ for each $i \in T_k$, and
(A2) $\{s^*_i | i \in \bigcup_{h=1}^{k} T_h\} \subseteq \bigcup_{h=1}^{k} T_h$.

Let $M_t := N\backslash \bigcup_{h=1}^{t-1} T_h$. Without loss of generality, we may assume that $M_t \neq \emptyset$. The property (A2) means that if every agent $i \in N\backslash M_t$ $(= \bigcup_{h=1}^{t-1} T_h)$ takes strategy s^*_i, then no agent $j \in M_t$ can get any commodity in $N\backslash M_t$ under the outcome function f.

Construct a strategy vector $s \in S^N$ that satisfies (i) $s_i \in B(i, M_t)$ for $i \in M_t$ with $f_i(s^*) \notin B(i, M_t)$, (ii) $s_i = s^*_i$ for $i \in M_t$ with $f_i(s^*) \in B(i, M_t)$, and (iii) $s_i = s^*_i$ for $i \in N\backslash M_t$.

Consider the case that there exists $j \in M_t$ with $s_j \notin M_t$. Then the set $\{j\}$ forms a singleton trading cycle in $f(s)$ with $f_j(s) = j$. Let $T_t := \{j\}$. Since $B(j, M_t) \subseteq M_t$, if $s_j \notin M_t$, then $s_j \notin B(j, M_t)$. By the definition of s, we must have $f_j(s^*) \in B(j, M_t)$ and $s_j = s^*_j \notin M_t$. We then see that $f_j(s^*) = j$. Therefore,

[4] By the definition of f, allocation $f(s)$ has such a trading cycle for *any* s.

$T_t = \{j\}$ is also a trading cycle in $f(s^*)$ with the properties that $f_j(s^*) \in B(j, M_t)$ and $\{s_i^* | i \in \bigcup_{h=1}^t T_h\} \subseteq \bigcup_{h=1}^t T_h$.

Next, consider the case that $s_i \in M_t$ for all $i \in M_t$. Then there is at least one trading cycle $T \subseteq M_t$ with property (4.3). Pick any such trading cycle and name it T_t. We show that $s_i = s_i^*$ for all $i \in T_t$. Suppose that $D := \{j \in T_t | \, s_j \neq s_j^*\}$ is nonempty. By the same argument as in Step 1, we have

$$f_j(s_D, s_{-D}^*) \in B(j, M_t) \text{ and } f_j(s_D, s_{-D}^*) \; P_j \; f_j(s^*) \text{ for each } j \in D.$$

Since agents in D cannot get any commodities in $N \backslash M_t$ as mentioned above, $f_j(s_D, s_{-D}^*)$ is the most preferred commodity agent j can obtain. Thus s_D is a self-enforcing strategy vector in $G(D, s^*)$ that makes each agent of D better off than s_D^*. This contradicts Coalition-Proofness of s^*. We must have $s_i = s_i^*$ for all $i \in T_t$. Therefore, T_t is also a trading cycle in $f(s^*)$ with the properties that $f_i(s^*) \in B(i, M_t)$ for all $i \in T_t$ and $\{s_i^* | i \in \bigcup_{h=1}^t T_h\} \subseteq \bigcup_{h=1}^t T_h$.

We have proved that $f(s^*)$ has trading cycles $T_1, \ldots, T_t$ such that properties (A1) and (A2) hold for each $k = 1, \ldots, t$.

By updating t by $t := t + 1$, repeat the argument above until the updated set M_t becomes empty. In a finite number of steps, set M_t becomes empty, and all the trading cycles in $f(s^*)$ are indexed in an order such as (4.2). This completes the proof. □

Acknowledgements

The author is most grateful to Myrna Wooders for her helpful comments and encouragement. This paper was completed while the author was visiting Stanford Universeity (1995-1997). He also thanks Masahiko Aoki at Stanford University.

References

[1] Aumann, R. [1959], *Acceptable Points in General Cooperative n-Person Games*, in Contributions to the Theory of Games (A. W. Tucker and R. D. Luce, eds.), Vol.4, Princeton University Press, Princeton, NJ.

[2] Bernheim, B. D., Peleg, B. and Whinston, M. D. [1987], *Coalition-Proof Nash Equilibria I. Concepts*, Journal of Economic Theory **42**, 1-12.

[3] Konishi, H., Le Breton, M. and Weber, S. [1997], *Equilibrium in a Model with Partial Rivalry*, Journal of Economic Theory **72**, 225-237.

[4] Ma, J. [1994], *Strategy-Proofness and the Strict Core in a Market with Indivisibilities*, International Journal of Game Theory **23**, 75-83.

[5] Roth, A. E. and Postlewaite, A. [1977], *Weak versus Strong Domination in a Market with Indivisible Goods*, Journal of Mathematical Economics **4**, 131-137.

[6] Roth, A. E. and Sotomayor, M. [1990], *Two-sided matching: a study in game-theoretic modeling and analysis*, Cambridge University Press, NY.

[7] Shapley, L. S. and Scarf, H. [1974], *On Cores and Indivisibilities*, Journal of Mathematical Economics **1**, 23-37.

[8] von Neumann, J. and Morgenstern, O. [1947], *Theory of Games and Economic Behavior*, Princeton University Press, Princeton, NJ.

[9] Wako, J. [1984], *A Note of the Strong Core of a Market with Indivisible Goods*, Journal of Mathematical Economics **13**, 189-194.

[10] Wako, J. [1991], *Some Properties of Weak Domination in an Exchange Market with Indivisible Goods*, The Economic Studies Quarterly **42**, 303-314.

Appendix

Presentations

at

The Fields Institute Workshop

New Directions in the Theory of Markets and Games

October 19 – 23, 1995
Toronto Ontario, Canada

Joseph Abou — Tight Free Interaction: A Class of Nash-Solvable Game Forms

Beth Allen — On the Existence of Core Allocations in a Large Economy with Incentive Compatibility Constraints

C. D. Aliprantis, O. Burkinshaw and K. C. Border
Market Economies

Robert M. Anderson and William R Zame
Edgeworth's Conjecture with Infinitely Many Commodities

Niels Anthonisen — On the Convergence of Simple Learning to Equilibrium in Games

Daniel Arce — Aumann's Micrceconomics? A Theory of Fiscal Pacts

Robert J. Aumann — Game Theory in The Real World: Hospital Interns, Talmudic Bankruptcy Law, Spiderwebs, Oil Lease Auctions, Formation of Governments and Other Stories

Lawrence Ausubel and Abijit Sen
Sequential Recontracting Under Incomplete Information

Erik Balder — A Unifying Approach to Existence of Nash Equilibria

Vicky Barham and Laura Razzolini
From Each According to their Capacity to Pay: Differentiated Pricing in Clubs

Dirk Bergemann — Learning and Commitment in Incentive Contracts

James Bergin and John Duggan
Non-Cooperative Foundations of the Core: An Implementation-Theoretic Approach

Volker Böhm — Equilibria for Exchange Economies and for Games of Exchange

Steven J. Brams — Modelling Free Choice in Games

Sandro Brusco — Perfect Bayesian Implementation in Economic Environments

Don Campbell and Jerry Kelly
A Trade-Off Result for the Gibbard-Sattherthwaite Theorem

Subir Chakrabarti — Pure Strategy Markov Equilibria in Stochastic Games with Many Players

John Conley — Fundamental Nonconvexities in Arrovian Markets and a Coasian Solution to the Problem of Externalities

David Cooper — Signalling, Adaptive Learning and Equilibrium Refinements in an Entry Limit Pricing Game

Luis Corchon and Carmen Herrero
A Decent Proposal

Ross Cressman — Should Cheaters be Punished? The Surprising Finds in the Buyer-Seller Game

Martin Cripps and Jonathan Thomas
The Folk Theorem in Repeated Games of Incomplete Information

Gabrielle Demange — Interim Design of Core Mechanisms

J. J. M. Derks and H. H. Haller
Weighted Nucleoli

Egbert Dierker and Birgit Grodal
The Price Normalization Problem in Imperfect Competition and the Objective of a Firm

Victor Domansky and Victoria Kreps
Explicit Solutions for a Class of Repeated Games with Incomplete Information

Irinel Dragan — On Some Relationships Between the Shapley Value and the Banzhaf Value

Theo Driessen	The Assignment Problem: A Description of Threats in Terms of Bargaining Ranges Within and Outside the Core of the Assignment Game
John Duggan	Sequentially Rational Implementation with Incomplete Information
Sjur Flåm	Learning Economic Equilibrium
Wulf Gaertner and Nick Baigent	Rationality and Norms: Never Choose the Uniquely Largest
Rod Garratt and Cheng-Zhong Qin	On a Market for Coalitions with Indivisible Agents and Lotteries
Andrzej Granas	Some General Geometric Principles in the Theory of KKM-Maps
Gordon Green	Market Institutions and Core Allocations
Thomas A. Gresik and Eric W. Bond	Information Sharing Among Asymmetrically Informed Governments
Birgt Grodal and Egbert Dierker	Maximization of Shareholder's Real Wealth
Peter Hammond	Multilaterally Strategy-Proof Mechanisms in Random Aumann-Hildebrand Macroeconomies
Sergiu Hart	A Preliminary Communication on Pre-Play Communication
Ebbe Hendon, Birgitte Sloth and Hans Jørgen Jacobsen	Adaptive Learning in Extensive Form Games and Sequential Equilibrium
Carmen Herrero	Personal Rights and Collective Ownership: The Rights-Egalitarian Solution
Lu Hong	Problem Solving by Teams of Heterogeneous Agents
Charles Horvath	On the Topological Choice Problem
Tatsuro Ichiishi	A Profit-Centre Game with Incomplete Information
Matthew Jackson and Ehud Kalai	Recurring Games and Social Learning

Mark Johnson — Algebraic Automata, Strategy Implementation and Economic Consistency

Michael Jones — Cones of Cooperation for Indefinitely Repeated, Generalized Prisoner's Dilemma Games

Hans Keiding and Peter Fristrup
Strongly Implementable Social Choice Correspondences and the Supernucleus

Marc Kilgour and Steven Brams
The Truel

Hideo Konishi, Shlomo Weber and Michel Le Breton
Equilibrium in a Model with Partial Rivalry

Van Koplin — Equitable Nonlinear Price Regulation and an Alternative Approach to Serial Cost Sharing

Alexander Kovalenkov and Myrna Holtz Wooders
On Nonemptiness of Approximate Cores of Games Without Side Payments

Chantale LaCasse and Ignatius Horstman
Dynamic Auctions with Learning

Ehud Lehrer — Breaking the Barriers of the Feasible Set on Repeated Games with Different Time Preferences

Roger Lagunoff and Akihiko Matsui
An "Anti-Folk" Theorem for a Class of Asynchroneously Repeated Games

Bart Lipman — Decision Theory Without Logical Omniscences Toward and Axiomatic Framework for Bounded Rationality

Juan Enrique Martinez-Legaz
A New Characterization of Totally Balanced Games

Kin Chung Lo — Equilibrium in Beliefs Under Uncertainty

Richard McLean — Customer Specific Aumann-Shapley Prices

Jean-Francois Mertens
Relative Utilitarianism

Antoni Meseguer and Myrna Holtz Wooders
NTU Games, Markets and a Social Choice Rule

Paulo Monteiro, Frank Page and Myrna Holtz Wooders
Arbitrage and Equilibrium

Hervé Moulin — Three Additive Methods to Share Joint Costs

Roger Myerson — Dual Extraction in Strategic-Form Games

Matjaz Omladic and Vesna Omladic
A Linear Alegebra Approach to Non-Transitive Expected Utility

Joseph Ostroy, L. Makowski and Uzi Segal
Perfect Competition as the Blueprint for Efficiency and Incentive Compatibility

Frank Page and Myrna Holtz Wooders
Asymmetric Information, the Efficienty Core and Stable Standards of Behaviour

Motty Perry — Implementation Without Common Prior

Clara Ponsati and Jozsef Sakovics
Mediation is Necessary for Dynamic Efficiency in Bargaining

Cheng-Zhong Qin — The Core of an Economy: Edgeworth vs. Walras

T. E. S. Raghavan — Algorithms for Locating the Nucleolus for Special Classes of Cooperative Games

James Ratliff — Repeated Games with Private Information about Duration

Lakshmi Raut — Aumann-Shapley Random Order Value and Haar Measure

Laura Razzolini and Rajat Deb
Auction-Like Mechanisms for Pricing Excludable Public Goods

Philip Reny — Local Payoff Security and the Existence of Nash Equilibrium in Discontinuous Games

Arthur Robson — Adaptive Behaviour, Idiosyncratic Risk and Aggregate Uncertainty

Bryan Routledge — Artificial Selection: Genetic Algorithms and Learning in a Rational Expectations Model

Jozsef Sakovics — Mirage Equilibria

Larry Samuelson and Ken Binmore
Evolutionary Drift and Equilibrium Selection

David Schmeidler, Elchanan Ben-Porath and Itzhak Gilboa
Social Welfare Under Uncertainty

Norman Schofield — The C-1 Topology on the Space of Preference Profiles and the Existence of a Continuous Preference Aggregator

Uzi Segal — Preferences Over Solutions to the Bargaining Problem

E. Somanathan — Evolutionary Stability of Pure-Strategy Equilibria in Finite Games

Tayfun Sonmez — Strategy-Proofness and Singleton Cores in Generalized Matching Problems

Sylvain Sorin — Strategic Market Games with Exchange Rates

Yeneng Sun and Ali Khan
Integrals of Set-Valued Functions with a Countable Range: Theory and Applications

Jeroen Swinkels and Wolfgang Pesendorfer
The Loser's Curse and Information Aggregation in Auctions

Rafael Tenorio and Linda Goldberg
Strategic Behaviour in Two-Sided Foreign Exchange Auctions

William Thomson — Axiomatic Analysis of Allocation Problems with Single-Peaked Preferences and Individual Endowments

Arja Turunen-Red — Harsanyi Aggregation, Arbitrage and the Core

Anne van den Nouweland
Consistency of Equilibria in Pure Exchange Economies and Public Good Economies

Antonio Villar — Classical Equilibrium with Increasing Returns

Karl Vind — Perfect Competition - Two Examples

Oscar Volij and Roberto Serrano
Axiomatizations of Neoclassical Concepts for Economies

Jun Wako — Coalition-Proofness of the Competitive Allocations in a Market with Indivisible Goods

Mark Walker, Jason Shachat and Jim Cox
An Experimental Test of Bayesian vs. Adaptive Learning

John Weymark	Strategy-Proofedness and the Tops-Only Property
Myrna Wooders, Frank Page and Paulo Monteiro	Arbitrage and Existence of Equilibrium with Unbounded Short Sales
Shinji Yamashige	Bayesian Approach with Finite Heirarchies of Beliefs: Bounded Rationality in Strategic Form Games
William Zame	Collateral and Default: A General Equilibrium View
Jiangang Zhao	The Endogenization of Coalition Structures in Normal Form TU Games